# Spline Smoothing and Nonparametric Regression

RANDALL L. EUBANK

*Department of Statistics*
*Southern Methodist University*
*Dallas, Texas*

MARCEL DEKKER, INC. New York and Basel

**Library of Congress Cataloging-in-Publication Data**

Eubank, Randall L.
Spline smoothing and nonparametric regression.

(Statistics, textbooks and monographs ; vol. 90)
Bibliography: p.
Includes index.
1. Regression analysis. 2. Nonparametric statistics.
3. Spline theory. I. Title. II. Series: Statistics, textbooks and monographs ; v. 90.
QA278.2.E93 1988 519.5'36 87-30599
ISBN 0-8247-7869-3





MARCEL DEKKER, INC.
270 Madison Avenue, New York, New York 10016

Current printing (last digit):
10 9 8 7 6 5 4 3 2 1

PRINTED IN THE UNITED STATES OF AMERICA

*To*

*LISA, MOTSI, LESTER, AND MARY KATHERINE*

# Preface

Statisticians have become increasingly aware that many questions which might be asked of a set of data can be answered equally well (or better) using nonparametric, rather than parametric, models. The extent of this realization is demonstrated by the rapid development of a subfield of statistics one might term nonparametric function estimation. This subject includes both nonparametric density estimation and nonparametric regression. The latter area is the topic of this book.

Nonparametric regression has its roots in the area of nonparametric density estimation. This is as it should be, since density estimation apparently came into focus as a topic of merit for statistical research at a somewhat earlier date (see Tapia and Thompson, 1978, for a chronology of research in density estimation). Nonparametric density estimation moved into the mainstream of statistical research in the nineteen sixties. There are now several books on the subject including Tapia and Thompson (1978), Prakasa Rao (1983), Devroye and Györfi (1985) and Silverman (1986).

Nonparametric regression has experienced much of its development within the last one and one-half decades. However, the field has developed and continues to develop rapidly. While controversies and unresolved issues remain, the area has now reached a level of development where systematic overviews in the form of texts and review articles have become feasible and worthwhile. My aim is to provide one such treatment of the topic which hopefully will prove useful for students, practitioners and researchers in the area.

This book evolved from a set of lecture notes I used for teaching a course on spline approximation in statistics at Southern Methodist University. These notes have continued to be revised and expanded during each subsequent presentation. The most recent of these was in a series of lectures given at Texas A & M University during the Fall of 1986.

In writing this text my objective was to satisfy two goals. I wanted to provide a treatment of the subject matter that could double as a text for a course on the subject (namely my own) and as an introductory treatment for more experienced statisticians who might benefit from a stepping stone into the literature. As a result, this book is not a textbook, *per se*. My aim was to take a middle ground and produce a flexible product that could be adapted to the needs of the user.

From a pedagogical viewpoint, the target audience for the text consists of second or third year students in a graduate program in Statistics. These students are assumed to have a good knowledge of linear models and regression but not necessarily any exposure to the concept of data smoothing. For that and other reasons I have kept the mathematical development at a fairly elementary level. Most derivations, proofs, etc. can be easily understood if one is acquainted with Taylor's Theorem and has a working knowledge of vector space and matrix theory. To fully appreciate the details of some developments in Chapters 3 and 6 a familiarity with certain results from real and Fourier analysis would also be helpful. However, knowledge of these latter two subjects is not essential. It is my belief that a reader who is unfamiliar with the tools of analysis can still appreciate the quality of results derived using such methods, provided they are willing to take certain things on faith.

Other people who will hopefully find this text useful are Statisticians who are like I was when I first became interested in splines and nonparametric regression. While I was quite enamoured by the topic, I found the process of sorting through the literature a formidable task. This text is designed to ease some of that burden by giving a fairly broad overview and working out the details for certain simple cases that exemplify general patterns or techniques. Literature citations are then given which point the way toward more in depth developments.

An outline of the contents of the book along with some thoughts on how it might be used for teaching purposes are as follows. Chapter 1 is intended to introduce the reader who is familiar primarily with parametric models to the nonparametric perspective. Chapter 2 then contains a general treatment of risk estimation, including such topics as unbiased risk estimation, and ordinary and generalized cross-validation. This chapter is structured so that it is accessible to a reader who knows nothing about nonparametric regression but is familiar with topics from linear regression such as ridge regression and variable selection. Knowledge of its content is presumed in all further chapters. If one desires to jump directly into smoothing after Chapter 1 they could skip Chapter 2 and go directly to Chapter 3 or 4 and then supplement using material from Chapter 2 whenever necessary. My personal preference is to cover the material in sequence. This is because the subject matter of Chapter 2 is of general interest and can be understood by a student, matching my description above, without the addition of the data smoothing issue. However, either approach seems workable and I have taught the material both ways.

Chapter 3 contains a study of Fourier series type estimators. Again, the idea is that a student who is comfortable with linear regression analysis will find these estimators a natural extension of familiar techniques to the nonparametric regression problem. Also, Fourier series estimators provide a convenient lead in, via the Dirichlet kernel, to the topic of kernel estimators treated in Chapter 4. Kernel estimators, being weighted averages, are easily understood by students. They represent an ideal setting where an appreciation for what asymptotic theory can (and cannot) tell us can be developed with a minimum of mathematical detail.

The heart of this text is Chapters 5 and 6 where smoothing splines and smoothing spline variants are studied. These estimators are somewhat more difficult to understand than kernel estimators. However, in my opinion, the payoff is worth the effort. The result is an extremely flexible estimation procedure with an intriguing connection to Bayesian estimation. My experience is that most students buy the estimation criterion immediately as a natural extension of polynomial regression and I have presented it from that viewpoint. Some of the work on smoothing spline bases in Chapter 5 and asymptotic theory in Chapter 6 can be given reduced treatment depending on the desired emphasis. The salient points of Chapter 6 (illustration of the adaptability of the spline smoothing concept, demonstration of the connection between smoothing splines and Fourier series and kernel estimators, and example applications) can be made without nearly as much detail as has been included. Finally, Chapter 7 represents a catch-all chapter where a number of other estimator types are mentioned and briefly discussed. The limited treatment given to these other estimators is a reflection of my own personal interest and should not be construed as a value judgement.

This book relies heavily on the work of Grace Wahba. It would not have been possible to write this particular text without her many contributions to the area. Her aid in writing the book by providing reprints, preprints, figures, etc. is also gratefully acknowledged. Thanks are also due to many others, including Dennis Cox, Ker Chau Li, John Rice, Paul Speckman, Theo Gasser, H. G. Müller and Florencio Utreras, for similar reasons and for their ground breaking work on nonparametric regression that has provided such a stimulating environment for study.

On a more personal level, I would like to thank my advisor, Manny Parzen, for giving me the reproducing kernel Hilbert space key that opened so much of the early literature on smoothing splines. I am grateful to Pat and Phil Smith for formally introducing me to spline functions. A few words of sage advice by Narayan Bhat are also gratefully acknowledged.

A number of friends, students and associates have read earlier drafts of the text and made helpful comments and suggestions. Among these I would particularly like to mention Darren Kline, Jeff Hart, Manny Parzen, Paul Speckman, Jyh-Jen Shiau and Grace Wahba. Detailed critiques by two anonymous reviewers have been most helpful in correcting a number of foolish errors on my part. I would also like to express my appreciation to the Office of Naval Research for supporting my research over the last seven years.

The manuscript was typed by Sheila Crain. Her patience in enduring endless changes, rewrites and revisions is demonstrated by the fact that I still live today. Valuable assistance with several of the examples and figures was provided by Kelly Cunningham and Tom Carmody. Thanks are due to L. Molinari and Paul Speckman for providing data and data analyses that were used in Chapters 4 and 7, respectively.

Finally, I must thank my wife Lisa and my second best friend Motsi for their unfailing support and love. Without their help and sacrifices this book could not have been written.

**Randall L. Eubank**

# Spline Smoothing and Nonparametric Regression

**STATISTICS: Textbooks and Monographs**



- Vol. 1: The Generalized Jackknife Statistic, *H. L. Gray and W. R. Schucany*
- Vol. 2: Multivariate Analysis, *Anant M. Kshirsagar*
- Vol. 3: Statistics and Society, *Walter T. Federer*
- Vol. 4: Multivariate Analysis: A Selected and Abstracted Bibliography, 1957-1972, *Kocherlakota Subrahmaniam and Kathleen Subrahmaniam* (out of print)
- Vol. 5: Design of Experiments: A Realistic Approach, *Virgil L. Anderson and Robert A. McLean*
- Vol. 6: Statistical and Mathematical Aspects of Pollution Problems, *John W. Pratt*
- Vol. 7: Introduction to Probability and Statistics (in two parts), Part I: Probability; Part II: Statistics, *Narayan C. Giri*
- Vol. 8: Statistical Theory of the Analysis of Experimental Designs, *J. Ogawa*
- Vol. 9: Statistical Techniques in Simulation (in two parts), *Jack P. C. Kleijnen*
- Vol. 10: Data Quality Control and Editing, *Joseph I. Naus* (out of print)
- Vol. 11: Cost of Living Index Numbers: Practice, Precision, and Theory, *Kali S. Banerjee*
- Vol. 12: Weighing Designs: For Chemistry, Medicine, Economics, Operations Research, Statistics, *Kali S. Banerjee*
- Vol. 13: The Search for Oil: Some Statistical Methods and Techniques, *edited by D. B. Owen*
- Vol. 14: Sample Size Choice: Charts for Experiments with Linear Models, *Robert E. Odeh and Martin Fox*
- Vol. 15: Statistical Methods for Engineers and Scientists, *Robert M. Bethea, Benjamin S. Duran, and Thomas L. Boullion*
- Vol. 16: Statistical Quality Control Methods, *Irving W. Burr*
- Vol. 17: On the History of Statistics and Probability, *edited by D. B. Owen*
- Vol. 18: Econometrics, *Peter Schmidt*
- Vol. 19: Sufficient Statistics: Selected Contributions, *Vasant S. Huzurbazar (edited by Anant M. Kshirsagar)*
- Vol. 20: Handbook of Statistical Distributions, *Jagdish K. Patel, C. H. Kapadia, and D. B. Owen*
- Vol. 21: Case Studies in Sample Design, *A. C. Rosander*
- Vol. 22: Pocket Book of Statistical Tables, *compiled by R. E. Odeh, D. B. Owen, Z. W. Birnbaum, and L. Fisher*
- Vol. 23: The Information in Contingency Tables, *D. V. Gokhale and Solomon Kullback*

**Vol. 24:** Statistical Analysis of Reliability and Life-Testing Models: Theory and Methods, *Lee J. Bain*
**Vol. 25:** Elementary Statistical Quality Control, *Irving W. Burr*
**Vol. 26:** An Introduction to Probability and Statistics Using BASIC, *Richard A. Groeneveld*
**Vol. 27:** Basic Applied Statistics, *B. L. Raktoe and J. J. Hubert*
**Vol. 28:** A Primer in Probability, *Kathleen Subrahmaniam*
**Vol. 29:** Random Processes: A First Look, *R. Syski*
**Vol. 30:** Regression Methods: A Tool for Data Analysis, *Rudolf J. Freund and Paul D. Minton*
**Vol. 31:** Randomization Tests, *Eugene S. Edgington*
**Vol. 32:** Tables for Normal Tolerance Limits, Sampling Plans, and Screening, *Robert E. Odeh and D. B. Owen*
**Vol. 33:** Statistical Computing, *William J. Kennedy, Jr. and James E. Gentle*
**Vol. 34:** Regression Analysis and Its Application: A Data-Oriented Approach, *Richard F. Gunst and Robert L. Mason*
**Vol. 35:** Scientific Strategies to Save Your Life, *I. D. J. Bross*
**Vol. 36:** Statistics in the Pharmaceutical Industry, *edited by C. Ralph Buncher and Jia-Yeong Tsay*
**Vol. 37:** Sampling from a Finite Population, *J. Hajek*
**Vol. 38:** Statistical Modeling Techniques, *S. S. Shapiro*
**Vol. 39:** Statistical Theory and Inference in Research, *T. A. Bancroft and C.-P. Han*
**Vol. 40:** Handbook of the Normal Distribution, *Jagdish K. Patel and Campbell B. Read*
**Vol. 41:** Recent Advances in Regression Methods, *Hrishikesh D. Vinod and Aman Ullah*
**Vol. 42:** Acceptance Sampling in Quality Control, *Edward G. Schilling*
**Vol. 43:** The Randomized Clinical Trial and Therapeutic Decisions, *edited by Niels Tygstrup, John M. Lachin, and Erik Juhl*
**Vol. 44:** Regression Analysis of Survival Data in Cancer Chemotherapy, *Walter H. Carter, Jr., Galen L. Wampler, and Donald M. Stablein*
**Vol. 45:** A Course in Linear Models, *Anant M. Kshirsagar*
**Vol. 46:** Clinical Trials: Issues and Approaches, *edited by Stanley H. Shapiro and Thomas H. Louis*
**Vol. 47:** Statistical Analysis of DNA Sequence Data, *edited by B. S. Weir*
**Vol. 48:** Nonlinear Regression Modeling: A Unified Practical Approach, *David A. Ratkowsky*
**Vol. 49:** Attribute Sampling Plans, Tables of Tests and Confidence Limits for Proportions, *Robert E. Odeh and D. B. Owen*
**Vol. 50:** Experimental Design, Statistical Models, and Genetic Statistics, *edited by Klaus Hinkelmann*
**Vol. 51:** Statistical Methods for Cancer Studies, *edited by Richard G. Cornell*
**Vol. 52:** Practical Statistical Sampling for Auditors, *Arthur J. Wilburn*
**Vol. 53:** Statistical Signal Processing, *edited by Edward J. Wegman and James G. Smith*
**Vol. 54:** Self-Organizing Methods in Modeling: GMDH Type Algorithms, *edited by Stanley J. Farlow*
**Vol. 55:** Applied Factorial and Fractional Designs, *Robert A. McLean and Virgil L. Anderson*
**Vol. 56:** Design of Experiments: Ranking and Selection, *edited by Thomas J. Santner and Ajit C. Tamhane*
**Vol. 57:** Statistical Methods for Engineers and Scientists. Second Edition, Revised and Expanded, *Robert M. Bethea, Benjamin S. Duran, and Thomas L. Boullion*
**Vol. 58:** Ensemble Modeling: Inference from Small-Scale Properties to Large-Scale Systems, *Alan E. Gelfand and Crayton C. Walker*
**Vol. 59:** Computer Modeling for Business and Industry, *Bruce L. Bowerman and Richard T. O'Connell*
**Vol. 60:** Bayesian Analysis of Linear Models, *Lyle D. Broemeling*

**ADDITIONAL VOLUMES IN PREPARATION**

# Contents

$X_n = o_p(Y_n)$ means that $X_n/Y_n \xrightarrow{P} 0$.

$(x)_+ = \max\{x,0\}$ .

$\ln x$ = natural logarithm of x .

* is used to indicate complex conjugate (or complex conjugate transpose for a matrix).

$g^{(m)}$ = mth derivative of a function g; g' and g'' are usually used to indicate first and second derivatives.

$||g|| = \{\int_a^b g(t)^2 dt\}^{\frac{1}{2}} = L_2[a,b]$ norm of a function g.

EX denotes the mathematical expectation of a random variable X.

P(A) or Prob(A) stands for the probability of some event A.

$Z_\alpha$ = 100(1 - α) percentage point of the standard normal distribution.

Vectors are always column vectors and are signified by an underbar. Thus $\underline{t} \in \mathbb{R}^p$ indicates that $\underline{t}$ is a p-dimensional column vector.

$\underline{t}'$ = transpose of $\underline{t}$.

$|\underline{t}| = (\underline{t}'\underline{t})^{\frac{1}{2}}$ = Euclidean norm of a vector $\underline{t}$.

$\underline{1}$ = vector of all ones. The dimension will be clear from context.

$A = \{a_{ij}\}_{\substack{i=1,p \\ j=1,k}}$ = a p × k matrix with $a_{ij}$ as the element in the ith row and jth column.

# Notation

The following is a list of some of the important notation and terminology employed throughout the text.

order m: usually refers to a polynomial (or polynomial segment) and means that the polynomial degree is at most $m - 1$.

functional = a mapping from some set of functions into the real line.

$\{x \in X\colon P(x)\}$ = the set of all elements in X having the property $P(x)$.

$\mathbb{R}^p$ = Euclidean p-space.

$g(x) = O(f(x))$ as $x \to a$, some constant, means that there exists a finite constant M (possibly depending on g, f and a) such that $|g(x)| \le M|f(x)|$ as $x \to a$.

$g(x) = o(f(x))$ as $x \to a$ means that $g(x)/f(x) \to 0$ as $x \to a$.

$g(x) \sim f(x)$ as $x \to a$ means that $g(x)/f(x) \to 1$ as $x \to a$.

$X_n \xrightarrow{p} X$ signifies that a sequence of random variables $X_n$ converges in probability to the random variable X.

$X_n \xrightarrow{d} X$ means that $X_n$ converges in distribution to X.

$X_n = O_p(Y_n)$ indicates that $X_n/Y_n$ is bounded in probability.

$A'$ = transpose of $A$.

$0$ = matrix of all zeros. The dimension will always be clear from context.

$I$ = identity matrix. Sometimes $I_r$ will be used to indicate the $r \times r$ identity if the dimension is not clear from context.

$\mathrm{diag}(a_1, \ldots, a_p)$ = $p \times p$ diagonal matrix with diagonal elements $a_1, \ldots, a_p$.

$A^j$ = $j$-fold product of a symmetric matrix $A$.

$\mathrm{tr}A$ = trace of the matrix $A$.

$E\underline{X}$ - mathematical expectation of a random vector $\underline{X}$. Expectation is taken component wise. Thus, if $\underline{X}$ is a p-vector, $E\underline{X} = (EX_1, \ldots, EX_p)'$.

$\mathrm{Var}\ \underline{X} = E(\underline{X} - E\underline{X})(\underline{X} - E\underline{X})'$ = the variance-covariance matrix for a random vector $\underline{X}$.

$$\frac{\partial^{\alpha_1+\cdots+\alpha_p} g(\underline{t})}{\partial t_1^{\alpha_1} \cdots \partial t_p^{\alpha_p}} = \alpha_1 + \cdots + \alpha_p \text{ order partial derivative of } g$$

with iterated derivatives of order $\alpha_i$ in the variable $t_i$.

$d\underline{t} = dt_1 \ldots dt_p$ .

# Spline Smoothing and Nonparametric Regression

# 1
# Introduction

## 1.1 Regression Analysis

When someone says they have performed regression analysis this usually conjures up visions of estimated regression coefficients, residual plots, etc. which have been obtained by fitting a linear model to a set of data. This is certainly one aspect of regression analysis. However, there are many other topics which also merit inclusion under this broad heading. This text focuses on a particular collection of such topics generally referred to as nonparametric regression.

To begin our study it will be helpful to restrict attention to a simple model that provides a convenient framework for the discussion of parametric and nonparametric regression. Our initial objective is to distinguish between the philosophies which underlie these two approaches to regression analysis.

Suppose that observations are taken on a continuous random variable Y at n predetermined values of a continuous independent variable t. Let $(t_j, y_j)$, $j=1,\ldots,n$, be the values of t and Y which result from this sampling scheme and assume that the $t_j$ and $y_j$ are related by the regression model

$$y_j = \mu(t_j) + \varepsilon_j, \quad j = 1,\ldots,n, \tag{1}$$

where the $\varepsilon_j$ are zero mean uncorrelated random variables with a common variance $\sigma^2$ and the $\mu(t_j)$ are the values of some unknown function $\mu$ at the points $t_1,\ldots,t_n$. The function $\mu$ is usually referred to as the *regression function* or *regression curve*. In this setting, the phrase "regression analysis" refers to methods for statistical inference about the regression function.

The determination of a suitable inferential methodology for model (1) will hinge on the assumptions it is possible to make about $\mu$. A *parametric regression model* assumes that the form of $\mu$ is known except for finitely many unknown parameters. More specifically, it is assumed that there is a vector of parameters $\underline{\beta} = (\beta_1, \ldots, \beta_p)' \in B$, some subset of $\mathbb{R}^p$, and a function $\mu(\cdot;\underline{\beta})$ such that $\mu(\cdot) = \mu(\cdot;\underline{\beta})$. It is therefore clear that in parametric regression models inference about $\mu$ is tantamount to inference about $\underline{\beta}$.

Parametric models can depend on the parameters in a linear and/or nonlinear fashion. Some examples of models where some of the parameters enter in a nonlinear fashion are

$$\mu(t) = \beta_1 t^{\beta_2}$$

$$\mu(t) = \beta_1 + \beta_2 \exp\{-\beta_3 t^{\beta_4}\}$$

and

$$\mu(t) = \beta_1 \cos\beta_2 t + \beta_3 \sin\beta_4 t .$$

In contrast, for a linear model there are known functions $x_1, \ldots, x_p$ such that

$$\mu(t) = \sum_{j=1}^{p} \beta_j x_j(t) \quad . \tag{2}$$

For instance, $x_1(t)$ might represent the temperature of the coolant in some production process on day t, $x_2(t)$ might be a measure of air flow, etc. Inference about the parameters in model (1) when (2) holds is generally referred to as *linear regression analysis*.

Regression analysis techniques for parametric models represent one approach for conducting inference about $\mu$. Using an appropriate estimation methodology (c.f. the Appendix) it is possible to utilize the data to estimate parameters and thereby obtain an estimate of $\mu$. The resulting estimate is a curve that has been selected from the family of curves allowed under the model and conforms to the data in some fashion.

There are, however, other methods of fitting curves to data. One collection of procedures that can be used for this purpose are nonparametric regression techniques. These methods give estimates of $\mu$ that allow great flexibility in the possible form of the regression curve and, in particular, make no assumptions of a parametric model. A *nonparametric regression model* generally only assumes that $\mu$ belongs to some infinite dimensional collection of functions. For example, $\mu$ may only be assumed to be differentiable or differentiable with a square integrable second derivative, etc. Assumptions of this type are concerned only with qualitative properties of $\mu$ and are to be contrasted with those used in a parametric model where a much greater level of specificity is required.

An important difference between parametric and nonparametric regression methodologies is their respective degree of reliance on the information about $\mu$ obtained from the experimenter and from the data. To construct a nonparametric regression model the experimenter generally chooses an appropriate function space to which $\mu$ is believed to belong. This choice is usually motivated by smoothness properties the regression function can be assumed to possess. The data is then utilized to determine an element of this function space that is representative of the unknown regression curve. In contrast, under a parametric model, the experimenter chooses one possible family of curves, from the collection of all curves, and inputs this choice into the inferential process. The information the data can supply concerning model development is restricted to what can be extracted from the data under this assumed parametric form. Consequently, nonparametric regression techniques rely more heavily on the data for information about $\mu$ than their parametric counterparts.

When appropriate, parametric models have some definite advantages. The corresponding inferential methods usually have nice efficiency properties. Also, the parameters generally have some physical meaning which makes them interpretable and of interest in their own right. Unfortunately, possibly because of these factors, parametric models are often used when there is little available information concerning the functional form of $\mu$. The experimenter assumes that $\mu$ has some ad hoc parametric representation with little or no knowledge concerning the accuracy of his or her assumption. If the assumed parametric model is grossly in error, the aforementioned advantages of the parametric approach will not obtain. Thus, there are few gains to be realized from using a poorly specified parametric form for $\mu$. In fact, the use of such a model can have exactly the opposite effect and is therefore a dangerous practice. (See Gasser et al., 1984 and Section 4.9 for an example.)

When there is little that is known about the regression function the bulk of the information about $\mu$ lies in the data rather than the person conducting the study or experiment. Thus, it seems reasonable to use inferential techniques which rely more heavily on the data in such instances. Consequently nonparametric regression techniques are ideally suited to problems of inference where the available knowledge about $\mu$ is of a limited nature.

Nonparametric regression techniques overcome the inherent difficulty with parametric techniques that the functional form of $\mu$ must be known. However, their use is not without its price. Generally, nonparametric estimators are less efficient than the parametric variety when the parametric model is valid.

For most parametric estimators the risk, or expected squared error of estimation, will decay to zero at a rate of $n^{-1}$. As we shall see, the corresponding rate for nonparametric estimators is usually $n^{-\delta}$ for some parameter $\delta \in (0,1)$ that depends on the smoothness of $\mu$. For example, if $\mu$ is twice differentiable, $n^{-4/5}$ is an often quoted rate. Thus, nonparametric regression techniques suffer a loss of efficiency when compared to parametric methods. However, this comparison is not entirely fair for if we knew the form of $\mu$ we would utilize a parametric approach to estimation in most cases. Nonparametric estimators become candidates for estimation of $\mu$ only when there is some question about an appropriate parametric form for $\mu$. In that case if an incorrect, ad hoc, parametric model is used, the $n^{-1}$ rate will no longer obtain (in fact, the estimator does not even converge to the true regression curve) and nonparametric regression techniques will tend to fare better than a parametric approach.

The result of a nonparametric regression analysis is a curve fitted to a set of data. Since this is produced without assuming a parametric form for $\mu$ there is some loss in the interpretability of estimators obtained in this fashion. Thus, for instance, there are no longer quantities such as estimated regression coefficients, that, e.g., would result from fitting a linear model, to be interpreted. However, the fitted curve itself is an estimate of the parameter $\mu$ and any functional of $\mu$ is also a parameter which can be estimated using an estimate of the regression curve. For example, if $\mu$ represents a growth curve then the point at which the velocity curve $\mu'$ attains its maximum may be of interest. This type of parameter can be estimated and interpreted from either a nonparametric or parametric viewpoint. Thus there are regression curve based parameters (i.e., functionals of $\mu$) about which inference can be made and interpretations can be drawn using either parametric or nonparametric techniques. In most cases where nonparametric regression is really appropriate it is parameters of this nature which are actually of interest anyway.

To summarize, parametric and nonparametric regression techniques represent two different approaches to the problem of regression analysis. Parametric methods require very specific, quantitative information from the experimenter about the form of $\mu$ that places restrictions on what the data can tell us about the regression function. Such techniques are most appropriate when theory, past experience and/or other sources are available that provide detailed knowledge about the process under study. In contrast, nonparametric regression techniques rely on the experimenter to supply only qualitative information about $\mu$ and let the data speak for itself concerning the actual form of the regression curve. These methods are best suited for inference in situations where there is little or no prior information available about the regression curve.

In conclusion it should be noted that even though parametric and nonparametric regression models represent distinctly different approaches to regression analysis, this does not mean that the use of one approach precludes the use of the other. Indeed, nonparametric regression techniques can be used to assess the validity of a proposed parametric model. Conversely, it may be that the form of a fitted regression curve obtained by nonparametric techniques will suggest an appropriate parametric model for use in future studies. Thus, nonparametric regression procedures may represent the final stage of data analysis or merely an exploratory or confirmatory step in the modeling process.

## 1.2 Nonparametric Regression

In this section we discuss the nonparametric regression problem from a general viewpoint and give an indication of the important issues which must be addressed for any particular solution. To begin, let us assume that model (1) is in effect and that $\mu \in W$, where W is some infinite dimensional collection of functions. The objective is inference about $\mu$ in the form of both point and interval estimation.

The specific choice of a point estimator for $\mu$ will depend on the choice made for W. Estimators which are suitable for particular function spaces are discussed in Chapters 3-7. For the present, however, we will leave W unspecified and attempt to draw some broad conclusions about the estimation problem.

First note that if only $\underline{\mu} = (\mu(t_1),\ldots,\mu(t_n))'$ is to be estimated, the corresponding response vector $\underline{y} = (y_1,\ldots,y_n)'$ gives one possible estimator. This estimators is unbiased, that is, $Ey_i = \mu(t_i)$, and has variance-covariance matrix $\sigma^2 I$. If observations are repeated at some values of t they would, of course, be grouped and their averages used to obtain more precise estimators. This does not alter the conclusions which follow so we will assume that the $t_j$ are distinct in subsequent discussions.

The vector of responses will not, in general, be a suitable estimator of $\underline{\mu}$. One reason for this is that $y_i$ has an unacceptably large squared error risk as an estimator of $\mu(t_i)$. We will always take risk to mean squared error risk. Thus, with this convention, the risk associated with $y_i$ is

$$E(\mu(t_i) - y_i)^2 = \sigma^2 \quad ,$$

so the quality of $y_i$ as an estimator does not improve as the sample size increases. We will see that one can construct estimators of $\mu(t_i)$ for which the risk will decay like $n^{-\delta}$ for some $\delta > 0$ when W has certain properties.

Another problem with the choice of $\underline{y}$ as an estimator is that it only provides an estimator of $\underline{\mu}$ and does not solve the problem of estimating the function $\mu$ at points other than the $t_i$. If the objective is to estimate the entire function $\mu$, it will no longer be possible to obtain an unbiased estimator since this would involve estimating more parameters (actually an infinite number of parameters) than there are data values.

If $\mu$ is believed to be smooth, then the observations at points $t_j$ near t should contain information about the value of $\mu$ at t. Thus, it should be possible to use something like a local average of the data near t to construct an estimator of $\mu(t)$. This approach should also result in an estimator of $\mu(t_j)$ with smaller variance and, if the bias that is introduced is not too large, smaller risk than $y_j$.

In this text we will focus exclusively on what are known as *linear estimators* of $\mu$. This type of estimator includes local averages as well as many other estimators of interest. A linear estimator of $\mu$ will have the form

$$\mu_\lambda(t) = \sum_{j=1}^{n} K(t,t_j;\lambda)y_j \quad , \tag{3}$$

where $K(\cdot,t_j;\lambda)$, $j = 1,..,n,$ is a collection of weight functions that depend on a parameter or, more generally, a set of parameters, $\lambda$. The reason (3) is called a linear estimator is because, given $\lambda$, it is a linear function of the responses $y_1,...,y_n$.

As an example of a linear estimator let $\lambda$ be an integral valued parameter and define $K(t,t_j;\lambda)$ to vanish unless $t_j$ is one of the $\lambda$ closest points to t among $t_1,...,t_n$. In this case $\mu_\lambda(t)$ is a type of local average of observations near t. There are many other possible weight functions. Some of these are discussed in the next section.

If the weights are all nonnegative and $\Sigma_{j=1}^n K(t,t_j;\lambda) = 1$, estimator (3) is actually a weighted average of the $y_j$. It is, in fact, a weighted least-squares estimator in the sense that it is the minimizer of the weighted squared error criterion

$$\sum_{j=1}^{n} K(t,t_j;\lambda)(y_j - \theta)^2 \tag{4}$$

with respect to $\theta$.

Estimators of the form (3) provide one class of estimators of $\mu$. Thus it is certainly possible to suggest potential estimators for the regression function. This leads us to a new issue. Given an estimator of $\mu$, what properties would we like the estimator to possess?

As previously noted, one cannot expect unbiasedness for a nonparametric regression estimator of $\mu$, if $\mu$ is just some arbitrary element of W. However, one would certainly want an estimator that will be nearly unbiased for large n over all the regression functions in W. . Related to this is the property of consistency. As n grows large any reasonable estimator of $\mu$ should converge to $\mu$ is some probabilistic sense.

Of particular interest for subsequent work is the performance of an estimator $\hat{\mu}$ as measured by the (average) risk

$$R(\hat{\mu}) = n^{-1} \sum_{j=1}^{n} E(\mu(t_j) - \hat{\mu}(t_j))^2 .$$

If $R(\hat{\mu})$ converges to zero as n becomes large this gives a form of squared-error consistency for $\hat{\mu}$. The rate at which a performance measure such as $R(\hat{\mu})$ decays to zero provides information about how effective $\hat{\mu}$ is as an estimator of $\mu$ and allows us to critique the performance of one estimator relative to another.

An estimator $\hat{\mu}$ will be said to be *asymptotically optimal* if, uniformly in $\mu \in W$, $R(\hat{\mu})$ decays like $n^{-\delta}$ for some $\delta > 0$ as n increases and no estimator can achieve a better rate of convergence. In the case of a parametric model one finds that $\delta = 1$ under standard regularity conditions. However, this is not true for nonparametric models and generally $\delta < 1$. Slower convergence rates are the price which is paid for making fewer assumptions about $\mu$ than in the parametric case. An important problem is the characterization of $\delta$ for certain important choices for W.

It should be noted that our definition of asymptotic optimality deals only with optimal rates of convergence. If $n^{-\delta}$ is the optimal rate, then we could consider all estimators with $R(\hat{\mu}) \sim Cn^{-\delta}$. One might then find an optimal estimator (or estimators) in this class by choosing the one (or ones) which make the asymptotic constant C as small as possible. While we will be primarily concerned only with rates of convergence in this text, the importance of the asymptotic constants associated with estimators obtaining optimal convergence rates should not be overlooked. In many cases of interest there will be a number of candidate estimators which are capable of obtaining the optimal convergence rates. Thus, useful performance comparisons among these estimators can be made by comparing their corresponding asymptotic constants to one another and, whenever possible, to an optimal constant.

To illustrate the types of convergence rates which might be expected we mention one important case where the best convergence rate for $R(\hat{\mu})$ is known. Suppose that W is the set of all functions $\mu$ on a finite interval [a,b] which, for some integer $m \geq 1$, have $m - 1$ absolutely continuous derivatives and an mth derivative that is square integrable and satisfies $\int_a^b \mu^{(m)}(t)^2 dt \leq \alpha$ for a specified $\alpha > 0$. Convergence rates for this choice of W have been studied by Speckman (1985) and Nussbaum (1985). Let L denote the set of all linear estimators of $\mu$. Speckman (1985) has shown that if the $t_j$ are generated by a positive, bounded design density, p, through the relation

$$\int_a^{t_j} p(t)dt = (2j - 1)/2n ,$$

then

$$\min_{\hat{\mu}\varepsilon L} \max_{\mu\varepsilon W} R(\hat{\mu}) = C(\alpha,m)n^{-2m/(2m+1)} , \tag{5a}$$

where

$$C(\alpha,m) = [\alpha(2m + 1)]^{1/(2m+1)} \left\{\sigma^2 m \int_a^b p(t)^{1/2m} dt/\pi(m + 1)\right\}^{2m/2m+1} . \tag{5b}$$

Nussbaum (1985) shows that, if the $t_j$ are uniformly spaced on [a,b], (5) still holds if we minimize over all estimators, rather than just linear estimators. Thus, for this particular W, the best uniform rate of convergence will be $n^{-2m/(2m+1)}$ and the constant $C(\alpha,m)$ plays the role of a Fisher information type bound. Similar results for related settings can be found in Farrell (1972), Stone (1980, 1982) and Ibragimov and Hasminski (1981).

By studying consistency and rates of convergence this tells us how an estimator behaves as a point estimator of the regression function. However, it is also of interest to construct interval estimates for $\mu$. Thus, we also need to know something about the probability distribution of an estimator. This problem is simplified if the estimator is asymptotically normal since standard tables can then be used and complicated distributional issues can be avoided. While not all estimators of $\mu$ can be expected to be asymptotically normal, the relation between estimators of the form (3) and averages leads one to anticipate that a central limit effect might hold for estimators of this type.

To this point discussion has centered on a general nonparametric estimator of $\mu$. In Chapters 3-7 we will focus in on particular estimators arising from special methods of choosing the weights in (3). Given a choice of the weight function, our attention turns to what properties are possessed by the resulting estimator. Apart from any other properties it may be possible to establish, the above discussion suggests that attention must also be directed toward questions of consistency, rates of convergence and asymptotic normality of the estimator.

## 1.3 Scope

In this final section we give a brief preview of the types of estimators to be studied in subsequent chapters. The discussion serves two purposes. First, it introduces the key estimators to be studied in the text and thereby provides other examples of linear estimators. Secondly, it provides concrete illustrations of the need for objective methods of assessing an estimator's performance from the data.

Our study of nonparametric estimators begins in Chapter 3 where it is assumed that the regression function can be represented as a linear combination of known functions $\{x_i\}_{i=1}^{\infty}$. Thus, there are coefficients $\{\beta_i\}_{i=1}^{\infty}$ such that $\mu = \Sigma_{i=1}^{\infty}\beta_i x_i$. This form for $\mu$ is reminiscent of the linear model (2) except that now there are infinitely many $\beta_i$. If, for example, the coefficient series is square summable, then the coefficients must eventually decay to zero. It therefore seems reasonable to assume that there is an integer $\lambda$ such that $\mu \doteq \Sigma_{i=1}^{\lambda}\beta_i x_i$. If this is so, $\mu$ might be viewed as following an approximate linear model and techniques for estimation from linear models such as those in the Appendix might be employed for inference about $\mu$. The resulting estimators for $\mu$ are termed series estimators.

One objective of Chapter 3 is to study the extent to which parallels of results from linear regression analysis can be developed for series estimators. There is, however, another problem which must be dealt with: the selection of $\lambda$. Since $\mu$ and, hence, the $\beta_j$ are unknown there is no way to know precisely what will constitute a good value for $\lambda$. Instead, one must generally rely on the data to aid in such determinations.

The series estimators studied in Chapter 3 are linear estimators. The parameter $\lambda$ for the weights in (3) is the number of $x_j$ used to construct an estimator. Unfortunately, the weights for series estimators generally involve the $t_j$ in a complicated fashion. A simpler method of choosing the weights in (3) is to use, for example, a symmetric, unimodal probability density function K on [-1,1] and, for some real number $\lambda > 0$, estimate $\mu$ by

$$\mu_\lambda(t) = \sum_{j=1}^{n} K(\lambda^{-1}(t - t_j))y_j / \sum_{j=1}^{n} K(\lambda^{-1}(t - t_j)). \tag{6}$$

This estimator is one illustration of what are called kernel estimators. These types of estimators are the subject of Chapter 4. Estimator (6) is a weighted average of the $y_j$. Since K is unimodal the heaviest weights will generally be assigned to observations near t, depending on how $\lambda$ is chosen. The parameter $\lambda$ is called the bandwidth and governs the amount of averaging which takes place. In order to use a kernel estimator one must first select an appropriate value for the bandwidth.

Chapters 5-6 focus on inference about $\mu$ based on smoothing spline estimators. These estimators can be viewed as a generalization of polynomial regression estimators. If, for example, we assume that $\mu$ belongs to the set of all continuously differentiable functions on [0,1] with square integrable second derivatives, then an appropriate smoothing spline estimator of $\mu$ would be the function $\mu_\lambda$ which minimizes the penalized least-squares criterion

$$n^{-1}\sum_{j=1}^{n}(y_j - f(t_j))^2 + \lambda\int_0^1 f''(t)^2 dt \ , \quad \lambda \geq 0 \ , \tag{7}$$

over all functions f in this set. The resulting estimator provides a generalization of the linear regression estimator for $\mu$. The parameter $\lambda$ in (7) is also the parameter for the weights in (3). In this case $\lambda$ determines the influence of the roughness penalty $\int_0^1 f''(t)^2 dt$ and thereby governs the smoothness of the estimator. For this reason it is usually termed the smoothing parameter. Estimation of $\mu$ by smoothing splines will require the selection of an appropriate value for $\lambda$.

A variety of other possible nonparametric estimators of $\mu$ are discussed in Chapter 7. They stem from varying motivations. However, like the other estimators discussed above, there is a common thread which connects them. Each estimator depends on a parameter or parameters which must be selected in order to conduct inference. Thus, we are led to wonder what constitutes an optimal choice for these parameters and if there are techniques which employ the data to effectively guide their selection in practice. These subjects are the essence of the next chapter which provides a general discussion of how to assess the quality of an estimator and presents techniques that utilize the data to select an optimal estimator from a class of estimators. However, the subject matter in Chapter 2 is of much broader applicability than just to nonparametric regression. Consequently, we will study this topic from a rather general viewpoint with applications to the estimators discussed above being detailed in subsequent chapters.

# 2
# What Is a Good Estimator?

## 2.1 Performance Criteria

In subsequent chapters we will study several families of estimators for the regression function. Each family is indexed by a parameter (often called a smoothing parameter). The selection of an estimator from a given family requires the choice of a value for the index parameter. While this can be done subjectively, an objective choice will usually be preferred. The present chapter provides an overview of several techniques which have been found useful for smoothing parameter selection. Their usefulness is not restricted to nonparametric regression problems and they can, in fact, be used to solve a wide variety of estimation problems. Some of these will be studied in the examples and exercises.

To provide a general development that is also suitable for subsequent work let us assume that a vector of observations, $\underline{y} = (y_1,\ldots,y_n)'$, has been obtained following the model

$$\underline{y} = \underline{\mu} + \underline{\varepsilon}, \tag{1}$$

where $\underline{\mu} = (\mu_1,\ldots,\mu_n)'$ is an unknown mean vector and $\underline{\varepsilon} = (\varepsilon_1,\ldots,\varepsilon_n)'$ is a vector of zero mean, uncorrelated random errors with common variance $\sigma^2$. Suppose now that we are considering a class of estimators for $\underline{\mu}$, $C(\Lambda) = \{\underline{\mu}_\lambda : \lambda \in \Lambda\}$, with $\Lambda$ representing some index set. As seen below, the index parameter $\lambda$ may be scalar, vector or even set valued. To simplify things we will also assume that the elements of $C(\Lambda)$ are linear estimators. This means that for each $\lambda$ there is a $n \times n$ matrix $H(\lambda)$ such that

$$\underline{\mu}_\lambda = H(\lambda)\underline{y}.$$

Unless stated otherwise, $H(\lambda)$ will be assumed to be symmetric and positive semi-definite.

The problem to be considered is selection of a best estimator of $\underline{\mu}$ from among the elements of $\{\underline{\mu}_\lambda: \lambda \in \Lambda\}$. The word "best" in our problem statement has been left open for interpretation and, until it is given a precise meaning, there is clearly no way to formulate a solution. Thus, to effect an objective choice for an estimator we must quantify what is meant by "best". Of course, one person's idea of "best" may not coincide with another's so there is no universally accepted measure of an estimator's performance. Nonetheless, there are certain criteria which are widely accepted and used. We will focus on three of these.

Define the *loss* in estimating $\underline{\mu}$ by

$$L(\lambda) = n^{-1} \sum_{j=1}^{n} (\mu_j - \mu_{\lambda j})^2 , \tag{2}$$

where $\mu_{\lambda j}$ is the jth element of $\underline{\mu}_\lambda$. $L(\lambda)$ is just the squared Euclidean distance between $\underline{\mu}$ and $\underline{\mu}_\lambda$ (apart from the factor $n^{-1}$). It therefore represents a natural measure of the closeness of $\underline{\mu}$ and $\underline{\mu}_\lambda$. Associated with the loss is its average or expected value

$$R(\lambda) = EL(\lambda) , \tag{3}$$

which is called the *risk*. Both $L(\lambda)$ and $R(\lambda)$ provide assessments of an estimator's performance. Smaller values of the criteria are indicative of better estimators.

Another measure that is closely related to (3) is the *prediction risk* (sometimes called the mean squared error of prediction). The idea behind the prediction risk is as follows. Suppose we are planning to observe n new observations $\underline{y}^* = \underline{\mu} + \underline{\varepsilon}^*$, with $\underline{\mu}$ the same vector of means as in (1) and $\underline{\varepsilon}^*$ a vector of random errors with zero means that are uncorrelated with each other and with $\underline{\varepsilon}$ and have common variance $\sigma^2$. To evaluate the performance of $\underline{\mu}_\lambda$ as a predictor of future observations we could then use

$$P(\lambda) = E\, n^{-1} \sum_{j=1}^{n} (y_j^* - \mu_{\lambda j})^2 \quad . \tag{4}$$

The prediction risk in (4) is closely related to R(λ). In fact

$$P(\lambda) = \sigma^2 + R(\lambda) \tag{5}$$

(Exercise 1). Thus, an estimator that minimizes the risk, R(λ), will also minimize P(λ) and conversely.

Ideally one would select an estimator by choosing λ to minimize the risk (equivalently P(λ)) or the loss. The resulting estimator would provide a *best* estimator of $\underline{\mu}$. Unfortunately, neither L(λ), R(λ) nor P(λ) can actually be computed without knowing $\underline{\mu}$. Thus, in practice, the criterion of interest must be estimated from the data and its estimate then minimized with respect to λ to obtain an estimate of the best estimator for $\underline{\mu}$. Of the three criteria, R(λ) and P(λ) are the most mathematically tractable since, unlike L(λ), they are not stochastic. Consequently, most of the discussion which follows will concentrate on estimating the risk and the prediction risk.

The loss, risk and prediction risk provide three possible measures of performance for an estimator of $\underline{\mu}$. There are, of course, many others. For example, instead of squared errors we might use absolute values. However, squared errors are generally easier to deal with from a mathematical viewpoint and, given no compelling reason to use absolute values, will often be preferred for this reason.

If only one of the means, $\mu_j$, is of interest, then another possible measure is, e.g.,

$$E(\mu_j - \mu_{\lambda j})^2 \ .$$

In contrast to this latter measure, R(λ) and P(λ) provide global measures of the performance of $\underline{\mu}_\lambda$, that is, they give an indication of how $\underline{\mu}_\lambda$ behaves as an estimator over all the elements of $\underline{\mu}$.

In the context of nonparametric regression the use of a global performance measure seems desirable. Recall that in that setting we have (c.f. model (1.1))

$$y_i = \mu(t_i) + \varepsilon_i \ , \quad i = 1,\ldots,n,$$

with $\mu$ some unknown regression function. This corresponds to model (1) with $\mu_i = \mu(t_i)$. Thus, $R(\lambda)$ or $P(\lambda)$ provides a description of the efficacy of an estimator of the regression function over the entire interval of observations. There are other closely related measures of performance with similar properties. If $\mu_\lambda(\cdot)$ is an estimator of the regression function $\mu(\cdot)$, then we might consider measures such as the *integrated loss*

$$IL(\lambda) = \int_a^b (\mu(t) - \mu_\lambda(t))^2 dt,$$

or *integrated risk*

$$IR(\lambda) = \int_a^b E(\mu(t) - \mu_\lambda(t))^2 dt \ .$$

If the $t_j$ are uniformly spaced points over $[a,b]$, $L(\lambda)$ and $R(\lambda)$ are just quadrature formulae for $IL(\lambda)$ and $IR(\lambda)$, respectively. More generally, if the $t_j$ are unequally spaced we might wish to modify $IL(\lambda)$ and $IR(\lambda)$ by integrating against a weight function to reflect the spacing of the design.

At this point some examples of estimation problems which fit into the framework of model (1) will be helpful in clarifying and solidifying the concepts and notation. Two examples are discussed which, while not from nonparametric regression, may be more familiar to the reader.

**Example 1: Model Selection.** Suppose that $y_1,\ldots,y_n$ are obtained as responses to values of p input variables $X_1,\ldots,X_p$. Thus, $y_j$ is the output obtained from input vector

$$\underline{x}_j = (x_{j1},\ldots,x_{jp})' \ , \quad j = 1,\ldots,n.$$

Our objective is to develop a scheme for prediction of future responses to various input configurations.

Assume that the response variable Y is linearly related to the input variables so that $\mu_j = \underline{x}_j'\underline{\beta}$ for some coefficient vector $\underline{\beta} = (\underline{\beta}_1, \ldots, \underline{\beta}_p)'$. To estimate $\underline{\beta}$ we can use the least-squares estimator $\hat{\underline{\beta}}$ obtained by minimizing

$$n^{-1} \sum_{j=1}^{n} (y_j - \underline{x}_j'\underline{\beta})^2$$

with respect to $\underline{\beta}$ (c.f. the Appendix). If $X' = [\underline{x}_1, \ldots, \underline{x}_n]$ has rank p, $\hat{\underline{\beta}}$ is given explicitly by

$$\hat{\underline{\beta}} = (X'X)^{-1}X'\underline{y}.$$

Our estimate of $\underline{\mu}$ is then

$$\hat{\underline{\mu}} = X\hat{\underline{\beta}} = X(X'X)^{-1}X'\underline{y}. \tag{6}$$

By using $\hat{\underline{\mu}}$ one obtains an unbiased estimator of $\underline{\mu}$. However, as seen in the next section, the use of an unbiased estimator need not lead to an estimator with minimal prediction risk. Thus, define our set $\Lambda$ to consist of all possible subsets of the indices of the variables $X_1, \ldots, X_p$. In general, there will be $2^p$ such subsets. For example, when p = 2, $\Lambda$ consists of the sets {1}, {2}, {1,2} and the empty set, $\phi$.

Corresponding to each $\lambda \in \Lambda$ (except $\lambda = \phi$) is a design matrix

$$X_\lambda = \{x_{jr}\}_{\substack{j=1,n \\ r\in\lambda}} \quad . \tag{7}$$

For instance, if p = 2 and $\lambda = \{2\}$, $X_\lambda = (x_{12}, \ldots, x_{n2})'$ and, if $\lambda = \{1,2\}$, $X_\lambda$ and X, the design matrix for the full set of predictors, coincide.

To estimate $\underline{\mu}$ we can now choose from the class of estimators $C(\Lambda) = \{\underline{\mu}_\lambda : \lambda \in \Lambda\}$ with

$$\underline{\mu}_\lambda = X_\lambda \underline{\beta}_\lambda = X_\lambda (X_\lambda' X_\lambda)^{-1} X_\lambda' \underline{y} \quad , \qquad \lambda \in \Lambda, \tag{8}$$

and the convention that $\underline{\mu}_\lambda = \underline{0}$ when $\lambda = \phi$. For a given $\lambda$, $\underline{\mu}_\lambda$ is recognized as the least-squares estimator of $\underline{\mu}$ corresponding to the predictor variable subset whose indices are in $\lambda$. Note that $H(\lambda)$ in this setting is, from (8),

$$H(\lambda) = X_\lambda (X_\lambda' X_\lambda)^{-1} X_\lambda' \quad , \tag{9}$$

which is easily seen to be symmetric and nonnegative definite.

As an example of a situation where variable selection methods are of interest we consider a study of housing rent data for 455 households in Mariposa County, Arizona, that was reported in Gunst and Mason (1980). The data consists of a response variable, monthly rent for an unfurnished dwelling unit, and 17 additional predictor variables which are believed to be related to housing rent. One aspect of the study was concerned with developing a predictive equation for housing rents. Due to the costs and difficulty of maintaining information on the predictor variables it was also important to determine if some subset of the predictor variables could be substituted for the entire set of predictors with comparable (or better) results from a prediction standpoint.

To make the presentation somewhat more manageable we will concentrate here on predicting housing rent, Y, using only four of the predictor variables in the housing rent study: namely,

$X_2$ - a measure of school quality and lack of transportation for the census tract where the dwelling unit is located,

$X_3$ - a dichotomous variable which takes on the values 1 or 0 according as whether or not the length of residence is 5 to 10 years,

$X_4$ - the logarithm of the total floor area of the dwelling unit,

and

$X_5$ - a measure of average surface area and structural quality, 0 being worst and 3 the best.

There is also a fifth predictor, $X_1$, which is defined to be unity for all observations. Some summary statistics for Y and $X_2,\ldots,X_5$ are presented in Table 2.1.

**Table 2.1 Summary Measures for the Housing Rent Data**

| Variable | Mean | Standard Deviation |
|---|---|---|
| Y | 129.8718 | 44.3970 |
| $X_2$ | .1171 | 1.0906 |
| $X_3$ | .1077 | .3103 |
| $X_4$ | 6.2794 | .3539 |
| $X_5$ | 2.2973 | .6364 |

**Source: Gunst and Mason (1980).**

Let us now assume that housing rents follow a linear model in $X_1,\ldots,X_5$, i.e., that Y follows the model

$$Y = \beta_1 + \beta_2X_2 + \beta_3X_3 + \beta_4X_4 + \beta_5X_5 + \varepsilon$$

$$= \sum_{j=1}^{5} \beta_jX_j + \varepsilon ,$$

with $\varepsilon$ denoting a random error term. The question to be addressed is "Can an optimal prediction equation for housing rents be found involving $X_1,\ldots,X_5$?" To answer this question we must first decided on what is to be meant by optimal. If a particular type of dwelling unit is to be studied with known or specified values $x_2^0,\ldots,x_5^0$ for $X_2,\ldots,X_5$, then a particular mean value $\mu_0 = \beta_1 + \Sigma_{j=2}^{5}\beta_jx_j^0$ is of interest. In that case, one might define the least-squares estimator $\mu_{\lambda 0}$, based on some subset of $X_1,\ldots,X_5$, as being optimal if it minimized $(\mu_0 - \mu_{\lambda 0})^2$, or perhaps, $E(\mu_0 - \mu_{\lambda 0})^2$ over all subsets of interest, $\lambda$, of the predictor variables. It seems unlikely in this setting that a particular type of housing unit will be the focal point of interest. Instead, one would anticipate that a least-squares predictive equation which performed well over all types of dwelling units in the study would be preferred. Thus, it is reasonable to define an optimal predictive equation for this problem as one which corresponds to the minimizer of the loss (2) or, possibly, the predictive risk (5).

Given that the selection of a variable subset to minimize (2) or (5) is the objective, one must now ask how this can be accomplished. If the true Y means were available for each of the sample points, it would be a simple matter to compute $L(\lambda)$ and find the optimal set of predictors. However, since only sample information is available, the true y means are unknown and hence neither the loss nor risk are computable. Thus the choice of a variable subset must be guided by the data, possibly through sample estimates of the predictive risk in (5). Consequently, it becomes necessary to postpone further consideration of the optimal subset question for this data until the next section where estimators of $P(\lambda)$ are developed.

**Example 2: Ridge Regression.** Consider now a related problem where we again have predictor variables $X_1,\ldots,X_p$ which are linearly related to Y and are to be used in estimating $\underline{\mu}$. However, in some cases the X variables are so closely related that the X matrix is nearly rank deficient (i.e., has rank less than p). This has some unpleasant consequences for estimator variances and can cause difficulties when attempting to solve the normal equations $X'X\underline{\beta} = X'\underline{y}$ from which (6) derives. This and other factors led Hoerl and Kennard (1970a,b) to propose what is called a *ridge regression estimator* for $\underline{\beta}$ which is defined by

$$\underline{\beta}_\lambda = (X'X + \lambda I)^{-1}X'\underline{y} \quad , \qquad \lambda \geq 0 \ .$$

The ridge estimator of $\underline{\mu}$ is then obtained by the relation

$$\underline{\mu}_\lambda = X\underline{\beta}_\lambda = X(X'X + \lambda I)^{-1}X'\underline{y} \quad . \tag{10}$$

Ridge estimators have a variety of optimality properties, some of which are explored in the exercises.

As seen from (10), ridge estimators involve a scalar parameter $\lambda$ that is often called the *ridge parameter*. Thus, to choose an estimator, a value of $\lambda$ must first be selected. Note that the choice $\lambda = 0$ gives $\underline{\mu}_0 = \hat{\underline{\mu}}$.

We now see that ridge estimators fit into the framework set out above. More precisely, ridge regression corresponds to the case $\Lambda = \{\lambda: 0 \leq \lambda < \infty\}$ and $C(\Lambda) = \{\underline{\mu}_\lambda: \lambda \in \Lambda\}$ with $\underline{\mu}_\lambda$ defined by (10) and $H(\lambda) = X(X'X + \lambda I)^{-1}X'$.

There are several generalizations of ridge estimators. For example, we might consider an estimator of the form

$$\underline{\mu}_\lambda = X(X'X + V(\lambda))^{-1}X'\underline{y}$$

where, eg., $V(\lambda) = \mathrm{diag}(\lambda_1,\ldots,\lambda_p)$, $\lambda_j > 0$. In this case $\Lambda$ can be viewed as the set of all p-vectors with positive coordinates. More generally still, $V(\lambda)$ could be some positive semi-definite matrix. Estimators of this type are called *generalized ridge estimators*.

## 2.2 Estimating P(λ) and R(λ)

In the previous section we considered three measures of the performance of an estimator of $\underline{\mu}$: the loss, the risk, and the prediction risk. In this section we concentrate our attention on obtaining estimators of the latter two measures.

As before, let $C(\Lambda) = \{\underline{\mu}_\lambda: \lambda \in \Lambda\}$ be a class of linear estimators for $\underline{\mu}$ with $\underline{\mu}_\lambda = H(\lambda)\underline{y}$ and consider estimation of the prediction risk defined in (4). The naive estimator of $E(y_j^* - \mu_{\lambda j})^2$ in this case is the square of the residual (or error) $y_j - \mu_{\lambda j}$. Thus, one might estimate $P(\lambda)$ by the average of the squared errors

$$MSE(\lambda) = n^{-1} \sum_{j=1}^{n} (y_j - \mu_{\lambda j})^2$$

(the abbreviation MSE stands for mean squared error).

A few words of caution about notation and terminology are probably in order at this point. First, we should note that while $y_j - \mu_{\lambda j}$ represents the error in predicting $y_j$ by $\mu_{\lambda j}$, one should not confuse this usage of the word "error" with our previous usage associated with the $\varepsilon_j$ in (1). The difference will always be clear from the context. Secondly, the use of MSE($\lambda$) to indicate the average of the squared residuals is not entirely standard. In some circles MSE is used to signify what we have called the risk and denoted by R. Our choice of notation is motivated by classical analysis of variance where MSE denotes the error sum of squares divided by degrees of freedom. However, to simplify matters, a divisor of n is always used since it is often unclear how to parallel the concept of degrees of freedom in many of the problems we consider.

The question that must now be asked is "How does MSE($\lambda$) perform as an estimator of P($\lambda$)?" An answer can be obtained by computing the expected value of MSE($\lambda$) and comparing it to P($\lambda$).

To evaluate EMSE($\lambda$) note that

$$\begin{aligned}\mathrm{MSE}(\lambda) &= n^{-1}(\underline{y} - \underline{\mu}_\lambda)'(\underline{y} - \underline{\mu}_\lambda)\\ &= n^{-1}\underline{y}'(I - H(\lambda))^2\underline{y}.\end{aligned}$$

Thus, an application of our Lemma A.1 from the Appendix with $\underline{z} = \underline{y}$, $\underline{\theta} = \underline{\mu}$ and $\Sigma = \sigma^2 I$ gives

$$\begin{aligned}\mathrm{EMSE}(\lambda) &= n^{-1}\underline{\mu}'(I - H(\lambda))^2\underline{\mu} + n^{-1}\sigma^2\mathrm{tr}[(I - H(\lambda))^2]\\ &= n^{-1}\underline{\mu}'(I - H(\lambda))^2\underline{\mu} + \sigma^2 + n^{-1}\sigma^2\mathrm{tr}[H(\lambda)^2]\\ &\quad - 2n^{-1}\sigma^2\mathrm{tr}[H(\lambda)] \ . \qquad (11)\end{aligned}$$

In contrast,

$$P(\lambda) = \sigma^2 + R(\lambda) = \sigma^2 + n^{-1}\sum_{j=1}^{n} E(\mu_j - \mu_{\lambda j})^2$$

$$= \sigma^2 + n^{-1}E(\underline{\mu} - \underline{\mu}_\lambda)'(\underline{\mu} - \underline{\mu}_\lambda)$$

$$= \sigma^2 + n^{-1}\ \underline{\mu}'(I - H(\lambda))^2\underline{\mu} + n^{-1}\sigma^2 tr[H(\lambda)^2] \ . \tag{12}$$

The last equality is obtained by noting that $\underline{\mu}_\lambda = H(\lambda)\underline{y} = H(\lambda)\underline{\mu} + H(\lambda)\underline{\varepsilon}$, and by using Lemma A.1 with $\underline{z} = \underline{\varepsilon}$, $\underline{\theta} = \underline{0}$ and $\Sigma = \sigma^2 I$.

Comparing (11) and (12) we see that MSE($\lambda$) is a biased estimator of P($\lambda$) with bias equal to $-2n^{-1}\sigma^2 tr[H(\lambda)]$. Assuming that $\sigma^2$ is known, an unbiased estimator of P($\lambda$) is therefore given by

$$\hat{P}(\lambda) = MSE(\lambda) + 2n^{-1}\sigma^2 tr[H(\lambda)] \ . \tag{13}$$

Similarly, an unbiased estimator of the risk when $\sigma^2$ is known is

$$\hat{R}(\lambda) = \hat{P}(\lambda) - \sigma^2 \ . \tag{14}$$

The estimator $\hat{R}(\lambda)$ is closely related to a general criterion for selecting $\lambda$ proposed by Mallows (1973). His criterion, often termed $C_L$, is

$$M(\lambda) = \frac{n}{\sigma^2} MSE(\lambda) - n + 2tr[H(\lambda)] \ .$$

Note that $M(\lambda) = n\hat{R}(\lambda)/\sigma^2$ and, hence, $\hat{R}(\lambda)$ and M($\lambda$) will be minimized at the same value of $\lambda$. Thus $\hat{R}(\lambda)$ and M($\lambda$) are interchangable measures for the purpose of selecting $\lambda$. The reason for Mallows' use of M($\lambda$), rather than $\hat{R}(\lambda)$, apparently stems from graphical considerations pertinent to the variable selection problem where it was first utilized.

Let us pause for a moment to examine the risk in more detail. Note from (12) that

$$R(\lambda) = n^{-1}\underline{\mu}'(I - H(\lambda))^2\underline{\mu} + n^{-1}\sigma^2 tr[H(\lambda)^2]$$

$$= n^{-1}\sum_{j=1}^{n} (\mu_j - E\mu_{\lambda j})^2 + n^{-1}\sum_{j=1}^{n} Var(\mu_{\lambda j}) \ ,$$

where $Var(\mu_{\lambda j})$ is the variance $E(\mu_{\lambda j} - E\mu_{\lambda j})^2$. This shows that the risk breaks into two components: namely,

$$b^2(\lambda) = n^{-1} \sum_{j=1}^{n} (\mu_j - E\mu_{\lambda j})^2$$

and

$$v(\lambda) = n^{-1} \sum_{j=1}^{n} \operatorname{Var}(\mu_{\lambda j}) \; .$$

The first component, $b^2(\lambda)$, is a measure of the estimator's bias, while the second is an indication of the estimator's variance. If the elements of $C(\Lambda)$ are all unbiased for $\underline{\mu}$, minimization of $R(\lambda)$ (or $P(\lambda)$) is tantamount to selecting an estimator which minimizes $v(\lambda)$. Generally, the estimators in $C(\Lambda)$ will not all be unbiased, however, and minimization of $R(\lambda)$ will usually entail a balancing of the (squared) bias component and the variance component. Note that even when $C(\Lambda)$ contains an unbiased estimator it may not represent the optimal estimator (Exercise 4). Thus making $b^2(\lambda)$ vanish, when that is possible, is not always the best strategy and it is possible for a biased estimator to have a smaller risk than an unbiased one (c.f. Exercise 4).

If $\sigma^2$ is known, $\hat{P}(\lambda)$ and $\hat{R}(\lambda)$ can be used directly to evaluate the performance of $\underline{\mu}_\lambda$. However, this will generally not be the case and, as a result, $\sigma^2$ will usually require estimation in (13) and (14).

In some cases the dependence of (13) and (14) on $\sigma^2$ is not a serious problem. For example, in the model selection problem there are many estimators of $\sigma^2$. If $\#(\lambda)$ denotes the number of elements in $\lambda$, then $n\text{MSE}(\lambda)/(n - \#(\lambda))$ provides an estimator of $\sigma^2$ for each $\lambda \in \Lambda$. A common approach, in this setting, is to use the estimator of $\sigma^2$ corresponding to $\lambda = \{1,\ldots,p\}$, i.e., the estimator obtained when all variables are used in estimating $\underline{\mu}$.

There is, of course, one case in which no estimator of $\sigma^2$ is needed for estimation of $P(\lambda)$, even when $\sigma^2$ is unknown. This occurs when $\text{tr}[H(\lambda)] = 0$ for all $\lambda \in \Lambda$. Estimators of this type are called *nil-trace estimators*. They will be discussed in more detail in the next section.

Most estimators do not have the nil-trace property, however, and there are many situations where there are no obvious estimators of $\sigma^2$. This is particularly true of the nonparametric regression problems we will study which do not have as much structure as variable selection and ridge regression problems. Thus $\hat{P}(\lambda)$ (and $\hat{R}(\lambda)$) is not an entirely satisfactory estimator. Consequently, we now consider two estimators of the prediction risk that do not require estimation of $\sigma^2$.

One alternative estimator of $P(\lambda)$ that does not require estimation of $\sigma^2$ can be motivated by further examination of $MSE(\lambda)$. In using $MSE(\lambda)$ to estimate $P(\lambda)$ we used the $y_j - \mu_{\lambda j}$ as estimates of future prediction errors. Since $\mu_{\lambda j}$ was constructed using the data $y_1, \ldots, y_n$ we would anticipate that $\mu_{\lambda j}$ would do better in predicting $y_j$ than some future response $y_j^*$. As a result, we would expect $MSE(\lambda)$ to be smaller on the average than $P(\lambda)$. This is, in fact, the case. As seen from (11) and (12), $MSE(\lambda)$ tends to underestimate $P(\lambda)$ by a factor of $2\sigma^2 tr[H(\lambda)]/n$.

An intuitive approach to correcting $MSE(\lambda)$ for underestimation of $P(\lambda)$ would involve replacement of $(y_j - \mu_{\lambda j})^2$ by some quantity which gives a better estimate of $E(y_j^* - \mu_{\lambda j})^2$. To give some insight into what follows, consider what could be done if we had multiple observations for each mean. As a simple example, suppose we had observed

$$y_{kj} = \mu_j + \varepsilon_{kj}, \quad k = 1,2, \quad j = 1,\ldots,n.$$

In words, suppose that instead of one observation for each mean we had two. Then, an estimate of $\underline{\mu}$ could be obtained using only $\underline{y} = (y_{11}, \ldots, y_{1n})'$ and $E(y_j^* - \mu_{\lambda j})^2$ could be estimated using $(y_{2j} - \mu_{\lambda j})^2$. Since $y_{2j}$ was not used in computing $\mu_{\lambda j}$ we are no longer assessing the prediction properties of the estimator by the data used for its construction. Consequently, $n^{-1}\Sigma_{j=1}^{n}(y_{2j} - \mu_{\lambda j})^2$, with $\mu_{\lambda j}$ based only on the $y_{1r}$, $r=1,\ldots,n$, would be expected to give a better estimator of $P(\lambda)$ than $MSE(\lambda)$. This is, in fact, the case and it can be shown (Exercise 15) that this particular estimator is unbiased for $P(\lambda)$.

Of course, we will not generally have multiple observations for each $\mu_j$. Also, when we do, it will usually be preferable to include them in the estimation process rather than holding them back for validation of the estimator. However, the idea of using a subsample of the data for estimating the prediction error suggests another approach that is applicable in certain situations.

In some cases it is possible to compute a parallel of, or approximation to, the estimator $\mu_{\lambda j}$ that does not involve the jth observation. In situations such as this the data can be effectively broken into n subsamples of size n - 1 and the observation held out of each subsample used for estimation of the prediction error. Before providing a general formulation it will be helpful to consider an example where this type of development is actually possible.

For the variable selection problem it is possible to delete the point $(\underline{x}_j', y_j)$ from consideration and, for any given variable subset, recompute the coefficient estimates. More precisely, for a variable subset with indices in $\lambda$, let $X_{\lambda(j)}$ denote the matrix with elements

$$x_{kr}; \quad k = 1,\ldots,j-1,j+1,\ldots,n, \qquad r \in \lambda,$$

and let $\underline{y}_{(j)} = (y_1,\ldots,y_{j-1},y_{j+1},\ldots,y_n)'$. Then

$$\underline{\beta}_{\lambda(j)} = (X_{\lambda(j)}' X_{\lambda(j)})^{-1} X_{\lambda(j)}' \underline{y}_{(j)}$$

is the least-squares estimate of the coefficients for the variables $X_r$, $r \in \lambda$, computed without the use of $(\underline{x}_j', y_j)$. An estimate of $\mu_j$ is provided by

$$\mu_{\lambda(j)} = \underline{x}_{\lambda j}' \underline{\beta}_{\lambda(j)}$$

with $\underline{x}_{\lambda j}$ the vector of the values for the X variables with indices in $\lambda$ that corresponds to the jth response. The estimator $\mu_{\lambda(j)}$ is directly related to $\mu_{\lambda j}$ in the sense that both are least-squares estimates of $\mu_j$ involving the predictors indexed by $\lambda$. The only difference is that one involves $(\underline{x}_j', y_j)$ while the other does not. In fact, the connection is stronger than this since it can be shown that

$$\mu_{\lambda(j)} = \mu_{\lambda j} - h_{jj}(\lambda)(y_j - \mu_{\lambda j})/(1 - h_{jj}(\lambda)) \ , \tag{15}$$

where $h_{jj}(\lambda)$ is the jth diagonal element of $H(\lambda)$ (see, e.g., Gunst and Mason 1980, Chapter 7 and Exercise 5). Similar conclusions hold for ridge regression (Exercise 5).

Returning to the general case, assume that $\mu_{\lambda(j)}$ represents a parallel of $\mu_{\lambda j}$ that is computed without the jth observation; estimator (15) is one example. Then, the estimator $\mu_{\lambda(j)}$ is constructed from a subsample of size n - 1 taken from the original data. If the jth observation is treated like an additional observation or holdout sample, our previous discussion would suggest using $(y_j - \mu_{\lambda(j)})^2$ to estimate $E(y_j^* - \mu_{\lambda j})^2$ and estimating $P(\lambda)$ by

$$CV(\lambda) = n^{-1} \sum_{j=1}^{n} (y_j - \mu_{\lambda(j)})^2 \ . \tag{16}$$

$CV(\lambda)$ is called the *cross-validation mean squared error* and selection of $\lambda$ through minimization of (16) is called *cross validation (CV)*. The quantity $nCV(\lambda)$ is sometimes called the prediction sum of squares or PRESS (Allen 1974).

Cross-validation is a method of reusing the data in a fashion which simulates what would be done if repeated observations were available. However, since there is actually only one observation for each $\mu_j$ it is no surprise that $CV(\lambda)$ is generally biased for $P(\lambda)$ (c.f. Exercise 16). (It is however unbiased for a prediction risk that is related to $P(\lambda)$; See Section 3.) Stone (1974) and Geisser (1975) have considered more elaborate cross-validation schemes than the ones discussed here. An overview of various applications for CV is provided by Titterington (1985). For a comparison of cross validation with other sample reuse techniques, see Efron (1982).

Another useful method for selecting $\lambda$ is *generalized cross validation* (GCV). Assuming that $\mathrm{tr}[H(\lambda)] \le n$, the GCV criterion is defined by

$$GCV(\lambda) = MSE(\lambda)/(n^{-1}\mathrm{tr}[I - H(\lambda)])^2 \quad . \qquad (17)$$

The GCV criterion was first proposed by Craven and Wahba (1979) for use in the context of a particular nonparametric regression problem. In other work Wahba (1977b,c, 1984) and Golub, Heath and Wahba (1979) have shown that GCV can be used to solve a wide variety of problems involving selection of a parameter to minimize $R(\lambda)$ or $P(\lambda)$.

In most cases of interest GCV is closely related to CV. An illustration of this is provided by the variable selection problem where we find, using (16), that

$$CV(\lambda) = n^{-1} \sum_{j=1}^{n} (y_j - \mu_{\lambda(j)})^2 = n^{-1} \sum_{j=1}^{n} \left[\frac{y_j - \mu_{\lambda j}}{1 - h_{jj}(\lambda)}\right]^2$$

and, hence, that

$$GCV(\lambda) = n^{-1} \sum_{j=1}^{n} (y_j - \mu_{\lambda(j)})^2 \left\{\frac{1 - h_{jj}(\lambda)}{n^{-1}\mathrm{tr}[I - H(\lambda)]}\right\}^2 \quad .$$

Thus, in this case, GCV is essentially a weighted version of CV.

While the previous example illustrates that GCV and CV may be related it also shows that CV is not a special case of GCV. Thus the title generalized cross validation is somewhat of a misnomer. Nevertheless, we can still interpret $GCV(\lambda)$ as being an attempt to account for the downward bias of $MSE(\lambda)$ for $P(\lambda)$ through multiplication by the factor $(n^{-1}\mathrm{tr}[I - H(\lambda)])^{-2}$ which is necessarily greater than or equal to one. However the best motivation for criterion (17) is probably provided by the so called GCV Theorem first established by Craven and Wahba (1979). (See, also Wahba, 1977c and Golub, Heath and Wahba, 1979 for other variants of the theorem.) A general form of the GCV Theorem can be stated as follows.

**Theorem 2.1. (The GCV Theorem).** Let

$$\tau_j(\lambda) = n^{-1}\mathrm{tr}[H(\lambda)^j], \qquad j = 1,2 ,$$

and assume that $\tau_1(\lambda) \le 1$. Then,

$$|EGCV(\lambda) - P(\lambda)|/R(\lambda) \le g(\lambda) , \tag{18}$$

where

$$g(\lambda) = [2\tau_1(\lambda) + \tau_1(\lambda)^2/\tau_2(\lambda)]/(1 - \tau_1(\lambda))^2. \tag{19}$$

Before proving Theorem 1 let us pause to interpret its content. If $g(\lambda)$ is small, then Theorem 1 has the consequence that the distance between $P(\lambda)$ and $EGCV(\lambda)$ is small relative to the intrinsic accuracy measure $R(\lambda)$. This can be roughly interpreted to mean that, in this case, $GCV(\lambda)$ is nearly an unbiased estimator of $P(\lambda)$.

There are, in fact, many cases where $g(\lambda)$ will be small, particularly for larger sample sizes. For example, in the variable selection problem

$$\tau_1(\lambda) = n^{-1}\mathrm{tr}[H(\lambda)] = \#(\lambda)/n \le p/n$$

and, since $H(\lambda)$ is idempotent, $\tau_2(\lambda) = \tau_1(\lambda)$. Thus,

$$g(\lambda) = 3\tau_1(\lambda)/(1 - \tau_1(\lambda))^2 \le 3\frac{p}{n} /(1 - \frac{p}{n})^2 = 3\frac{p}{n} + O((\frac{p}{n})^2) .$$

Since p is fixed, it follows that $g(\lambda)$ will be small for large n. The ridge regression problem admits similar conclusions (Exercise 7).

*Proof of Theorem 2.1.* Using Lemma A.1 we can show that

$$\begin{aligned} EGCV(\lambda) &= n^{-1}E\underline{y}'(I - H(\lambda))^2\underline{y}/(1 - \tau_1(\lambda))^2 \\ &= n^{-1}\{\underline{\mu}'(I - H(\lambda))^2\underline{\mu} + \sigma^2\mathrm{tr}[(I - H(\lambda))^2]\}/(1 - \tau_1(\lambda))^2 . \end{aligned}$$

Thus, in view of (5) and (12),

$$EGCV(\lambda)-P(\lambda) = \frac{R(\lambda)}{(1 - \tau_1(\lambda))^2} + \sigma^2 \frac{1 - 2\tau_1(\lambda)}{(1 - \tau_1(\lambda))^2} - (R(\lambda) + \sigma^2)$$

$$= \frac{\tau_1(\lambda)(2 - \tau_1(\lambda))}{(1 - \tau_1(\lambda))^2} R(\lambda) - \sigma^2 \frac{\tau_1(\lambda)^2}{(1 - \tau_1(\lambda))^2} .$$

This implies that

$$|EGCV(\lambda) - P(\lambda)|/R(\lambda) = \left| \frac{\tau_1(\lambda)(2 - \tau_1(\lambda))}{(1 - \tau_1(\lambda))^2} - \frac{\sigma^2 \tau_1^2(\lambda)}{R(\lambda)(1 - \tau_1(\lambda))^2} \right|$$

$$\leq \frac{\tau_1(\lambda)(2 - \tau_1(\lambda))}{(1 - \tau_1(\lambda))^2} + \frac{\tau_1(\lambda)^2/\tau_2(\lambda)}{(1 - \tau_1(\lambda))^2} \leq g(\lambda).$$

In obtaining the last two inequalities we used the fact that $R(\lambda) \geq \sigma^2\tau_2(\lambda)$ and that $\tau_1(\lambda) \leq 1$ which entails that $1 \leq 2 - \tau_1(\lambda) \leq 2$. □

So we now have available three criteria for selecting a good value of $\lambda$ from the data: $\hat{P}(\lambda)$ (and $\hat{R}(\lambda)$), $CV(\lambda)$ and $GCV(\lambda)$. The strategy is to choose one (or more) of these criteria and use its minimizer $\hat{\lambda}$ as an estimate of the minimizer of $P(\lambda)$ (and $R(\lambda)$). The choice of which criterion to choose will depend on the situation and will be governed by whether or not a value for $\sigma^2$ is known (or easily estimated) and the available computational resources. If $\sigma^2$ is unknown and no reasonable estimate is available, GCV or CV can be used since they do not require estimation of the error variance. As a rule the use of CV will involve more computational labor than GCV since its use entails recomputation of estimators from data subsets.

In some cases it may be feasible to compute two or even all three of our risk estimates. This is a sound practice and should be used whenever possible. By computing two or more of the measures we obtain some basis for comparison and, as a result, are less likely to be misled by an aberration associated with a particular measure. Also, when all the risk estimators are minimized by the same value of $\lambda$ this should strengthen our belief that a good choice has been made for $\lambda$.

In all of our previous discussions we have treated $\lambda$ as a nonrandom or deterministic variable. Thus, $R(\lambda)$ and $P(\lambda)$ are nonrandom functions of the nonrandom variable $\lambda$ while $L(\lambda)$, $MSE(\lambda)$, $\hat{R}(\lambda)$, $\hat{P}(\lambda)$, $GCV(\lambda)$ and $CV(\lambda)$ are all random functions of a deterministic variable $\lambda$. However, once we begin to consider the value $\hat{\lambda}$ which minimizes one of our criteria (i.e., $\hat{P}(\lambda)$, $CV(\lambda)$ or $GCV(\lambda)$) an additional level of stochastisticity is added; for $\hat{\lambda}$ is data dependent and therefore random. There are cases where we may want to study $L(\hat{\lambda})$, $P(\hat{\lambda})$ or $R(\hat{\lambda})$ and, hence, some confusion may arise as to whether or not the argument for the loss or risk is random or deterministic. To avoid such difficulties we will employ the convention that the symbol $\lambda$ corresponds to a nonrandom variable while special notation such as $\hat{\lambda}$ or $\tilde{\lambda}$ signifies the random case. A good rule of thumb to follow is that *the value of the index parameter under consideration should be assumed nonrandom unless indicated otherwise.*

**Example 1 (continued).** To illustrate the use of the various criteria for estimating $P(\lambda)$ and $R(\lambda)$ described in this section we return again to the housing rent study discussed in Section 1. Recall that the objective was to find an optimal least-squares predictive equation involving some subset of the five predictor variables $X_1, \ldots, X_5$.

We have chosen to include the constant in all predictive equations. Under this constraint there are $2^4 = 16$ possible predictor combinations involving the variables with index subsets $\{1\}$, $\{1,2\}$, $\{1,3\}$, $\{1,4\}$, $\{1,5\}$, $\{1,2,3\}$, $\{1,2,4\}$, $\{1,2,5\}$, $\{1,3,4\}$, $\{1,3,5\}$, $\{1,4,5\}$, $\{1,2,3,4\}$, $\{1,2,3,5\}$, $\{1,2,4,5\}$, $\{1,3,4,5\}$ and $\{1,2,3,4,5\}$. The collection of these subsets makes up the index set $\Lambda$. We now wish to find a variable subset which estimates the minimizer of $P(\lambda)$ over $\Lambda$.

As estimates of the minimizer of $P(\lambda)$ both the minimizers of $\hat{P}(\lambda)$ and $GCV(\lambda)$ will be considered. In the case of $\hat{P}(\lambda)$ an estimator of $\sigma^2$ is required. For this purpose the associated least-squares estimator of $\sigma^2$ corresponding to the least-squares estimator of $\mu$ based on all five predictor variables (c.f. (A.10) of the Appendix) will be used. Its value is found to be $\hat{\sigma}^2 = 711$ for this data.

Again let $\#(\lambda)$ denote the number of elements in $\lambda$. Then, due to the fact that $\mathrm{tr}H(\lambda) = \#(\lambda)$ for the variable selection problem, we find that our predictive risk estimators have the simple representations

$$\hat{P}(\lambda) = MSE(\lambda) + 2\hat{\sigma}^2\#(\lambda)/n$$

and

$$GCV(\lambda) = MSE(\lambda)/(1 - \#(\lambda)/n)^2.$$

As a result, these quantities can both be easily computed using the standard output of almost any linear regression software package. Table 2.2 gives a summary of the values of $\hat{P}(\lambda)$ and $GCV(\lambda)$ over all $\lambda \in \Lambda$ that was obtained by making use of this latter observation.

**Table 2.2 Values of MSE($\lambda$), GCV($\lambda$) and $\hat{P}(\lambda)$ for the Housing Cost Study**

| Variable Subset | MSE($\lambda$) | GCV($\lambda$) | $\hat{P}(\lambda)$ |
|---|---|---|---|
| $X_1$ | 1966.8 | 1975.5 | 1969.9 |
| $X_1,X_2$ | 1955.8 | 1973.1 | 1962.0 |
| $X_1,X_3$ | 1920.8 | 1937.8 | 1927.1 |
| $X_1,X_4$ | 1304.4 | 1315.9 | 1310.7 |
| $X_1,X_5$ | 1278.1 | 1289.4 | 1284.3 |
| $X_1,X_2,X_3$ | 1911.0 | 1936.5 | 1920.4 |
| $X_1,X_2,X_4$ | 1298.2 | 1315.5 | 1307.6 |
| $X_1,X_2,X_5$ | 1277.3 | 1294.3 | 1286.7 |
| $X_1,X_3,X_4$ | 1259.3 | 1276.1 | 1268.7 |
| $X_1,X_3,X_5$ | 1248.4 | 1265.0 | 1257.8 |
| $X_1,X_4,X_5$ | 733.5 | 743.3 | 742.9 |
| $X_1,X_2,X_3,X_4$ | 1253.9 | 1276.2 | 1266.4 |
| $X_1,X_2,X_3,X_5$ | 1247.8 | 1270.0 | 1260.3 |
| $X_1,X_2,X_4,X_5$ | 733.5 | 746.6 | 746.0 |
| $X_1,X_3,X_4,X_5$ | 703.3 | 715.8 | 715.8 |
| $X_1,X_2,X_3,X_4,X_5$ | 703.2 | 718.9 | 718.8 |

Examination of the values in Table 2.2 reveals that both $\hat{P}(\lambda)$ and $GCV(\lambda)$ point toward $\{X_1, X_3, X_4, X_5\}$ as the best predictor subset, i.e., both criteria choose $\hat{\lambda} = \{1,3,4,5\}$ as an estimate of the minimizer of $P(\lambda)$. The second best choice involves all five predictors while the third best choice consists of the variables corresponding to $\hat{\lambda}$ with the exclusion of $X_3$. From a practical viewpoint, there is not a substantial difference between our risk estimates for any of these three variable subsets. Thus, one would probably use $X_1$, $X_3$, $X_4$, $X_5$ to construct a predictive equation or, if information on $X_3$ (length of residence) is difficult or costly to obtain, perhaps just $X_1$, $X_4$ and $X_5$ would be used.

## 2.3 Other Properties of CV and GCV

The motivation for the GCV and CV criteria introduced in the previous section is not nearly as direct as that for $\hat{P}(\lambda)$ in (13). In fact, the reader may not be entirely convinced of their importance until later chapters where precise statements about their optimality properties can be made. One difficulty in interpreting $CV(\lambda)$ and $GCV(\lambda)$ is that, until this point, we have not indicated precisely what they estimate. Thus, in this section we attempt to shed more light on CV and GCV by first showing that the CV and GCV criteria provide unbiased estimates of the prediction risk for a class of nil-trace estimators of $\underline{\mu}$ that approximates $C(\Lambda)$. We then show that GCV is closely related to an unbiased risk estimate derived by Stein (1981). The results we present are based on work by Li (1983, 1985).

Recall from Section 2 that a nil-trace estimator of $\underline{\mu}$ is one for which $tr[H(\lambda)] = 0$. In this case we saw that $EMSE(\lambda) = P(\lambda)$ and, hence, no estimator of $\sigma^2$ was needed to obtain an unbiased estimate of the prediction risk.

Now define a class of estimators $\tilde{C}(\Lambda)$ obtained from $C(\Lambda)$ by

$$\tilde{\underline{\mu}}_\lambda = \tilde{H}(\lambda)\underline{y}, \qquad \lambda \in \Lambda,$$

where

$$\tilde{H}(\lambda) = (1 + \alpha(\lambda))H(\lambda) - \alpha(\lambda)I$$

with

$$\alpha(\lambda) = tr[H(\lambda)]/tr[I - H(\lambda)],$$

and $H(\lambda)$ corresponds to $\underline{\mu}_\lambda \in C(\Lambda)$. When $\alpha(\lambda)$ is small $\tilde{\underline{\mu}}_\lambda$ provides an approximation to $\underline{\mu}_\lambda$. For example in the variable selection problem $\alpha(\lambda) = \#(\lambda)/(n - \#(\lambda)) \leq p/n + O((p/n)^2)$ and will typically be small.

Observe that $tr[\tilde{H}(\lambda)] = 0$. Thus $\tilde{C}(\Lambda) = \{\tilde{\underline{\mu}}_\lambda : \lambda \in \Lambda\}$ is a class of nil-trace estimators for $\underline{\mu}$. Now note that

$$\begin{aligned} \tilde{MSE}(\lambda) &\equiv n^{-1}\sum_{j=1}^{n}(y_j - \tilde{\mu}_{\lambda j})^2 \\ &= n^{-1}\underline{y}'(I - \tilde{H}(\lambda))^2\underline{y} \\ &= n^{-1}\underline{y}'(I - H(\lambda))^2\underline{y}/(n^{-1}tr[I - H(\lambda)])^2 \\ &= GCV(\lambda). \end{aligned}$$

In words, this says that the MSE for $\tilde{\underline{\mu}}_\lambda$ is the GCV criterion for $\underline{\mu}_\lambda$. From the nil-trace property of $\tilde{\underline{\mu}}_\lambda$ it follows that $\tilde{MSE}(\lambda)$ is unbiased for the prediction risk for $\tilde{\underline{\mu}}_\lambda$. Thus $GCV(\lambda)$ represents an unbiased estimator of the prediction risk for an estimator, $\tilde{\underline{\mu}}_\lambda$, which approximates the estimator of interest, $\underline{\mu}_\lambda$. In this particular case it follows from Li (1983) that, under reasonable conditions, $\tilde{\underline{\mu}}_\lambda$ and $\underline{\mu}_\lambda$ are asymptotically equivalent estimators of $\underline{\mu}$.

A similar property to that stated for GCV holds for CV: namely, CV is just the MSE for a class of nil-trace estimators for $\underline{\mu}$. Verification of this statement is left for the exercises. (See Exercises 10 and 11).

Another view of GCV can be obtained by considering the Stein estimate of $\underline{\mu}$ corresponding to $\underline{\mu}_\lambda$ that is defined by

$$\bar{\underline{\mu}}_\lambda = \underline{y} - \sigma^2(\underline{y} - \underline{\mu}_\lambda)/\underline{y}'B(\lambda)\underline{y} , \qquad (20)$$

where

$$B(\lambda) = \{tr[I - H(\lambda)]I - 2(I - H(\lambda))\}^{-1}(I - H(\lambda))^2$$

and it is assumed that the largest eigenvalue of $I - H(\lambda)$ is less than half the trace of $I - H(\lambda)$. The quantity $\sigma^2/\underline{y}'B(\lambda)\underline{y}$ is called a shrinkage factor.

A simple illustration of $\bar{\underline{\mu}}_\lambda$ can be obtained from the model selection problem. In this case one finds that $\underline{y}'B(\lambda)\underline{y} = (n - \#(\lambda))s^2(\lambda)/(n - \#(\lambda) - 2)$, where $s^2(\lambda) = \underline{y}'(I - X_\lambda(X_\lambda'X_\lambda)^{-1}X_\lambda')\underline{y} /(n - \#(\lambda))$ is the least-squares estimator of $\sigma^2$ for the model with variables corresponding to $\lambda$. If the model is correct i.e., we have chosen the correct $\lambda$, then $\sigma^2/\underline{y}'B(\lambda)\underline{y}$ should be near one and, hence, $\bar{\underline{\mu}}_\lambda$ will be close to $\mu$. Conversely, if $\lambda$ is incorrect one might expect $s^2(\lambda)$ to be large relative to $\sigma^2$ and then $\bar{\underline{\mu}}_\lambda$ shrinks $\underline{\mu}_\lambda$ towards $\underline{y}$ to guard against model violations.

Assuming normal errors in (1), various consistency and optimality properties of $\bar{\underline{\mu}}_\lambda$ have been established by Li and Hwang (1984). For our purposes, we are more interested in Stein's unbiased risk estimator (SURE) associated with $\bar{\underline{\mu}}_\lambda$ which, assuming $\sigma^2$ is known, is defined by

$$SURE(\lambda) = \sigma^2 - \sigma^4 MSE(\lambda)/(\underline{y}'B(\lambda)\underline{y})^2. \tag{21}$$

Stein (1981) shows that in the case of normal errors SURE($\lambda$) is an unbiased estimate of the risk

$$\bar{R}(\lambda) = n^{-1}\sum_{j=1}^{n} E(\mu_j - \bar{\mu}_{\lambda j})^2 .$$

Now suppose that the largest eigenvalue of $I - H(\lambda)$ is small relative to its trace. Then,

$$B(\lambda) = (tr[I - H(\lambda)])^{-1}[I - \frac{2}{tr[I - H(\lambda)]}(I - H(\lambda))]^{-1}(I - H(\lambda))^2$$

$$\doteq (I - H(\lambda))^2/tr[I - H(\lambda)] . \tag{22}$$

Substituting this approximation into (21) gives

$$\mathrm{SURE}(\lambda) \doteq \sigma^2 - \sigma^4 \mathrm{MSE}(\lambda)/(n\mathrm{MSE}(\lambda)/\mathrm{tr}[I - H(\lambda)])^2$$

$$= \sigma^2 - \sigma^4/\mathrm{GCV}(\lambda) \ . \qquad (23)$$

Thus, minimizing Stein's unbiased risk estimate and minimizing $\mathrm{GCV}(\lambda)$ are essentially equivalent in this case. So by minimizing $\mathrm{GCV}(\lambda)$ we can obtain an optimal value of $\lambda$ for $\bar{\underline{\mu}}_\lambda$.

Li (1983) uses the approximation (21) to obtain a simplified version of (20) and establishes certain consistency properties for the resulting estimator of $\underline{\mu}$ and the estimator (23) when $\lambda$ is selected by GCV. He is also able to show that $\underline{\mu}_{\hat{\lambda}}$, with $\hat{\lambda}$ the GCV estimate of $\lambda$, is consistent in the sense that $L(\hat{\lambda})$ converges to zero in probability, under certain restrictions. We will explore these results in more detail in subsequent chapters. For now, it suffices to note that i) minimizing $\mathrm{GCV}(\lambda)$ is roughly equivalent to minimizing Stein's unbiased risk estimate for the Stein estimator (20) that is related to $\underline{\mu}_\lambda$ and ii) $\underline{\mu}_{\hat{\lambda}}$, with $\hat{\lambda}$ selected by GCV, is a consistent estimator of $\underline{\mu}$ in certain cases.

## 2.4 Other Criteria

There are a variety of other criteria which have appeared in the literature for data driven estimator selection. In this final section we discuss a few of these criteria, and relate them to those we have already studied.

For simplicity we restrict attention to the variable selection problem for the remainder of this discussion. In this case there are numerous criteria which have been suggested for selecting an optimal $\lambda$. A partial list of some of these is

$$\mathrm{AIC}(\lambda) = n\log\mathrm{MSE}(\lambda) + 2\#(\lambda) \ , \qquad (24)$$

$$\mathrm{FPE}(\lambda) = (n + \#(\lambda))\mathrm{MSE}(\lambda)/(n - \#(\lambda)) \ , \qquad (25)$$

$$S(\lambda) = (n + 2\#(\lambda))\mathrm{MSE}(\lambda), \qquad (26)$$

and

$$U(\lambda) = MSE(\lambda)\frac{n}{n - \#(\lambda)}(1 + \#(\lambda)/(n - 1 - \#(\lambda))). \tag{27}$$

These all have obvious generalizations to the problem of selecting $\lambda$ for the more general estimation problem discussed in Section 2.1.

Both AIC and FPE were proposed by Akaike (1970, 1973, 1974) and derive from information theory and time series considerations. The abbreviation AIC stands for information criterion A, thereby allowing for future criteria to be named BIC, etc. (There is, in fact, a BIC; See Akaike, 1978). The notation FPE stands for final prediction error and the criterion was suggested by prediction considerations for autoregressive processes. The $U(\lambda)$ criterion was proposed by Hocking (1976) and Thompson (1978). Its motivation is tied to the distributional properties of conditional prediction errors in regression models. Criterion (26) is due to Shibata (1981).

The list given above is by no means complete. There are many other criteria. Some of these can be found in work by Parzen (1974, 1977), Schwarz (1978), Suigiura (1978), Atkinson (1978), Hannan and Quinn (1979), Young (1982) and Atilgan (1984). Atkinson (1980, 1981) gives an excellent discussion of the relationships between some of these criteria and studies their asymptotic behavior for models with only *finitely* many parameters. Shibata (1981) and Li (1987) also discuss the asymptotic properties of certain criteria under the alternative assumption that the true model contains an *infinite* number of parameters. The conclusions regarding optimality of the criteria appear to be quite different for these two cases (c.f., Stone, 1979). Other asymptotic treatments can be found in work by Breiman and Friedman (1983), who study the $U(\lambda)$ statistic, and Stone (1977) who shows that AIC and CV are asymptotically equivalent in a certain sense.

All four criteria (24)-(27) are closely related to GCV. To see this let us first note that, for the variable selection problem,

$$\mathrm{GCV}(\lambda) = \mathrm{MSE}(\lambda)/(1 - \#(\lambda)/n)^2 \ .$$

Now, assuming that $p/n \to 0$ as $n \to \infty$, observe that, by use of the Maclaurin series for $\exp\{x\}$, we have

$$\begin{aligned} \exp\{\mathrm{AIC}(\lambda)/n\} &= \mathrm{MSE}(\lambda)\exp\{2\#(\lambda)/n\} \\ &= \mathrm{MSE}(\lambda)/(\exp\{-\#(\lambda)/n\})^2 \\ &\doteq \mathrm{MSE}(\lambda)/(1 - \#(\lambda)/n)^2 = \mathrm{GCV}(\lambda). \end{aligned}$$

To relate $\mathrm{GCV}(\lambda)$ to criteria (25) and (26) we use the geometric series which allows us to see that

$$\frac{n + \#(\lambda)}{n - \#(\lambda)} = \frac{1 + \#(\lambda)/n}{1 - \#(\lambda)/n} \doteq (1 - \#(\lambda)/n)^{-2}$$

and

$$(n + 2\#(\lambda)) = n(1 + 2\#(\lambda)/n) \doteq n/(1 - \#(\lambda)/n)^2.$$

Also, as $(n - 1)/n \sim 1$ as $n \to \infty$ ,

$$\begin{aligned} \frac{n}{n - \#(\lambda)}(1 + \#(\lambda)/(n - 1 - \#(\lambda))) &\doteq (1 + \frac{\#(\lambda)/n}{1 - \#(\lambda)/n})/(1 - \#(\lambda)/n) \\ &= (1 - \#(\lambda)/n)^{-2} \ . \end{aligned}$$

Thus, (24)-(27) are all essentially variants of the GCV criterion with $(1 - \#(\lambda)/n)^{-2}$ replaced by some nearly equivalent factor. The reader should be cautioned however that this does not prove that $\mathrm{GCV}(\lambda)$ and (24)-(27) are asymptotically equivalent, but does indicate that all these criteria are closely related and will tend to have similar values when $\#(\lambda)/n$ is small.

**Example 1 (continued).** Table 2.3 below lists the values of criteria (24) - (27) for the housing rent data. Examination of the values in the table leads to the same conclusion that was reached in Section 2 concerning the three best predictive equations. In this particular example, all our criteria for variable selection, including $\hat{P}(\lambda)$ and $GCV(\lambda)$, appear to be very much in agreement concerning preference for the various possible prediction equations.

**Table 2.3 Values of Various Subset Selection Criteria for the Housing Cost Study**

| Variable Subset | AIC(λ) | FPE(λ) | S(λ) | U(λ) |
|---|---|---|---|---|
| $X_1$ | 3452.8 | 1975.5 | 898827.6 | 1975.5 |
| $X_1,X_2$ | 3452.2 | 1973.1 | 897712.3 | 1973.0 |
| $X_1,X_3$ | 3444.0 | 1937.8 | 881647.3 | 1937.8 |
| $X_1,X_4$ | 3267.9 | 1315.9 | 598719.6 | 1316.0 |
| $X_1,X_5$ | 3258.7 | 1289.4 | 586647.9 | 1289.4 |
| $X_1,X_2,X_3$ | 3443.7 | 1936.4 | 880971.0 | 1936.5 |
| $X_1,X_2,X_4$ | 3267.8 | 1315.4 | 598470.2 | 1315.5 |
| $X_1,X_2,X_5$ | 3260.4 | 1294.3 | 588835.3 | 1294.3 |
| $X_1,X_3,X_4$ | 3253.9 | 1276.0 | 580537.3 | 1276.1 |
| $X_1,X_3,X_5$ | 3250.0 | 1265.0 | 575512.4 | 1265.0 |
| $X_1,X_4,X_5$ | 3008.0 | 743.2 | 338143.5 | 743.3 |
| $X_1,X_2,X_3,X_4$ | 3254.0 | 1276.1 | 580555.7 | 1276.3 |
| $X_1,X_2,X_3,X_5$ | 3251.8 | 1270.0 | 577777.7 | 1270.2 |
| $X_1,X_2,X_4,X_5$ | 3010.0 | 746.5 | 339610.5 | 746.6 |
| $X_1,X_3,X_4,X_5$ | 2990.9 | 715.8 | 325627.9 | 715.8 |
| $X_1,X_2,X_3,X_4,X_5$ | 2992.8 | 718.8 | 326988.0 | 718.9 |

## Exercises

1. Verify identity (5).
2. Establish that the ridge estimator of $\underline{\beta}$ is optimal in the sense of minimizing the penalized least-squares criterion

$$\sum_{j=1}^{n} (y_j - \underline{x}_j'\underline{\beta})^2 + \lambda\underline{\beta}'\underline{\beta} .$$

Discuss the meaning of this criterion paying particular attention to what happens when λ is increased or decreased.

3. Consider the case where $\underline{y}$, conditional on $\underline{\beta}$, has an n-variate normal distribution with mean $X\underline{\beta}$ and variance-covariance matrix $\sigma^2 I$. Let $\underline{\beta}$ be p-variate normal with mean $\underline{0}$ and variance-covariance matrix $\sigma_s^2 I$. Show that the ridge estimator $\underline{\beta}_\lambda$ with $\lambda = \sigma^2/\sigma_s^2$ is the posterior mean of $\underline{\beta}$, i.e., the mean of $\underline{\beta}$ given $\underline{y}$. (This exercise shows that $\underline{\beta}_\lambda$ is optimal in the sense of being a Bayes estimate of $\underline{\beta}$ for this model.)

4. Establish the existence of biased estimators which have smaller risks than unbiased ones. [Hint: Consider the ridge regression setting and proceed as follows:

a) Let the singular value decomposition of X be X = UDV', where U is a n × n orthonormal matrix, V is a p × p orthonormal matrix and D is a n × p matrix whose diagonal entries are the square roots of the eigenvalues of X'X. Using this form for X obtain a simple expression for $R(\lambda)$ as a sum of two functions which are, respectively, monotone increasing and decreasing in $\lambda$.

b) Differentiate your expression for $R(\lambda)$ and show that there is a $\lambda_0$ such that $dR(\lambda)/d\lambda < 0$ whenever $\lambda < \lambda_0$.

c) Conclude from b) (Why?) that there is a $\lambda$ such that $R(\lambda) < R(0)$, where $R(0)$ is the risk for the unbiased estimator $\hat{\underline{\mu}}$.]

5. Establish a parallel of identity (15) for ridge regression. Derive (15) as a special case of your formula and use it to establish a relationship between CV and GCV in the ridge regression setting. [Hint: First consider the minimizer of

$$\sum_{\substack{k=1 \\ k\neq j}}^{n} (y_k - \underline{x}_k'\underline{\beta})^2 + (z - \underline{x}_j'\underline{\beta})^2 + \lambda\underline{\beta}'\underline{\beta} .$$

Denoting the minimizer by $\underline{\beta}_\lambda(z)$ show that the estimator $\underline{\beta}_{\lambda(j)}$, obtained by deleting the point $(\underline{x}_j', y_j)$ is $\underline{\beta}_\lambda(\mu_{\lambda(j)})$, where $\mu_{\lambda(j)} = \underline{x}_j'\underline{\beta}_{\lambda(j)}$. Now use the fact that $\underline{x}_j'\underline{\beta}_\lambda(z)$ is linear in z to see that $\mu_{\lambda(j)} = \mu_{\lambda j} + h_{jj}(\lambda)(\mu_{\lambda(j)} - y_j)$.]

6. For the ridge regression problem, as in Exercise 4 let X = UDV' and let W be the n × n unitary matrix which diagonalizes circulant matrices (see, e.g., Fuller, 1976, Chapter 4). W has

typical element $n^{-1/2} e^{2\pi ijk/n}$, j, k=0,...,n - 1 (note: $e^{ix}$ = cosx + isinx with $i^2 = -1$). Now transform from $\underset{\sim}{y}$ to $\tilde{\underset{\sim}{y}} = WU'\underset{\sim}{y}$ and note that

$$\tilde{\underset{\sim}{y}} = \tilde{\underline{\mu}} + \tilde{\underline{\varepsilon}} ,$$

where $\tilde{\underline{\mu}} = WDV'\underline{\beta}$ and $\tilde{\underline{\varepsilon}} = WU'\underline{\varepsilon}$.

a) Derive the ridge estimator of $\tilde{\underline{\mu}}$ and show that it has the form $\tilde{H}(\lambda)\tilde{\underset{\sim}{y}}$ with $\tilde{H}(\lambda)$ a circulant matrix. Comment on the implications of this fact.

b) Show that GCV is just CV in this transformed coordinate system.

[Golub, Heath and Wahba 1979]

**7.** Show that $|EGCV(\lambda) - P(\lambda)|/R(\lambda) \leq 3p/n + O((p/n)^2)$ for ridge regression.

**8.** Refer to Exercise 3. Argue that under the Bayesian formulation any estimator of $\lambda$ should be invariant under an orthogonal transformation of $\underset{\sim}{y}$ and/or $\underline{\beta}$. Show that GCV($\lambda$) has this property but CV($\lambda$) does not.

**9.** Refer to Exercise 3. Let E and $E_\beta$ denote expectations with respect to the distributions of $\underline{\varepsilon}$ and $\underline{\beta}$, respectively. Show that $E_\beta EL(\lambda)$ and $E_\beta EGCV(\lambda)$ are both minimized at $\lambda = \sigma^2/\sigma_s^2$.

[Golub, Heath and Wahba, 1979]

**10.** Suppose the estimators in C($\Lambda$) have the form $\underline{\mu}_\lambda = H(\lambda)\underset{\sim}{y}$ but $H(\lambda)$ is not symmetric. Derive analogs of (11)-(14) for this case.

**11.** Using Exercise 10 show that CV($\lambda$) is an unbiased estimator of the prediction risk for a class of nil-trace estimators of $\underline{\mu}$. Work out the form of these estimators explicitly for the variable selection problem.

**12.** Show that for the variable selection problem $\underset{\sim}{y}'B(\lambda)\underset{\sim}{y}$ in (20) is $(n - \#(\lambda))s^2(\lambda)/(n - \#(\lambda) - 2)$, where $s^2(\lambda) = \underset{\sim}{y}'(I - H(\lambda))^2\underset{\sim}{y}/(n - \#(\lambda))$.

**13.** Twelve individuals are randomly divided into three groups of four subjects each. The first group is allowed to follow their normal sleep patterns while groups 2 and 3 are deprived of sleep for twenty-four and forty-eight hours, respectively. The experimental subjects are then tested concerning their ability to do numerical calculations. The following data is the percentage of correct answers for the individuals on a test measuring arithmetic skills given at the end of the experimental period for each group.

| **Group 1: Normal Sleep** | **Group 2: Sleep Deprived 24 Hours** | **Group 3: Sleep Deprived 48 Hours** |
|---|---|---|
| 85 | 81 | 60 |
| 83 | 68 | 48 |
| 76 | 90 | 38 |
| 64 | 74 | 47 |

Let $\mu_1, \mu_2$ and $\mu_3$ denote the population means corresponding to the three experimental groups. Compute the GCV criterion and the four criteria (24)-(27) for the five models

1) $\mu_1 \neq \mu_2 \neq \mu_3$
2) $\mu_1 = \mu_2 \neq \mu_3$
3) $\mu_1 \neq \mu_2 = \mu_3$
4) $\mu_1 = \mu_3 \neq \mu_2$
5) $\mu_1 = \mu_2 = \mu_3$ .

According to your criteria, which model seems best? Which is second best, etc? Discuss the similarities and differences concerning model preference indicated by the five criterion functions.

**14.** Suppose n observations are classified according to whether they fall into C possible cells or categories. Let $n_i$ be the number of observations falling in the ith cell with

$$\sum_{i=1}^{C} n_i = n .$$

Let $p_i$ be the true probability of an individual belonging to cell i and define an unbiased estimator of $p_i$ by

$$\hat{p}_i = n_i/n .$$

In small samples the $\hat{p}_i$ may not perform well as estimators of the $p_i$. Thus, in some cases it is preferable to use an estimator which "smoothes" the $\hat{p}_i$ such as

$$p_{\lambda j} = \sum_{r=1}^{C} w_{jr}(\lambda)\hat{p}_r \quad ,$$

where

$$w_{jr}(\lambda) = \begin{cases} (1 - \lambda)/(C - 1) \;, & r \neq j \;, \\ \lambda \;, & r = j \;, \end{cases}$$

and $C^{-1} \leq \lambda \leq 1$. An optimal value of $\lambda$ is one that minimizes the risk

$$\begin{aligned} R(\lambda) &= E \sum_{j=1}^{C} (p_{\lambda j} - p_j)^2 \\ &= \sum_{j=1}^{C} (Ep_{\lambda j} - p_j)^2 + \sum_{j=1}^{C} \text{Var } p_{\lambda j} \quad . \end{aligned}$$

a) Show that there exists a value of $\lambda < 1$ such that $R(\lambda) < R(1)$. (The calculations are easiest to see when $C = 2$, i.e., we have binomial data, and you may wish to only work with this case.) Interpret the meaning of this result.

b) Show that $R(\lambda)$ is estimated unbiasedly by

$$\begin{aligned} &(n - 1)^{-1}\{[\sum_{j=1}^{C} n_j^2 - 1] + 2\lambda \\ &\quad - 2[\lambda \sum_{j=1}^{C} n_j^2 + (2(1 - \lambda)/(C - 1)) \sum_{j<r} n_j n_r]/n\} \\ &\quad + \{\alpha \sum_{j=1}^{C} n_j^2 + [2(1 - \alpha)/(C - 1)] \sum_{j<r} n_j n_r\}/n^2 \;, \end{aligned}$$

where $\alpha = \lambda^2 + (1 - \lambda)^2/(C - 1)$ .

c) Establish that your estimator of $R(\lambda)$ in b) is minimized by

$$\hat{\lambda} = \{\underline{n}'A\underline{n}[n + (n - 1)/(C - 1)] - n^2\}/\{[C(n - 1)/(C - 1)]\underline{n}'A\underline{n}\} ,$$

where $\underline{n} = (n_1,\ldots,n_C)'$ and $A = \{a_{ij}\}$ has elements

$$a_{ij} = -1/(C - 1), \quad i \neq j \quad ,$$

and

$$a_{ii} = 1 .$$

d) Verify that $\hat{\lambda}$ can be written in the alternative form

$$\hat{\lambda} = 1 - \{(C - 1)/[C(n - 1)]\}\{\frac{n^2}{\underline{n}'A\underline{n}} - 1\} .$$

Use this expression to conclude that $\hat{\lambda} \xrightarrow{P} 1$ as $n \to \infty$ at a rate of $n^{-1}$. This has been shown by Bowman (1980) to be a necessary and sufficient condition for the consistency of the estimator $\hat{p}_{\lambda j}$, $j = 1,\ldots,C$.
[Brown and Rundell, 1985. See, also, Titterington, 1985, for a general discussion of smoothing in this setting.]

**15.** Let

$$y_{kj} = \mu_j + \varepsilon_{kj}, \quad k = 1,2, \quad j = 1,\ldots,n,$$

with the $\varepsilon_{kj}$ being zero mean, uncorrelated random variable having common variance $\sigma^2$. Set $\underline{y}_i' = (y_{i1},\ldots,y_{in})$, $i = 1,2$, and define $\underline{\mu}_\lambda = H_\lambda \underline{y}_1 = (\mu_{\lambda 1},\ldots,\mu_{\lambda n})'$ and $\tilde{P}(\lambda) = n^{-1}\Sigma_{j=1}^n (y_{2j} - \mu_{\lambda j})^2$. Show that $E\tilde{P}(\lambda) = P(\lambda)$. [Hint: Use Lemma A.1 of the Appendix.]

**16.** Give an example of a situation where $CV(\lambda)$ is generally biased for $P(\lambda)$. [Hint: Try ridge regression or variable selection.]

**17.** Consider the case of ridge regression and, for simplicity, assume that $X'X = I$. Show that $R(\lambda)$ is minimized in this case by taking $\lambda = \sigma^2 p/\underline{\beta}'\underline{\beta}$. Verify that for this choice of $\lambda$ $EMSE(\lambda) = n^{-1}tr(I - H(\lambda))$ and that $R(\lambda) = \sigma^2 trH(\lambda)/n$. Comment on the implications of the identity concerning $MSE(\lambda)$ for the selection of an optimal $\lambda$. In particular, does it suggest yet another data driven method for choosing the ridge parameter.

Let $h_{ii}(\lambda)$ denote the ith diagonal element of $H(\lambda)$. Suppose we define a confidence interval for $\mu_i = \underline{x}_i'\underline{\beta}$ to be $\underline{x}_i'\underline{\beta}_\lambda \pm Z_{1-\alpha/2}[\sigma^2 h_{ii}(\lambda)]^{1/2}$, where $Z_{1-\alpha/2}$ is the $100(1-\alpha/2)$ percentage point of the standard normal distribution. In view of your identity for $R(\lambda)$ how might such an interval be expected to perform.

[Hall and Titterington, 1985].

**18.** The data set given in the following table is concerned with measurements for 33 black female applicants to a police department of a large city. The response variable is subject height (HEIGHT) while the 9 predictor variables are

SITHT - sitting height,
UARM - upper arm length,
FORE - forearm length,
HAND - hand length,
ULEG - upper leg length,
LLEG - lower leg length,
FOOT - foot length,
BRACH - a brachial index,

and

TIBIO - a tibio-femural index.

The variables have all been standardized by subtracting off their means and dividing by their respective standard deviations. Using these standardized values obtain a ridge estimator of the vector of regression coefficients. Select $\lambda$ using GCV and give a plot of the GCV function (as a function of $\lambda$).

| Sub. # | HEIGHT | SITHT | UARM | FORE | HAND | ULEG | LLEG | FOOT | BRACH | TIBIO |
|---|---|---|---|---|---|---|---|---|---|---|
| 1 | .2527 | 1.1887 | -.3374 | .0575 | .1311 | -.8961 | -.8129 | 1.0569 | .3648 | .1302 |
| 2 | 1.0704 | 1.6819 | .0198 | .8163 | -.2910 | .4379 | 1.0010 | .5856 | .6967 | .6827 |
| 3 | 1.2544 | .8093 | .7344 | 1.1199 | 2.2415 | .6158 | .2372 | 1.8424 | .6928 | -.5247 |
| 4 | 1.2953 | .5817 | -.8138 | .1334 | .0256 | .6158 | -.0014 | 1.0569 | .9607 | -.8274 |
| 5 | -1.4440 | -1.6187 | -.1588 | -.5496 | -1.1352 | -1.8744 | -.4788 | .8998 | -.3948 | 2.0908 |
| 6 | .2732 | .9990 | -.3374 | .7404 | .0256 | -.1401 | -.0014 | .7427 | 1.0127 | .1674 |
| 7 | -1.1987 | .2023 | -1.0519 | -.1702 | -.1855 | -1.0295 | -1.7199 | -.1999 | .9337 | -.9348 |
| 8 | .2936 | 1.1887 | -1.2901 | -.8531 | -1.1352 | -.3180 | -.7652 | -.1999 | .5269 | -.6042 |
| 9 | -1.1987 | -.7082 | -.7542 | -.7013 | -.2910 | -1.5186 | -1.4812 | .1143 | .0840 | .0972 |
| 10 | -.6263 | -1.6566 | -.0397 | -.1702 | .5532 | .2156 | -.2401 | .2714 | -.1610 | -.6196 |
| 11 | .5594 | 1.1508 | 1.4489 | -.5496 | -.2910 | .3490 | .5714 | 1.9995 | -1.9054 | .2513 |
| 12 | .5798 | -.8979 | 1.1512 | 1.5752 | .6587 | .4824 | .7623 | 1.2140 | .2252 | .3178 |
| 13 | -1.0965 | -1.5428 | -.8138 | -.5496 | -1.1352 | -1.1184 | -.4788 | -1.7710 | .2961 | .9163 |
| 14 | 1.1113 | .8852 | 1.0916 | 2.1822 | .8698 | .3490 | 1.4783 | .4285 | .8195 | 1.4176 |
| 15 | .3550 | 1.0370 | -1.0519 | .5887 | -.2910 | -.2291 | .1895 | -.1999 | 1.6818 | .5430 |
| 16 | .9069 | .0126 | .1389 | .5887 | .5532 | .1711 | .6668 | -.0428 | .3599 | .6156 |
| 17 | -1.7098 | -1.5049 | -.8138 | -1.8396 | -1.6628 | .8381 | -.7652 | -1.4568 | -.9594 | -2.0631 |
| 18 | -1.0147 | .3920 | .1985 | -1.9914 | -.9241 | -.1401 | -1.4812 | -1.6139 | -2.0822 | -1.7852 |
| 19 | -1.9551 | -1.3531 | -1.2306 | -1.0808 | -1.3462 | -1.9633 | -2.1495 | -1.2997 | .2329 | -.2072 |
| 20 | -.7081 | -.5564 | -1.7069 | -1.0808 | -.8186 | -1.6520 | -1.1471 | -.1999 | .7779 | .7751 |
| 21 | 1.1727 | .9611 | .9725 | .9681 | -.3965 | .3935 | .3804 | -.1999 | -.1362 | -.0532 |
| 22 | .6616 | -.6323 | .0794 | .4369 | 1.7139 | .4379 | 1.0965 | 1.8424 | .2813 | .8049 |
| 23 | -.2992 | .9611 | -.3970 | -.8531 | -.5020 | -.9850 | -.7652 | -.1999 | -.4380 | .3236 |
| 24 | .2527 | .5438 | .4962 | -1.3084 | .9753 | .3935 | .0463 | -.1999 | -1.7294 | -.4817 |
| 25 | 2.2152 | 1.5302 | 1.6871 | 1.5752 | .5532 | 1.2383 | 1.8602 | .4285 | -.2884 | .6973 |
| 26 | -.9738 | .0126 | -.5160 | -.7013 | .6587 | -.0067 | -.7652 | -.5141 | -.1693 | -1.1046 |
| 27 | .2936 | -.2529 | -1.1115 | .0575 | -.8186 | -.4959 | 1.1442 | .1143 | 1.2270 | 2.2051 |
| 28 | -.6876 | -.5564 | .2580 | .8922 | -.1855 | .1267 | .2372 | -.1999 | .5158 | .1178 |
| 29 | -.8512 | -.4805 | -1.1115 | -.1702 | -1.8738 | -.5848 | -.3833 | -.0428 | 1.0018 | .2750 |
| 30 | -.0539 | -.2150 | 1.5680 | -.1702 | .4477 | 2.1721 | .8578 | -1.6139 | -1.6810 | -1.6410 |
| 31 | .1914 | -1.1255 | 2.2825 | .4369 | 1.7139 | 1.1939 | .8100 | -.6712 | -1.7778 | -.5380 |
| 32 | .5389 | -.2150 | .7344 | -.7013 | 1.2918 | 1.6385 | .4759 | -.6712 | -1.4003 | -1.4812 |
| 33 | .5389 | -.8220 | .6748 | 1.2716 | .8698 | 1.2828 | 1.6215 | -1.2997 | .4316 | .3469 |
| **AVG** | 164.5637 | 85.5667 | 32.3667 | 28.0242 | 18.5758 | 41.3152 | 40.6030 | 6.0273 | 86.7713 | 96.0337 |
| **SD** | 4.8917 | 2.6359 | 1.6794 | 1.3178 | .9477 | 2.2489 | 2.0949 | .6365 | 4.3681 | 3.7800 |

**Source: Gunst and Mason (1980).**

# 3
# Function Spaces and Series Estimators

## 3.1 Introduction

Assume that we observe the values of a response variable Y at n predetermined values of an independent variable t. The resulting bivariate observations $(t_1,y_1),\ldots,(t_n,y_n)$ follow the model

$$y_j = \mu(t_j) + \varepsilon_j, \quad j = 1,\ldots,n \quad , \tag{1}$$

where $\underline{\varepsilon} = (\varepsilon_1,\ldots,\varepsilon_n)'$ is a vector of zero mean, uncorrelated, random errors having some common variance $\sigma^2$ and $\mu$ is an unknown regression function that we wish to estimate. In this chapter we encounter our first variety of nonparametric regression estimators known as orthogonal series estimators.

To motivate the type of estimators to be studied let us think, for the moment, about estimation of $\mu$ in a linear parametric model. In that case there are functions $x_1,\ldots,x_\lambda$ such that $\mu(t) = \Sigma_{j=1}^{\lambda}\beta_j x_j(t)$ for some set of regression coefficients $\beta_1,\ldots,\beta_\lambda$. Estimation of $\mu$ is, therefore, equivalent to estimation of the $\beta_j$. This can be accomplished in a variety of ways, least squares being one possibility.

The use of a linear model for $\mu$ assumes that the basic form of $\mu$ is known. This will frequently not be the case. When knowledge of $\mu$ is more limited one may prefer to broaden the class of functions to which $\mu$ is assumed to belong. For example, one may only be willing to assume that $\mu$ is smooth, in the sense of being continuous and differentiable, or some other assumptions of this nature.

When $\mu$ is assumed to belong to some general (infinite dimensional) function class (1) is called a *nonparametric regression model*. A model of this type may be closer to reality than a linear model. However, the improvement in realism of our model is not without its price, for it is no longer clear how $\mu$ should be estimated. It is therefore natural to ask if there is some parallel of the linear model for our more general set of assumptions. For example, if $\mu$ is only assumed to be continuous, can we express the regression function as a possibly infinite linear combination of some known set of functions? If the answer to this and other similar questions is in the affirmative, there are some obvious advantages. Foremost of these is the fact that now estimation of $\mu$ becomes equivalent to estimation of regression coefficients. Thus, we can attempt to adapt what is known from linear models to aid us in determining how to estimate more general regression functions.

As we shall see in the next section, there are many classes of functions whose elements admit what might be termed infinite linear models, i.e., for which it is possible to express $\mu(t) = \Sigma_{j=1}^{\infty}\beta_j x_j(t)$ for some set of functions $\{x_j\}$. In this instance one approach is to approximate $\mu$ by $\Sigma_{j=1}^{\lambda}\beta_j x_j$, for some integer $\lambda$, and then estimate the $\beta_j$'s to obtain an estimate of $\mu$. The result is what is known as a *series estimator*. The selection of an appropriate value of $\lambda$ then falls into the category of problems addressed in Chapter 2.

In the next section a short review of some relevant function space theory is given that will provide a framework for our work in Section 3. Then, in Section 3, orthogonal series estimators of $\mu$ are developed from a general point of view.

Sections 4 and 5 of this chapter are devoted to the study of two important special cases of orthogonal series estimators: classical Fourier series and polynomial regression estimators. These two estimator types result from the representation of $\mu$ as an infinite weighted sum of trigonometric functions and polynomials, respectively. Unfortunately, polynomial regression and classical Fourier series estimators, while often used in practice, generally do not perform as well as the other estimators of $\mu$ studied in Chapters 4-7. However, these other estimators have strong ties to polynomial regression and Fourier series estimators. Thus, one of the principal reasons for studying orthogonal series estimators is as a motivational tool that provides a point of departure into the work of subsequent chapters.

## 3.2 Some Function Space Theory

In the previous section a relaxation of standard parametric assumptions was made where the regression function was only assumed to lie in some general class of functions. In this section we discuss several function classes which play an important role in all future developments.

One of the most important function spaces is $L_2[a,b]$, the set of all square integrable functions on the interval $[a,b]$. For two functions $\mu_1, \mu_2 \in L_2[a,b]$ we define their *norm* by

$$||\mu_1|| = \{\int_a^b \mu_1(t)^2 dt\}^{1/2} \tag{2}$$

and *inner product* by

$$\langle\mu_1, \mu_2\rangle = \int_a^b \mu_1(t)\mu_2(t)dt. \tag{3}$$

Note that (3) is well defined due to the Cauchy-Schwarz inequality

$$|\langle\mu_1, \mu_2\rangle|^2 \le ||\mu_1||^2 ||\mu_2||^2 . \tag{4}$$

The space $L_2[a,b]$ represents a rich collection of functions. However, it is not entirely suited to our purpose. Two functions, $\mu_1$ and $\mu_2$, are viewed as identical in $L_2[a,b]$ if $||\mu_1 - \mu_2|| = 0$. Thus $\mu_1$ and $\mu_2$ are identical if they differ on sets of Lebesgue measure zero. This means that, strictly speaking, the elements of

$L_2[a,b]$ are equivalence classes of functions which are equal almost everywhere with respect to Lebesgue measure. Consequently, the evaluation of an element of $L_2[a,b]$ at points in $[a,b]$ is not a well defined operation. Some restrictions are therefore required on the types of regression functions that can be permitted in order to ensure that model (1) is mathematically valid. Since it is often reasonable to assume that $\mu$ in (1) is a smooth function, one way to accomplish this is to restrict attention to collections of functions that are smooth in the sense of satisfying various continuity and differentiability conditions.

There are several subsets of $L_2[a,b]$ which possess the types of properties we require in a function space. One of these is

$$C^m[a,b] = \{\mu\colon \mu^{(j)} \text{ is continuous, } \quad j=0,\ldots,m\} . \tag{5}$$

In words, $C^m[a,b]$ is the set of all functions on $[a,b]$ with m continuous derivatives. When m=0 we write $C^0[a,b] = C[a,b]$ for the set of all continuous functions on $[a,b]$. Obviously $L_2[a,b] \supset C[a,b] \supset C^1[a,b] \ldots$.

For some of our work it will not be necessary for $\mu$ to have m continuous derivatives. It will often suffice for $\mu^{(m)}$ to be square integrable. Thus, we define the mth order Sobolev space

$$W_2^m[a,b] = \left\{\mu\colon \begin{array}{l}\mu^{(j)} \text{ is absolutely continuous,} \\ j=0,\ldots,m-1, \text{ and } \mu^{(m)} \in L_2[a,b]\end{array}\right\} . \tag{6}$$

Note that $W_2^m[a,b] \supset C^m[a,b]$. When m=0 we have $W_2^0[a,b] = L_2[a,b]$.

If we restrict attention entirely to functions in $W_2^m[a,b]$ $(m \geq 1)$ or $C^m[a,b]$ $(m \geq 0)$ the ambiguity associated with point evaluation disappears; since if $\mu_1, \mu_2 \in C[a,b]$ and $\|\mu_1 - \mu_2\| = 0$ this means that $\mu_1 \equiv \mu_2$. Both $W_2^m[a,b]$ and $C^m[a,b]$ may be equipped with their own norms and studied independently of any connection with $L_2[a,b]$. On the other hand, the fact that they are subsets of $L_2[a,b]$ means that the properties of $L_2[a,b]$ functions described below will be applicable to the elements of $W_2^m[a,b]$ and $C^m[a,b]$ as well.

It is easy to verify that $L_2[a,b]$ is a vector space, i.e., if $\mu_1$, $\mu_2 \in L_2[a,b]$ then $\mu_1 + \mu_2 \in L_2[a,b]$, etc. Now, for finite dimensional vector spaces such as Euclidean p-space, $\mathbb{R}^p$, it is always possible to represent any element in the vector space as a linear combination of basis elements. One might expect an analogous property would hold for an infinite dimensional vector space such as $L_2[a,b]$. This is, in fact, the case. To make this result precise a few definitions are required.

**Definition 1.** Two functions $\mu_1$, $\mu_2 \in L_2[a,b]$ are said to be *orthogonal* if $\langle\mu_1,\mu_2\rangle = 0$. We denote this by $\mu_1 \perp \mu_2$.

**Definition 2.** A sequence of functions $\{x_j\}$ is said to be *orthonormal* if the $x_j$ are pairwise orthogonal and $||x_j|| = 1$ for all j.

**Definition 3.** A sequence of functions $\{x_j\}$ is said to be a *complete orthonormal system* (CONS) if $\mu \perp x_j$ for all j entails that $\mu \equiv 0$.

Recalling the definitions of orthogonality and orthonormality in $\mathbb{R}^p$, it is seen that Definitions 1 and 2 provide obvious parallels of these concepts for $L_2[a,b]$. Definition 3 states that $\{x_j\}$ is a CONS provided the only function orthogonal to all the $x_j$ is the zero function. There are many CONS's for $L_2[a,b]$. Two important examples are the following.

**Example 1: Complex Exponentials.** Consider the slightly more general case where $L_2[a,b]$ is allowed to contain complex as well as real valued functions. Thus any function $\mu \in L_2[a,b]$ will have real and imaginary parts $\mu_{Re}$ and $\mu_{Im}$, respectively. Both $\mu_{Re}$ and $\mu_{Im}$ are real valued functions and $\mu$ can be written as $\mu = \mu_{Re} +$

$i\mu_{Im}$ with $i^2 = -1$. The complex conjugate of $\mu$ is then defined by $\mu^* = \mu_{Re} - i\mu_{Im}$.

When complex valued functions are included the inner product and norm for $L_2[a,b]$ are defined by

$$\langle \mu_1, \mu_2 \rangle = \int_a^b \mu_1(t)\mu_2^*(t)dt \tag{7}$$

and

$$||\mu_1|| = \{\int_a^b |\mu_1(t)|^2 dt\}^{1/2} \quad , \tag{8}$$

where $|\mu_1(t)|^2 = \mu_1(t)\mu_1^*(t)$. The integration in (7) is done on a term by term basis with $\langle\cdot,\cdot\rangle$ being complex valued in general. By our definition of the modulus $|\mu_1(t)|$ it follows that (8) is real valued.

One CONS for $L_2[a,b]$ is provided by the complex exponentials. These functions are defined by

$$x_j(t) = (b-a)^{-1/2} e^{2\pi ijt/(b-a)} \quad , \quad j = 0, \pm 1, \ldots, \tag{9}$$

where $e^{ix} = \cos x + i\sin x$. The orthonormality property of the $x_j$ can be established directly. For verification of their completeness see Davis (1975, Chapter XII).

**Example 2. The Legendre Polynomials.** Another CONS for $L_2[a,b]$ can be derived from the polynomial functions $p_j(t) = t^j$, $j=0,1,\ldots$. One proceeds as follows. First take

$$x_0(t) = p_0(t)/||p_0|| \equiv 1/[b - a]^{1/2} \quad . \tag{10}$$

Now define the remaining functions recursively via the formula

$$x_j(t) = [p_j(t) - \sum_{k=0}^{j-1} \langle p_j, x_k \rangle x_k(t)]/||p_j - \sum_{k=0}^{j-1} \langle p_j, x_k \rangle x_k|| \quad . \tag{11}$$

The functions defined by (10)-(11) are known as the Legendre polynomials. It is easy to see that they are orthonormal. A proof that they are complete can be found in Davis (1975, Chapter XII).

The recursive method used to derive the Legendre polynomials is called the *Gram-Schmidt orthonormalization process.* As can be seen from (10)-(11) there is no reason to restrict its use to the polynomial functions. It can be used, more generally, to orthonormalize any sequence of functions.

The above examples serve to illustrate the existence of a CONS for $L_2[a,b]$. The following proposition states that any CONS actually gives us a basis for $L_2[a,b]$. Its proof can be found in any standard text concerning functional analysis, e.g., Luenberger (1969).

**Proposition 3.1.** Let $\{x_j\}_{j=1}^{\infty}$ be a CONS for $L_2[a,b]$ and for $\mu \in L_2[a,b]$ define

$$\beta_j = \langle \mu, x_j \rangle , \qquad j=1,\ldots. \tag{12}$$

Then $\Sigma_{j=1}^{\lambda} \beta_j x_j$ is a best approximation to $\mu$ in the sense that

$$||\mu - \sum_{j=1}^{\lambda} \beta_j x_j|| \le ||\mu - \sum_{j=1}^{\lambda} \alpha_j x_j||$$

for all $\underline{\alpha} = (\alpha_1,\ldots,\alpha_\lambda)' \in \mathbb{R}^\lambda$. Moreover

$$||\mu - \sum_{j=1}^{\lambda} \beta_j x_j||^2 \to 0$$

as $\lambda \to \infty$.

Proposition 3.1 states that the function sequence $\Sigma_{j=1}^{\lambda} \beta_j x_j$ converges to $\mu$ in $L_2[a,b]$ norm. We write this as

$$\mu(t) \sim \sum_{j=1}^{\infty} \beta_j x_j(t) \tag{13}$$

to indicate that $\mu$ and $\Sigma_{j=1}^{\infty} \beta_j x_j$ are identical functions when viewed as elements of $L_2[a,b]$. However, due to the properties of Lebesgue measure, this does not signify pointwise convergence in any sense; although that will also be true in most cases we consider.

As a result of Proposition 3.1 we can represent $\mu$ (in the $L_2[a,b]$ sense) as an infinite linear combination of the basis functions $\{x_j\}$. The optimal coefficients (12) used in the expansion (13) for $\mu$ are called *generalized Fourier coefficients* while the expansion itself is called a *generalized Fourier series* for $\mu$. The generalized Fourier coefficients can be shown to satisfy

$$\sum_{j=1}^{\infty} |\beta_j|^2 = ||\mu||^2 \quad , \tag{14}$$

which is known as *Parseval's relation*. It has the consequence that the generalized Fourier coefficients of $L_2[a,b]$ functions can be viewed as elements of the collection of all square summable sequences. The converse is also true; if $\{\beta_j\}$ is a square summable sequence then $\mu = \Sigma_{j=1}^{\infty}\beta_j x_j$ is in $L_2[a,b]$.

**Example 1 (continued).** Using Proposition 3.1 we see that if $\mu$ is any function in $L_2[a,b]$ then

$$\mu(t) \sim (b-a)^{-1} \sum_{j=-\infty}^{\infty} \beta_j e^{2\pi i j t/(b-a)} \quad , \tag{15a}$$

where

$$\beta_j = \int_a^b \mu(s) e^{-2\pi i j s/(b-a)} ds \quad . \tag{15b}$$

This is the classical Fourier series expansion for $\mu$ that spawned the work leading to the more general developments summarized in this section.

The functions we will be interested in will all be real valued. In this case (15) can be written in an alternative form. Since $\mu$ is real

$$\beta_j^* = \int_a^b \mu^*(t) e^{2\pi ijt/(b-a)} dt$$

$$= \int_a^b \mu(t) e^{2\pi ijt/(b-a)} dt = \beta_{-j} \quad .$$

Thus

$$(b-a)\mu(t) \sim \beta_0 + \sum_{j=1}^{\infty} (\beta_j e^{2\pi ijt/(b-a)} + \beta_{-j} e^{-2\pi ijt/(b-a)})$$

$$= \beta_0 + \sum_{j=1}^{\infty} (\beta_j e^{2\pi ijt/(b-a)} + \beta_j^* e^{-2\pi ijt/(b-a)}) \quad .$$

A little algebra using the definition of the complex exponentials then reveals that the Fourier series for $\mu$ can be expressed in terms of the sine and cosine functions as

$$(b-a)^{-1}[\beta_0 + \sum_{j=1}^{\infty} \{c_j \cos(2\pi jt/(b-a)) + s_j \sin(2\pi jt/(b-a))\}], \tag{16a}$$

where

$$c_j = 2 \int_a^b \cos(2\pi jx/(b-a))\mu(x)dx \tag{16b}$$

and

$$s_j = 2 \int_a^b \sin(2\pi jx/(b-a))\mu(x)dx \; . \tag{16c}$$

It is useful to remember that (16) holds since the properties of sines and cosines may be more familiar to some than those of the complex exponentials. However, the two forms (15) and (16) are equivalent, and hence, which one is used is primarily a matter of taste. Since the exponential functions provide more notationally compact expressions we will tend to use (15) in much of the sequel.

Throughout this section we have assumed t to be scalar valued. Similar developments can be pursued when t is vector valued. Some of these will be explored in the exercises.

## 3.3 Generalized Fourier Series Estimators

In the previous section we saw that given a CONS, $\{x_j\}$, any function in $L_2[a,b]$ had a formal generalized Fourier series expansion $\Sigma_{j=1}^{\infty}\beta_j x_j(t)$. Let us now assume that $\mu \in L_2[a,b]$ and that $\Sigma_{j=1}^{\lambda}\beta_j x_j$ converges to $\mu$ uniformly in t as $\lambda\to\infty$. (This is not as restrictive as it may seem. For example, the classical Fourier series expansion holds uniformly in t provided $\mu \in C^1[a,b]$, $\mu(a) = \mu(b)$ and $\mu'(a) = \mu'(b)$. See, e.g., Dym and McKean, 1972.) Under these assumpions on $\mu$, model (1) becomes

$$y_j = \sum_{k=1}^{\infty} \beta_k x_k(t_j) + \varepsilon_j, \quad j=1,\ldots,n. \tag{17}$$

Thus, the data follow a linear model with infinitely many unknown regression coefficients.

Since $\mu \in L_2[0,1]$ we know that $\Sigma_{j=1}^{\infty}\beta_j^2 < \infty$, so the $\beta_j$ must eventually decay to zero. Consequently, it seems reasonable to assume that there is an integer $\lambda$ such that $\mu \doteq \Sigma_{j=1}^{\lambda}\beta_j x_j$ and, hence,

$$y_i \doteq \sum_{j=1}^{\lambda} \beta_j x_j(t_i) + \varepsilon_i \quad , \qquad i = 1,\ldots,n.$$

This looks like a linear model. Consequently, it seems reasonable to attempt to apply inferential techniques for linear models to this setting.

To simplify what follows we will assume throughout the remainder of this chapter that, unless stated otherwise, the design points $t_1,\ldots,t_n$ are uniformly spaced over $[a,b]$, i.e., that

$$t_j = a + (b - a)(j - 1)/(n - 1) \ , \quad j = 1,\ldots,n.$$

If this is the case then, by analogy with the linear regression setting, we might estimate the unknown Fourier coefficients in our approximation to $\mu$ by the minimizers of the mean squared error

$$n^{-1}\sum_{j=1}^{n}(y_j - \sum_{k=1}^{\lambda}\alpha_k x_k(t_j))^2$$

with respect to $\alpha_1,\ldots,\alpha_\lambda$. When the $t_j$ are equally spaced this practice can be justified heuristicly by using the fact that $\beta_1,\ldots,\beta_\lambda$ minimize $||\mu - \Sigma_{k=1}^{\lambda}\alpha_k x_k||^2$ and that the mean squared error provides an estimate of this norm difference. If the $t_j$ are not equally spaced and are, instead, generated by a design density p through the relation

$$\int_a^{t_j} p(t)dt = (j - 1)/(n - 1) ,$$

then the mean squared error will estimate $\int_a^b(\mu(t) - \Sigma_{k=1}^{\lambda}\alpha_k x_k(t))^2 p(t)dt$ rather than the $L_2[0,1]$ norm difference. Thus either our estimation criterion or our definition of the $\beta_j$ must be adjusted to deal with this case. There are various ways to accomplish this: see, e.g., Rutkowski (1982), Rafajlowicz (1987) and the discussions in Section 3.5.2.

Define the design matrix by

$$X_\lambda = \{x_j(t_i)\}_{\substack{i=1,n\\ j=1,\lambda}} \quad . \tag{18}$$

If $X_\lambda$ has full column rank then the unique minimizer of the mean squared error is

$$\underline{\beta}_\lambda = (\beta_{\lambda 1},\ldots,\beta_{\lambda\lambda})' = (X_\lambda^* X_\lambda)^{-1}X_\lambda^*\underline{y}, \tag{19a}$$

where

$$\underline{y} = (y_1,\ldots,y_n)'$$

is the vector of observations and the "*" notation in (19a) indicates complex conjugate transpose. The resulting *generalized Fourier series estimator* of $\mu(t)$ is

$$\mu_\lambda(t) = \sum_{j=1}^{\lambda}\beta_{\lambda j}x_j(t) \ . \tag{19b}$$

To estimate the error variance we could then use

$$\mathrm{MSE}(\lambda) = n^{-1} \sum_{j=1}^{n} \left(y_j - \sum_{k=1}^{\lambda} \beta_{\lambda k} x_k(t_j)\right)^2$$

or

$$\sigma_\lambda^2 = n\mathrm{MSE}(\lambda)/(n - \lambda) \quad .$$

If $\beta_j = 0$, we do not need the function $x_j$ to approximate $\mu$. Thus, to ascertain if $x_j$ should be used in fitting the data we might examine the statistic (c.f. the Appendix)

$$Z = \beta_{\lambda j}/[\sigma_\lambda^2 C_{\lambda,jj}]^{1/2} \quad ,$$

where $C_{\lambda,jj}$ is the jth diagonal element of

$$C_\lambda = (X_\lambda^* X_\lambda)^{-1} \quad .$$

This statistic is patterned after the usual statistic used for testing hypotheses about the parameters in a linear model. Similar considerations suggest using an interval estimate for the value of $\mu$ at t of the form

$$\mu_\lambda(t) \pm Z_{\alpha/2}[\sigma_\lambda^2 \underline{x}_\lambda(t)^* C \underline{x}_\lambda(t)]^{1/2} \quad ,$$

with $\underline{x}_\lambda(t) = (x_1(t),\ldots,x_\lambda(t))'$ and $Z_{\alpha/2}$ the upper $\alpha/2$ percentage point of the standard normal distribution. Diagnostic analysis for the estimator could be conducted using the techniques for linear regression discussed in the Appendix.

The strategy we have just outlined for inference about $\mu$ is intuitively appealing. However, problems can arise with this approach. For example, there is no assurance that the estimator of $\mu$ constructed in this fashion is even consistent. Inference of any kind is of doubtful value unless this is the case.

Define the loss and risk for estimating $\mu$ with $\mu_\lambda$ by

$$L_n(\lambda) = n^{-1} \sum_{j=1}^{n} (\mu(t_j) - \mu_\lambda(t_j))^2 \tag{20}$$

and

$$R_n(\lambda) = EL_n(\lambda)$$

$$= n^{-1} \sum_{j=1}^{n} (\mu(t_j) - E\mu_\lambda(t_j))^2 + \sigma^2\lambda/n \quad .$$

Let us work with $R_n(\lambda)$ for the present. We will say that $\mu_\lambda$ is consistent if $R_n(\lambda) \to 0$ as $n\to\infty$.

The first term in the risk is the sum of the squared biases. Intuitively, this should decay as $\lambda$ grows since by increasing $\lambda$ we are including more and more terms in our approximation to $\mu$. Thus, for $R_n(\lambda)$ to converge to zero we would anticipate the necessity that $\lambda \to \infty$ as $n \to \infty$.

The second term in the risk, $\sigma^2\lambda/n$, corresponds to the variance of the estimator $\mu_\lambda$. This term grows with $\lambda$ so that $R_n(\lambda)$ is bounded away from zero unless $\lambda/n \to 0$. This gives the conclusion that consistency of $\mu_\lambda$ should require $\lambda$ to grow at a rate smaller than n.

To actually establish consistency one must now attempt to characterize the asymptotic behavior of $n^{-1}\Sigma_{j=1}^{n}(\mu(t_j) - E\mu_\lambda(t_j))^2$. It is difficult to get very far here without making some explicit assumptions about the elements in the CONS. However, our intuition would suggest that its behavior should depend on the unestimated Fourier coefficients and the ability of the $\beta_{\lambda j}$ to estimate the $\beta_j$.

Shibata (1981) gives results which are relevant to the consistency of least-squares estimators for model (17). Li (1984a) focuses on the existence of consistent estimators for some functional of the $\beta_j$, e.g., the value of $\mu$ at a fixed point t. He gives necessary and sufficient conditions for the existence of a consistent estimator and actually constructs an estimator which is consistent. Other consistent estimators that are closely related to estimator (19b) are studied by Rutkowski (1982) and Rafajlowicz (1987).

Suppose now that we have decided to use an estimator of the form (19) to estimate $\mu$. To determine which estimator to use it becomes necessary to select a value of $\lambda$. The choice of $\lambda$ will typically govern how smooth or wiggly the estimator will be. Usually, larger values of $\lambda$ will result in estimators that come closer to interpolation of the $(t_j, y_j)$ than smaller values and, hence, will tend to be more wiggly. In particular, since we have assumed that $X_\lambda$ has full column rank in (19), the value $\lambda=n$ corresponds to the interpolation case with $\mu_\lambda(t_j) = y_j$, $j=1,\ldots,n$.

The problem of selecting an optimal $\lambda$ is similar to the variable selection problem discussed in Chapter 2 and can be treated using the methodology developed there. For any particular value of n an optimal $\lambda$ is one which minimizes $L_n(\lambda)$. To estimate the minimizer of $L_n(\lambda)$ we can then use any one of the criteria studied in Chapter 2. For example, the GCV estimate of the optimal $\lambda$ is $\hat{\lambda}_n$ obtained by minimizing

$$GCV_n(\lambda) = n^{-1} \sum_{j=1}^{n} (y_j - \mu_\lambda(t_j))^2/(1 - \lambda/n)^2 \quad . \tag{22}$$

Under certain conditions the GCV method of selecting $\lambda$ is (asymptotically) optimal as indicated by the following theorem due to Li (1987).

**Theorem 3.1 (Li, 1987).** Assume that

i) the $\varepsilon_i$ are independent and identically distributed with $E\varepsilon_1^4 < \infty$ and there exists a constant K such that for any $c \geq 0$

$$\sup_x P(x - c \leq \varepsilon_1 \leq x + c) \leq K c \quad , \tag{23}$$

ii) $\inf_{\lambda \leq n} nR_n(\lambda) \to \infty$ as $n \to \infty$ ,

and

iii) $\inf_{\lambda \leq n} L_n(\lambda) \xrightarrow{P} 0$ .

Then, if $\hat{\lambda}_n$ is the minimizer of (22),

$$L_n(\hat{\lambda}_n)/ \inf_{\lambda \leq n} L_n(\lambda) \xrightarrow{P} 1. \tag{24}$$

Theorem 3.1 states that, under certain restrictions, the GCV estimator, $\hat{\lambda}_n$, and the optimal value of $\lambda$ behave the same for sufficiently large n in terms of the corresponding values for the loss. Alternatively, (24) can be viewed as saying that $L_n(\hat{\lambda}_n)$ will tend toward the minimal loss as $n \to \infty$. Similar results can be established for some of the other criteria for selecting $\lambda$ discussed in Chapter 2: See Li (1987).

The conditions required for Theorem 3.1 merit some discussion. Condition i) is concerned with the random errors and is self-explanatory except for (23). If, for example, the $\varepsilon$'s have a bounded density, $f_\varepsilon$, then

$$P(x - c \le \varepsilon \le x + c) = \int_{x-c}^{x+c} f_\varepsilon(t)dt$$

$$\le [\sup_t f_\varepsilon(t)]\ 2c$$

and (23) is satisfied. So (23) is a rather mild restriction.

Condition ii) states that the optimal rate of convergence of $R_n(\lambda)$ to zero must be slower than $n^{-1}$. Condition iii) merely requires the existence of an estimator that is consistent in the sense that the minimum loss decays to zero. To establish this we need only show that there exists a sequence $\lambda_n$ such that $L_n(\lambda_n) \overset{p}{\to} 0$ (Exercise 14). A somewhat stronger condition which implies condition iii) is that $\inf_{\lambda \le n} R_n(\lambda) \to 0$ as $n \to \infty$ (Exercise 4). The latter condition is often easier to verify in practice.

In the next section we examine the Fourier series estimator $\mu_\lambda$ in the special case of the complex exponential CONS. We will see that it is possible to establish consistency for $\mu_\lambda$ in this case and one can justify, asymptotically, the plan of attack discussed above for diagnostic and inferential analysis under this choice for a CONS. The details will be representative of, although undoubtably simpler than, what one might expect to encounter for some general orthonormal basis.

### 3.4 Classical Fourier Series Estimators

In this section we discuss the properties of the particular type of series estimator which results when the complex exponentials are used for a CONS. The resulting estimators correspond to the classical concept of a Fourier series. For that reason we will refer to them as *classical Fourier series estimators* or, when the context is clear, merely as Fourier series estimators.

Classical Fourier series estimators were, at one time, popular in the field of time series analysis. In some respects this is still true today; although modern time series methods employ more sophisticated, related techniques.

The stimulus for the use of Fourier methods by time series analysts was their recognition that, when observing data over time, some aspects of real world phenomena tended to exhibit cycles or periodicities. In some cases the cycles were often while in other cases they were less frequent. For example, retail sales experience yearly cycles (e.g., at Christmas) as well as longer term upward and downward trends. Thus, they began to think about analyzing the frequency composition of a time series.

When thinking of periodic behavior perhaps the first functions which come to mind are the sines and cosines. Thus, it was natural for time series analysts to consider the use of regression models involving sines and cosines to describe time series behavior. The idea was to use linear combinations of functions such as 1 and $\{\cos(2\pi jt), \sin(2\pi jt)\}$, $j=1,\ldots,\lambda$, for some nonnegative integer $\lambda$, to model a series. The high frequency (repeats often) aspects of the series should be picked up by the higher frequency elements of the model: i.e., $\sin(2\pi jt)$ and $\cos(2\pi jt)$ for large $j$. The low frequency components of the series, such as trends, would then be modeled by the constant and other low frequency terms.

For the problems we will be interested in, t may or may not represent time. Nonetheless insight into the properties of Fourier series estimators can be gained through analogy with the time series setting. We may think of $\mu$ in model (1) as representing a trend or low frequency component while the random errors, since they will tend to oscillate rapidly about the regression curve, can be thought of as high frequency in nature. To estimate $\mu$ (the low frequency component) we should therefore fit a linear model involving the first few sine and cosine terms. In view of (16), the result is recognized as a (classical) Fourier series estimator of $\mu$.

In the analysis which follows it will be useful to make some simplifying assumptions which make the notation more managable and aid in avoiding certain technical difficulties. Specifically, we hereafter assume that

$$[a,b] = [0,1] \tag{25a}$$

and, hence, that the $t_j$ are uniformly spaced on $[0,1]$, i.e., that

$$t_j = (j - 1)/n, \qquad j=1,\ldots,n. \tag{25b}$$

### 3.4.1 Form of the Estimator

In this section we will derive various forms for the classical Fourier series estimator, $\mu_\lambda$, under the assumption that (25) holds. These will be valuable in the sequel and also provide further insight into how Fourier series estimators actually work.

Let

$$x_j(t) = e^{2\pi ijt}, \qquad j = 0, \pm 1, \ldots,$$

and, for some integer $\lambda \geq 0$, consider fitting the function $\Sigma_{j=-\lambda}^{\lambda}\beta_j x_j(t)$ to the data by the method of least squares. As noted in Section 3, the estimates of the $\beta_j$ are

$$\underline{\beta}_\lambda = (\beta_{\lambda-\lambda},\ldots,\beta_{\lambda\lambda})' = (X_\lambda^* X_\lambda)^{-1} X_\lambda^* \underline{y}, \tag{26}$$

where

$$X_\lambda = \{e^{2\pi ijr/n}\}_{\substack{r=0,n-1\\ j=-\lambda,\lambda}} \quad .$$

This can be simplified somewhat through the use of the identity

$$n^{-1}\sum_{r=0}^{n-1} e^{2\pi ijr/n}e^{-2\pi ikr/n} = \begin{cases} 1, & j = k\ , \\ 0, & j \neq k\ . \end{cases} \tag{27}$$

As a result of (27) we see that $X_\lambda^* X_\lambda = nI$. Thus, (26) becomes

$$\beta_{\lambda j} = n^{-1}\sum_{r=1}^{n} y_r e^{-2\pi ij(r - 1)/n} \quad , \ j=-\lambda,\ldots,\lambda. \tag{28}$$

Our estimator of $\mu$ is therefore seen to be

$$\mu_\lambda(t) = \sum_{j=-\lambda}^{\lambda} \beta_{\lambda j} e^{2\pi ijt} \quad . \tag{29}$$

Notice that under this design the $\beta_{\lambda j}$ do not depend on $\lambda$. We will however continue to use the $\lambda$ subscript for notational continuity.

The benefits of assumption (25b) are apparent from (28)-(29). When the $t_j$ are evenly spaced over $[0,1]$ we see that the exponential functions exhibit an orthogonality property (27), when summed over $t_1,\ldots,t_n$, that parallels what transpires when they are integrated over $[0,1]$. In addition, $\beta_{\lambda j}$ is recognized as a direct estimate of the jth coefficient in the Fourier expansion for $\mu$.

Using developments similar to those required to establish (16), $\mu_\lambda$ can also be expressed in terms of sines and cosines. The result is

$$\mu_\lambda(t) = \beta_{\lambda 0} + \sum_{j=1}^{\lambda} (c_{\lambda j}\cos(2\pi jt) + s_{\lambda j}\sin(2\pi jt)) \quad ,$$

where

$$c_{\lambda j} = \frac{2}{n}\sum_{r=1}^{n} y_r\cos(2\pi j(r - 1)/n)$$

and

$$s_{\lambda j} = \frac{2}{n} \sum_{r=1}^{n} y_r \sin(2\pi j(r - 1)/n) .$$

This shows that $\mu_\lambda$ is obtained by regression of $\underset{\sim}{y}$ on the sine and cosine functions as discussed earlier.

There is still another useful expression for $\mu_\lambda$ which can be obtained by first writing

$$\mu_\lambda(t) = \sum_{j=-\lambda}^{\lambda} \beta_{\lambda j} e^{2\pi ijt}$$

$$= n^{-1} \sum_{r=1}^{n} y_r \sum_{j=-\lambda}^{\lambda} e^{2\pi ij(t-t_r)} .$$

Now use the identity

$$\sum_{j=-\lambda}^{\lambda} e^{2\pi ijx} = \sin[\pi(2\lambda + 1)x]/\sin\pi x \equiv K_\lambda(x) \tag{30}$$

to obtain

$$\mu_\lambda(t) = n^{-1} \sum_{r=1}^{n} y_r K_\lambda(t - t_r) . \tag{31}$$

The function defined in (30) is called the *Dirichlet kernel.* It is a periodic function with period one that is well defined for non-integral arguments. If u is an integer we define $K_\lambda(u) = (2\lambda + 1)$.

Note that

$$\sum_{r=1}^{n} K_\lambda(t - t_r) = n.$$

Thus, equation (31) has the implication that $\mu_\lambda(t)$ is a weighted average (if we allow for negative weights) of the $y_r$ with weights obtained from the Dirichlet kernel.

Graphs of the Dirichlet kernel are shown in Figure 3.1 for several values of $\lambda$. We see that as $\lambda$ grows larger the function

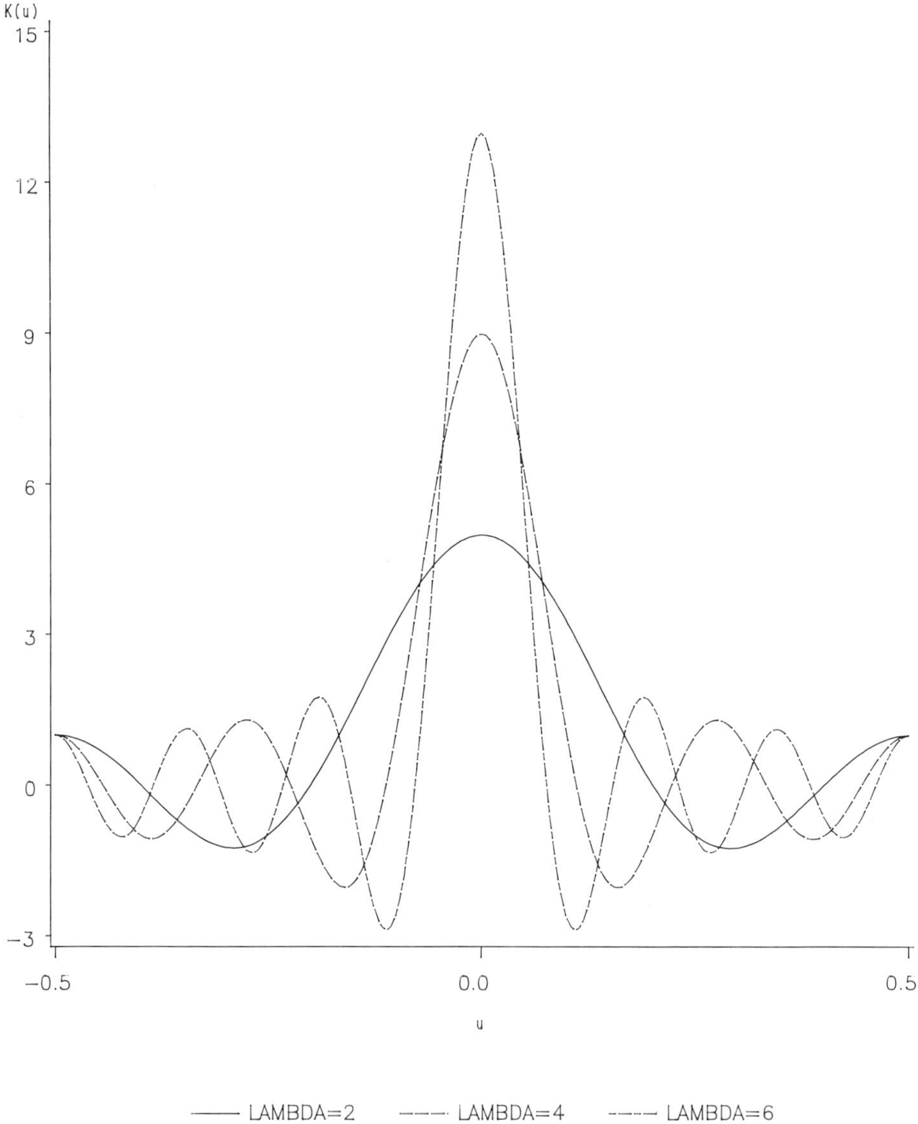


**Figure 3.1** Examples of the Dirichlet kernel.

becomes more peaked. This has the consequence that, for larger $\lambda$, the larger weights in computing $\mu_\lambda(t)$ will be placed on observations near t.

The kernel also has a rather displeasing property in that as $\lambda$ grows large it oscillates more and more rapidly away from its peak. This means that the estimator is adding in or subtracting off contributions from observations far from the point of estimation and doing so in an increasingly rapid fashion as $\lambda$ grows large. One might hope that the contributions from these oscillations will tend to knock each other out, on the average, in which case $\mu_\lambda(t)$ is essentially a local average of observations near t. The role of $\lambda$ is then to determine how dependent the estimator is on local behavior of the data. Increasing $\lambda$ would correspond to increasing local dependence and result in a more wiggly estimator.

In light of the previous discussion an obvious modification to $\mu_\lambda$ would be to replace the Dirichlet kernel in (31) with a function that is better behaved. One result of doing so is the kernel estimators of $\mu$ studied in the next chapter.

### 3.4.2 Some Asymptotic Theory

The simple expression for $\mu_\lambda$ provided by (28)-(29) makes it possible to obtain a simple form for the risk, $R_n(\lambda)$, as well. In this section we exploit this fact to investigate the asymptotic behavior of the risk as n and $\lambda$ grow large. We will see results similar to those obtained here in subsequent work.

Define the risk for estimating $\mu$ with $\mu_\lambda$ by

$$R_n(\lambda) = n^{-1} \sum_{r=0}^{n-1} E|\mu(r/n) - \mu_\lambda(r/n)|^2 .$$

Our objective will be to see how $R_n(\lambda)$ behaves as a function of n and $\lambda$ under certain restrictions on $\mu$. We begin by assuming that $\mu \in W_{2,per}^2[0,1]$, where $W_{2,per}^2[0,1] = \{\mu : \mu \in W_2^2[0,1]$ and $\mu(0)$

$= \mu(1),\ \mu'(0) = \mu'(1)\}$. Thus $W^2_{2,per}[0,1]$ represents a class of smooth, periodic functions. It is a well known fact (Davis, 1975, Chapter XII) that the Fourier series for $\mu$ converges to $\mu$ uniformly (in t) under this assumption.

Since $\operatorname{tr} X_\lambda(X^*_\lambda X_\lambda)^{-1}X^*_\lambda = n^{-1}\operatorname{tr}X^*_\lambda X_\lambda = 2\lambda + 1$, our work in Chapter 2 can be used to see that

$$R_n(\lambda) = n^{-1}\sum_{r=0}^{n-1} |\mu(r/n) - E\mu_\lambda(r/n)|^2 + \sigma^2(2\lambda + 1)/n. \tag{32}$$

Since the sum in (32) is nonnegative it follows immediately that $R_n(\lambda)$ is bounded away from zero unless $\lambda/n \to 0$ as $n\to\infty$. Thus, a necessary condition for the risk to approach zero is that the number of terms included in the Fourier series estimator has smaller order than the sample size. More precise statements along these lines will require study of the squared bias component $n^{-1}\Sigma_r|\mu(r/n) - E\mu_\lambda(r/n)|^2$ which is where we now direct our attention.

Let us begin by deriving a convenient expression for $\mu(t) - E\mu_\lambda(t)$. First note that, because $\mu$ is represented uniformly by its Fourier series, this means in particular that

$$\mu(k/n) = \sum_{j=-\infty}^{\infty} \beta_j e^{2\pi ijk/n}\quad , \ k=0,\ldots,n - 1\ , \tag{33}$$

where

$$\beta_j = \int_0^1 \mu(s)e^{-2\pi ijs}\, ds.$$

The expectation of $\mu_\lambda(t)$ depends on $E\beta_{\lambda j}$, $j = -\lambda,\ldots,\lambda$. Thus taking expectations and using (33) gives

$$E\beta_{\lambda j} = n^{-1}\sum_{k=0}^{n-1} \mu(k/n)e^{-2\pi ijk/n}$$

$$= \sum_{r=-\infty}^{\infty} \beta_r\{n^{-1}\sum_{k=0}^{n-1} e^{-2\pi ik(j-r)/n}\}$$

$$= \beta_j + \sum_{r\neq 0} \beta_{j+nr} \quad . \tag{34}$$

(In establishing (34) we have also used the fact that $e^{2\pi ik} = \cos 2k\pi + i\sin 2k\pi = 1$.) Combining (33) and (34) results in the expression

$$\mu(t) - E\mu_\lambda(t) = \sum_{|j|>\lambda} \beta_j e^{2\pi ijt} - \sum_{|j|\leq\lambda} \{ \sum_{r\neq 0} \beta_{j+nr} \} e^{2\pi ijt} \quad . \tag{35}$$

Notice that (34) gives a precise expression for the bias of $\beta_{\lambda j}$ in estimating $\beta_j$: namely,

$$\text{Bias } \beta_{\lambda j} = E\beta_{\lambda j} - \beta_j$$

$$= \sum_{r\neq 0} \beta_{j+nr} \quad .$$

As $n\to\infty$ this expression will involve only Fourier coefficient with indices of large magnitude. Consequently, we would expect the bias of $\beta_{\lambda j}$ to be negligible for large n. Under certain assumptions, this is actually the case as will be seen below. As a result, inference about the $\beta_j$ can be conducted in a similar manner to that of linear regression analysis. A demonstration of this is given in Section 3.4.5.

A useful form for $R_n(\lambda)$ results if we substitute (35) into (32) and use (27) along with a manipulation similar to that required for (34) to obtain

$$R_n(\lambda) = \sum_{|j|>\lambda} |\beta_j|^2 + \sigma^2(2\lambda + 1)/n$$

$$+ \sum_{|j|>\lambda} \beta_j \sum_{r\neq 0} \theta^*_{j+nr} \ , \tag{36}$$

where

$$\theta_j = \begin{cases} \beta_j & , \quad |j| > \lambda \ , \\ \beta_j - E\beta_{\lambda j} & , \quad |j| \leq \lambda \ , \end{cases}$$

and we have assumed that $2\lambda + 1 < n$. This expression reveals how $R_n(\lambda)$ depends on the Fourier coefficients of $\mu$. In particular, we can use Parseval's relation to see that $\Sigma_{|j|>\lambda}|\beta_j|^2$ is the $L_2[0,1]$ norm of the error from approximating $\mu$ by the $\lambda$th partial sum of its Fourier series, $\Sigma_{j=-\lambda}^{\lambda}\beta_j e^{2\pi ijt}$. For n sufficiently large the last term in (36) will generally be negligible relative to the other two. Thus, asymptotically, selection of an optimal $\lambda$ will essentially correspond to a balancing of the stochastic error or variance term, $\sigma^2(2\lambda + 1)/n$, and the purely deterministic error from approximating $\mu$ by the $\lambda$th partial sum of its Fourier series.

To study the asymptotic behavior of $R_n(\lambda)$ it now becomes clear that we must make some assumptions about the Fourier coefficients for $\mu$. It is apparent from Parseval's relation that, for example, $|\beta_j|^2 \to 0$ as $j \to \infty$. However, more precise statements concerning the rate of decay of the $|\beta_j|$ can be made for smooth functions. For example, if $\mu \in W_{2,per}^2[0,1]$, its second derivative is in $L_2[0,1]$ and integration by parts reveals that

$$\begin{aligned}\beta_j &= \int_0^1 \mu(t)\, e^{-2\pi ijt}dt \\ &= -(2\pi j)^{-2}\int_0^1 \mu''(t)e^{-2\pi ijt}dt \\ &\equiv -(2\pi j)^{-2}\beta_j'' \quad .\end{aligned}$$

Since

$$||\mu''||^2 = \sum_{j=-\infty}^{\infty} |\beta_j''|^2 = \sum_{j=-\infty}^{\infty} (2\pi j)^4|\beta_j|^2 < \infty$$

it follows that $|\beta_j|^2 = o(j^{-4})$. Thus, we know at least that the squared moduli of the Fourier coefficients for all functions in $W_{2,per}^2[0,1]$ must decay faster than $j^{-4}$.

One possible way to model the Fourier coefficients for $\mu$ is to assume that the $|\beta_j|^2$ decay algebraically, i.e., that

$$|\beta_j|^2 \sim C|j|^{-5-\delta} \tag{37}$$

for some positive constants C and $\delta$. If the $\beta_j$ satisfy this condition it follows from calculations above that $||\mu''|| < \infty$. By combining this fact with results from calculus concerning the continuity and differentiability of the limit function of a uniformly convergent sequence (see, e.g., Taylor and Mann, 1972, Chapter 20) it is readily verified that condition (37) suffices to ensure that $\mu \in W^2_{2,per}[0,1]$.

Under condition (37) the magnitude of the bias of $\beta_{\lambda j}$ behaves like a constant multiple of

$$\sum_{r \neq 0} |j + nr|^{-(5+\delta)/2} .$$

To deal with this sum it suffices to estimate the size of $\Sigma^{\infty}_{r=1}(nr \pm j)^{-(5+\delta)/2}$. This can be accomplished by using an integral estimate to obtain

$$\begin{aligned}\sum_{r=1}^{\infty} (nr \pm j)^{-(5+\delta)/2} &\leq \int_1^{\infty}(nx \pm j)^{-(5+\delta)/2}dx \\ &= (2/(3+\delta))n^{-1}(n \pm j)^{-(3+\delta)/2} \\ &= O(n^{-(5+\delta)/2}) \qquad (38)\end{aligned}$$

uniformly in $|j| \leq (n - 1)/2$. A simple calculation using (38) shows that the last term in (36) is $o(n^{-(5+\delta)/2)})$. An integral estimate can now be used along with (37) to see that as $\lambda \to \infty$

$$\sum_{|j|>\lambda} |\beta_j|^2 \sim 2C_1\lambda^{-4-\delta} . \qquad (39)$$

for $C_1 = C/(4 + \delta)$.

Combining the above results with (36) gives an asymptotic expression for $R_n(\lambda)$. If $\lambda \to \infty, n \to \infty$ and $\lambda/n \to 0$ then

$$R_n(\lambda) \sim 2C_1\lambda^{-4-\delta} + \sigma^2 2\lambda/n . \qquad (40)$$

To determine the asymptotically optimal rate of growth for $\lambda$ we can differentiate (40) and equate the result to zero to obtain

$$\lambda_{opt} \propto n^{1/(5+\delta)} \quad . \tag{41}$$

When (41) is substituted back into (40) this shows that

$$R_n(\lambda_{opt}) = O(n^{-(4+\delta)/(5+\delta)}) \ . \tag{42}$$

There are several important conclusions to be drawn from (41) and (42). The first is that $\mu_\lambda$ is a consistent estimator of $\mu$ provided only that $n, \lambda \to \infty$ and $\lambda/n \to 0$. The second is that the optimal rate of decay of the risk is slower than $n^{-1}$. This is in contrast to what is usually true for parametric models and can be viewed as the price that is paid for increasing the realism of our modeling assumptions. Also, we see from (41) that the number of terms included in the Fourier series estimator should grow much slower than the sample size. To give some indication of how much slower, it may be useful to note, for example, that $100^{1/5} = 2.5$, $1000^{1/5} = 4$, $10000^{1/5} = 6.3$ and $100{,}000^{1/5} = 10$. This, of course, is not to say that if $n = 1000$ we should use approximately 4 terms in our series estimator since this disregards the constant of proportionality in (41) and the value of $\delta$ in (37).

It is of interest to compare the rate of convergence for $\mu_\lambda$ to the optimal rate discussed in Chapter 1. It was noted there that the best uniform rate for an estimator of $\mu \in W_2^2[0,1]$ was $n^{-4/5}$. Our assumptions are somewhat different than those under which this result applies since we are dealing with a subset of $W_2^2[0,1]$: namely, the set of all functions in $W_{2,per}^2[0,1]$ whose Fourier coefficients decay algebraicly, i.e., for which (37) holds. However, for any function of this type the Fourier series estimators with $\lambda = O(n^{1/(5+\delta)})$ attains a rate of $n^{-(4+\delta)/(5+\delta)}$ for the risk. The best uniform rate for a Fourier series estimator over this subset of $W_2^2[0,1]$ is, therefore, $\inf_{\delta>0} n^{-(4+\delta)/(5+\delta)} = n^{-4/5}$, which is identical to the best rate on $W_2^2[0,1]$. While this is encouraging, we will see momentarily that Fourier series estimators are not efficient if $\mu$

does not belong to $W_{2,per}^2[0,1]$. This is one reason that the estiators discussed in later chapters are usually regarded as preferable to series type estimators.

Let us now alter our assumption about $\mu$ slightly to see what effect this has on $R_n(\lambda)$. Suppose, for example, that $\mu$ does not satisfy the boundary conditions for functions in $W_{2,per}^2[0,1]$. Specifically, suppose that $\mu'(0) \neq \mu'(1)$. In this case we have

$$\beta_j = (\mu'(1) - \mu'(0))/(2\pi j)^2 - (2\pi j)^{-2}\beta_j'' \quad .$$

Thus, $|\beta_j|^2 \sim (\mu'(1) - \mu'(0))^2/(2\pi j)^4$ by the Riemann-Lebesgue lemma. If we now mimic our previous arguments using this new behavior for the $\beta_j$ we find that the optimal $\lambda$ is proportional to $n^{1/4}$ and the optimal risk is $O(n^{-3/4})$. The important information to be gleaned from this analysis is that the behavior of $\mu$ at the *boundary* affects the risk *globally* and can result in slower rates of convergence. We will see this phenomenon again with other types of estimators.

In the above discussion we have restricted attention to the case where $\mu$ is twice differentiable. More generally, it might be assumed that $\mu \in W_{2,per}^m[0,1]$, the subset of $W_2^m[0,1]$ whose elements satisfy the boundary conditions $\mu^{(j)}(0) = \mu^{(j)}(1)$, $j=0,\ldots,m-1$. Similar conclusions can be drawn in this case that will be explored in the exercises.

Results such as (41) and (42) provide useful qualitative information about the estimator $\mu_\lambda$. However, they are not particularly useful from a practical viewpoint since, for example, the optimal $\lambda$ obtained from (40) involves the constants $C_1$ and $\delta$ in (37) that depend on the unknown function $\mu$. Thus, (41) is of questionable value for telling us how to select a value of $\lambda$ for estimation of $\mu$ in a particular data set. In addition, even if $\lambda_{opt}$ were known it is only an asymptotic quantity and would not give us the minimum risk for finite n. Thus we now indicate how the data can be used to guide the selection of a value for $\lambda$.

### 3.4.3 Selection of λ

To select a value of $\lambda$ for use with a given data set we might use the GCV criterion (22). To establish the optimality of GCV in this case we can show that the conditions of Theorem 3.1 are satisfied.

Our results in the previous section (see also Exercise 4) have the consequence that if $\mu \in W_{2,per}^{m}[0,1]$ and its Fourier coefficients decay algebraically then $\inf_{\lambda \le n} nR_n(\lambda) \to \infty$ and $\inf_{\lambda \le n} R_n(\lambda) \to 0$. As noted in the comments after Theorem 1 this is sufficient to insure that conditions ii) and iii) of the theorem hold. Thus, if condition i) can be assumed to hold, Theorem 3.1 is applicable to classical Fourier series estimators and we may conclude that for large n GCV should provide a nearly optimal value of $\lambda$ in the sense of (24).

### 3.4.4 Asymptotic Distribution Theory

Equation (31) has the implication that $\mu_\lambda(t)$ is a weighted sum of uncorrelated random variables. One might, therefore, expect that some type of central limit theorem should hold for Fourier series estimators. In this section we explore that possibility and investigate the probabilistic behavior of $\mu_\lambda$ as $n,\lambda \to \infty$. Throughout we will assume that the $\varepsilon$'s are independent and identically distributed with finite $(2 + \alpha)$th absolute moments for some $\alpha > 0$.

The tool we require for derivation of the asymptotic distribution of $\mu_\lambda(t)$ is the Liapunov central limit theorem (see, e.g., Serfling, 1980 for a proof) which may be stated as follows.

**Liapunov Theorem.** Let $\{V_{jn}: 1 \le j \le n;\ n = 1,2,\ldots\}$ be a triangular array of random variables with independent random variables within rows. Define

$$\sigma_n = \{\sum_{j=1}^{n} E(V_{jn} - EV_{jn})^2\}^{1/2}$$

and, for some $\theta > 2$, let

$$C_n(\theta) = \sum_{j=1}^{n} E|V_{jn} - EV_{jn}|^{\theta} \quad .$$

Then, if

$$C_n(\theta)/\sigma_n^{\theta} \to 0 \quad \text{as } n \to \infty,$$

the asymptotic distribution of $\Sigma_{j=1}^{n}(V_{jn} - EV_{jn})/\sigma_n$ is standard normal.

The advantage of the Liapunov Theorem is that it does not require the variables in the sum $\Sigma_{j=1}^{n}(V_{jn} - EV_{jn})$ to be identically distributed. This is precisely the setting we are in, for, using (31), it is seen that

$$\mu_{\lambda}(t) - E\mu_{\lambda}(t) = \frac{1}{n}\sum_{j=1}^{n}(y_j - Ey_j)K_{\lambda}(t - t_j)$$

$$= \frac{1}{n}\sum_{j=1}^{n}\varepsilon_j K_{\lambda}(t - t_j) \ ,$$

with $K_{\lambda}(\cdot)$ defined by (30). Thus, $\mu_{\lambda}(t) - E\mu_{\lambda}(t)$ has the form $\Sigma_{j=1}^{n}(V_{jn} - EV_{jn})$ with independent but not identically distributed $V_{jn}$ defined by

$$V_{jn} = \varepsilon_j K_{\lambda}(t - t_j)/n \ . \tag{43}$$

Now note that with $V_{jn}$ defined by (43) we have $EV_{jn} = 0$ and

$$EV_{jn}^2 = \sigma^2|K_{\lambda}(t - t_j)|^2/n^2 \quad ,$$

where, recall, $\sigma^2$ is the common variance of the $\varepsilon_j$. This means that

$$\sigma_n^2 = \sum_{j=1}^{n} EV_{jn}^2 = \sigma^2 n^{-2}\sum_{j=1}^{n}|K_{\lambda}(t - t_j)|^2$$

$$= \sigma^2 n^{-2} \sum_{j=1}^{n} \sum_{r=-\lambda}^{\lambda} \sum_{k=-\lambda}^{\lambda} e^{2\pi i r(t-t_j)} e^{-2\pi i k(t-t_j)}$$

$$= \sigma^2(2\lambda + 1)/n,$$

as a result of (27). Similarly, using (27) and the fact that $|K_\lambda(u)| \le (2\lambda + 1)$, we find

$$\sum_{j=1}^{n} E|V_{jn}|^{2+\alpha} = n^{-(2+\alpha)} E|\varepsilon_1|^{2+\alpha} \sum_{j=1}^{n} |K_\lambda(t - t_j)|^{2+\alpha}$$

$$< O((\lambda/n)^{1+\alpha}).$$

An application of the Liapunov Theorem with $\theta = 2 + \alpha$ then reveals that $\Sigma_{j=1}^{n} V_{jn}/\sigma_n$ will be asymptotically standard normal provided

$$\sum_{j=1}^{n} E|V_{jn}|^{2+\alpha}/[\sigma^2(2\lambda + 1)/n]^{1+\alpha/2} = O((\lambda/n)^{\alpha/2})$$

converges to zero as $n \to \infty$. We summarize this result in the following theorem.

**Theorem 3.2.** Suppose that $\lambda, n \to \infty$ in such a way that $\lambda/n \to 0$. Then, for any fixed t, $n^{1/2}(\mu_\lambda(t) - E\mu_\lambda(t))/\sigma(2\lambda + 1)^{1/2}$ converges in distribution to a standard normal random variable.

Theorem 3.2 is not all that one might want in this setting. It would be more satisfying if $n^{1/2}(\mu_\lambda(t) - \mu(t))/\sigma(2\lambda + 1)^{1/2}$ was asymptotically standard normal. This is true too under certain restrictions. The key lies in writing

$$n^{1/2} \frac{\mu_\lambda(t) - \mu(t)}{\sigma(2\lambda + 1)^{1/2}} = n^{1/2} \frac{\mu_\lambda(t) - E\mu_\lambda(t)}{\sigma(2\lambda + 1)^{1/2}} + n^{1/2} \frac{E\mu_\lambda(t) - \mu(t)}{\sigma(2\lambda + 1)^{1/2}} \quad . \tag{44}$$

The first term on the right hand side of (44) is asymptotically standard normal as a result of Theorem 3.2. If the second term converges to zero, an application of Slutsky's theorem (see, e.g., Serfling, 1980) gives the desired result.

From work in Section 3.4.2 we know that

$$|\mu(t) - E\mu_\lambda(t)| = |\sum_{|j|>\lambda} \beta_j e^{2\pi ijt} - \sum_{|j|\le\lambda} (\sum_{r\ne 0} \beta_{j+rn}) e^{2\pi ijt}|$$

$$\le \sum_{|j|>\lambda} |\beta_j| + \sum_{|j|\le\lambda} |\sum_{r\ne 0} \beta_{j+rn}|$$

$$= O(\lambda^{-(3+\delta)/2}) + O(\lambda n^{-(5+\delta)/2}) \tag{45}$$

provided (37) holds. If $\lambda/n \to 0$ the first term dominates so that

$$n^{1/2} |\mu(t) - E\mu_\lambda(t)|/(2\lambda + 1)^{1/2} \le n^{1/2} O(\lambda^{-(4+\delta)/2}) \tag{46}$$

uniformly in t. This gives the following corollary.

**Corollary 3.1.** Suppose that $\mu \in W^2_{2,per}[0,1]$ with Fourier coefficients satisfying (37). Then, for fixed t, $n^{1/2}(\mu_\lambda(t) - \mu(t))/\sigma(2\lambda + 1)^{1/2}$ has an asymptotic standard normal distribution provided n, $\lambda\to\infty$ in such a manner that $n\lambda^{-(4+\delta)} \to 0$.

It is interesting to note that the asymptotically optimal $\lambda$ described by (41) does not satisfy the conditions of the corollary. Thus, to insure asymptotic normality for the estimator centered at $\mu(t)$, $\lambda$ should be allowed to grow large at something faster than the optimal rate. This is not very useful from a practical viewpoint since it provides only a very rough guideline on how to choose $\lambda$ in a manner which makes a normal approximation valid for $\mu_\lambda(t) - \mu(t)$.

The reason for interest in results such as Corollary 3.1 concerns interval estimation of $\mu(t)$. When Corollary 3.1 holds it has the implication that an approximate $100(1 - \alpha)\%$ confidence interval for $\mu(t)$ is provided by

$$\mu_\lambda(t) \pm Z_{\alpha/2}\sigma\ ((2\lambda + 1)/n)^{1/2} \tag{47}$$

with $Z_{\alpha/2}$ the upper $\alpha/2$ percentage point of the standard normal distribution. This assumes, of course, that $\lambda$ has been chosen correctly and that $\sigma$ is known. Alternative approaches to interval estimation might be based on jackknife or bootstrap procedures (see Efron 1982).

The problem with confidence intervals based on estimators using the optimal $\lambda$ is that $\lambda_{opt}$ provides a balancing of variance and bias. This has the consequence that the variance of the estimator (equivalently, the length of the confidence interval) is shrinking too fast relative to the estimator's bias. Thus one needs to bump up the variance or length of the interval in some reasonable fashion to overcome this difficulty. One approach which has been suggested for selecting $\lambda$ to insure the validity of (47) is to choose $\lambda$ by GCV, which should give an estimate of the optimal $\lambda$ in (41), and then use a slightly larger value of $\lambda$ for construction of the interval. This approach has its problems since no one knows how much larger to choose $\lambda$ and, secondly, larger values of $\lambda$ (than the GCV estimate) will generally result in inferior estimators of $\mu$. Another approach might be to use an estimate of the risk, rather than the variance, in constructing an interval estimate. This has intuitive appeal from the standpoint that the risk is a better measure of the error in $\mu_\lambda$ as an estimator of $\mu$ than the estimator's variance. Use of a risk estimate will also tend to give wider intervals than the use of the estimator's variance thereby counteracting some of the bias difficulties associated with series estimators. However, if the optimal $\lambda$ is used, one can still argue that intervals constructed using risk estimates will not be centered at $\mu(t)$; so problems concerning coverage probabilities persist even in this case.

From a linear regression viewpoint it might also be of interest to construct interval estimates for the $\beta_j$, $|j| \le \lambda$, the true Fourier coefficients of $\mu$ that are estimated in $\mu_\lambda$. Using an argument similar to the one required for Theorem 3.2 one finds that $n^{1/2}(\beta_{\lambda j} - \beta_j)/\sigma$ has an asymptotic standard normal distribution (Exercise 9). This has the implication that tests, etc., concerning the $\beta_j$ can be conducted as if $\mu$ actually followed a linear model, provided the sample size is sufficiently large.

If $\sigma$ is unknown in (47) it can be estimated. The simplest estimator arises when there are repeated observations at each $t_j$. If, for example, we have observations

$$y_{rj} = \mu(t_j) + \varepsilon_{rj} , \quad r = 1,\ldots,n_j, \quad j=1,\ldots,n,$$

it is easy to see that

$$(N - n)^{-1} \sum_{j=1}^{n} \sum_{r=1}^{n_j} (y_{rj} - \bar{y}_j)^2,$$

with $\bar{y}_j = n_j^{-1} \Sigma_{r=1}^{n_j} y_{rj}$ and $N = \Sigma_{j=1}^{n} n_j$, is an unbiased estimator of $\sigma^2$. When there are no repeated observations consistent estimators can be constructed using observation differences. For example, when n is even, and $\mu$ is sufficiently smooth

$$\hat{\sigma}^2 = n^{-1} \sum_{j=1}^{n/2} (y_{2j} - y_{2j-1})^2 \tag{48}$$

provides a consistent estimator of $\sigma^2$ (Exercises 15).

Due to the similarity between Fourier series and linear regression estimators another candidate for an estimator of $\sigma^2$ is

$$\sigma_\lambda^2 = \sum_{j=1}^{n} (y_j - \mu_\lambda(t_j))^2/(n - (2\lambda + 1)) . \tag{49}$$

To study the consequences of using this estimator we draw on results from Section 2.2. Using the notation of that section we can write $\sigma_\lambda^2 = \text{MSE}(\lambda)/(1 - (2\lambda + 1)/n)$. Thus, using (2.11) - (2.12),

$$E\sigma_\lambda^2 = (1 - (2\lambda + 1)/n)^{-1}[R_n(\lambda) + \sigma^2 - 2\sigma^2(2\lambda + 1)/n]$$

$$\sim R_n(\lambda) + \sigma^2 - 2\sigma^2(2\lambda + 1)/n$$

$$\sim \sigma^2,$$

provided $\lambda$, $n\to\infty$ in such a way that $\lambda/n \to 0$. Consequently, $\sigma_\lambda^2$ is asymptotically unbiased for $\sigma^2$.

If, for example, the $\varepsilon$'s have a normal distribution $\sigma_\lambda^2$ will actually be consistent for $\sigma^2$. This is because, under normality, we can apply Lemma A.1 to obtain

$$\operatorname{Var} \sigma_\lambda^2 = (n - (2\lambda + 1))^{-2}\{2\sigma^4(n - 3(2\lambda + 1)) + \sigma^2 nR_n(\lambda)\}$$

$$= O(n^{-1})$$

provided $\lambda$ grows at a rate which insures that $R_n(\lambda) \to 0$ as $n \to \infty$.

### 3.4.5 An Example

To illustrate the concepts discussed in Sections 3.4.1-3.4.4 it will be helpful to consider a numerical example. The data we examine represent the voltage drops in the battery of a guided missile motor during its flight (Montgomery and Peck, 1982). A plot of the observed voltage drops as a function of time is shown in Figure 3.2. (The actual data can be found in Exercise 14.) Observations were taken at half second intervals between take-off and 20 seconds of flight. In the plot the time axis has been rescaled to conform with (25).

To estimate the regression function we intend to fit this data with an estimator of the form

$$\mu_\lambda(t) = \beta_{\lambda 0} + \sum_{j=1}^{\lambda} (c_{\lambda j}\cos(2\pi jt) + s_{\lambda j}\sin(2\pi jt)),$$

i.e., a classical Fourier series estimator. The first step is to select an appropriate value of $\lambda$. To accomplish this we compute the GCV criterion over several values of $\lambda$. Note that in this case

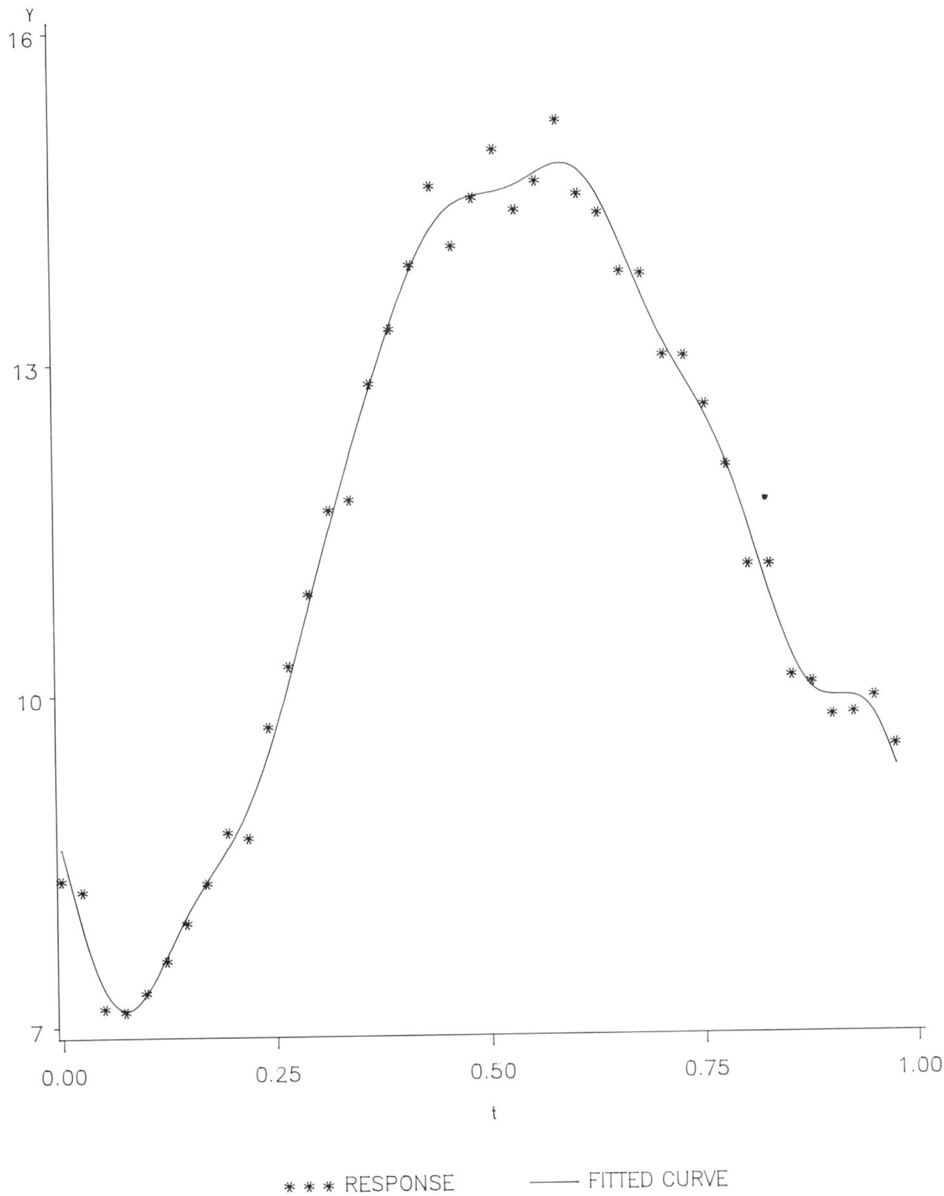


**Figure 3.2** Voltage drop data and fitted curve with lambda = 6.

$$GCV(\lambda) = MSE(\lambda)/(1 - (2\lambda + 1)/n)^2.$$

Table 3. 1 gives the values of GCV(λ) for λ between 1 and 10. The minimum occurs at $\hat{\lambda} = 6$ suggesting that we use $\mu_6$ to estimate μ. The corresponding estimator (49) for $\sigma^2$ is found to be $\sigma_6^2$ = .062.

Now that a suitable value for λ has been selected we should examine the nature of the fit to the data. A first step in this regard involves examination of fit diagnostics such as those used in linear regression analysis. A brief summary of diagnostic methodology used in linear regression is given in the Appendix.

**Table 3.1. GCV(λ) for the Voltage Drop Data**

| λ | GCV(λ) |
|---|---|
| 0 | 6.731 |
| 1 | .235 |
| 2 | .149 |
| 3 | .139 |
| 4 | .122 |
| 5 | .103 |
| 6 | .090 |
| 7 | .097 |
| 8 | .114 |
| 9 | .127 |
| 10 | .154 |

Application of the diagnostic tools discussed in the Appendix to the voltage drop data entails detection of points with high leverage, large residuals and, most importantly, significant influence. Leverage, in this case, presents no problem since the hat matrix is found to be constant down the diagonal. One finds that no observations have studentized residuals which exceed two in magnitude. A plot of the residuals shown in Figure 3.3 reveals no obvious patterns or trends, and the value of the runs test statistic is not statistically significant, which further supports this conclusion. Finally the largest Cook's distance value is .141, well below the rough bound provided by the 10th

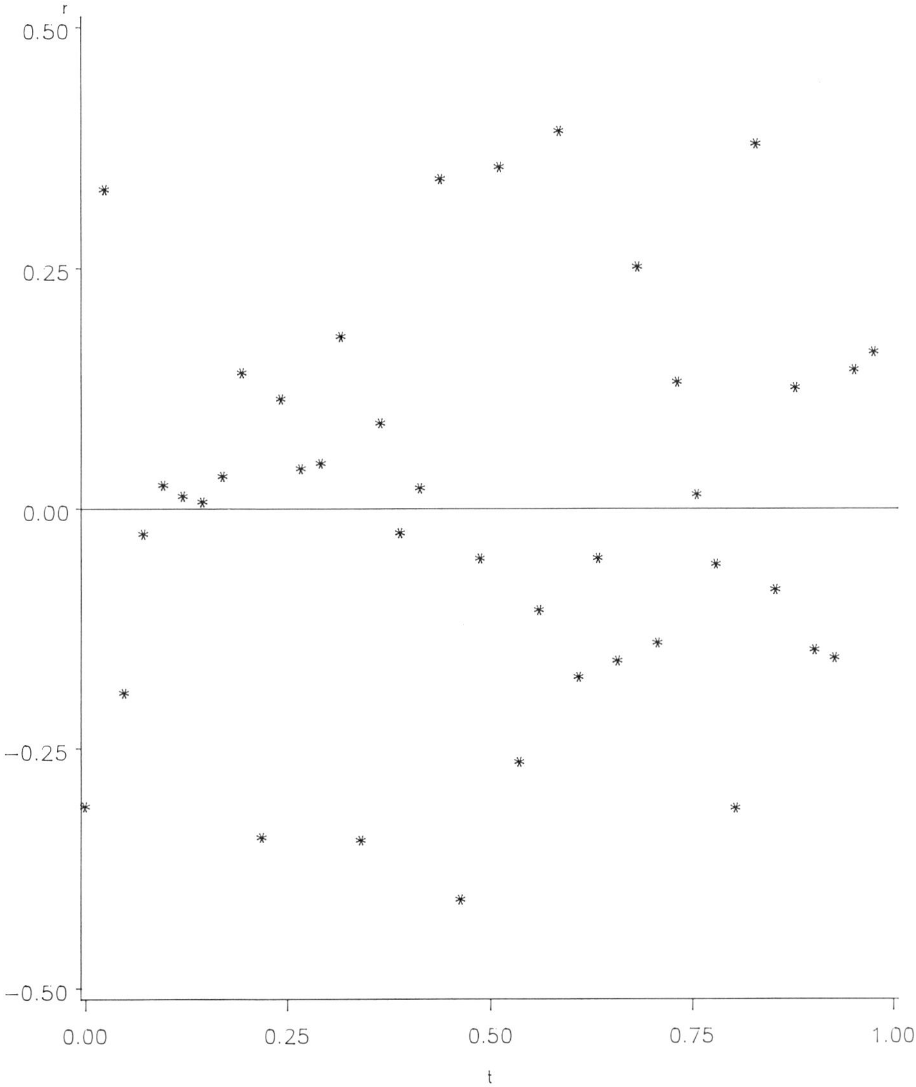


**Figure 3.3** Residuals for voltage drop data.

percentile of an F-distribution with 12 numerator and 28 denominator degrees of freedom (which is around .49). We leave the verification of the above statements to the reader. The conclusion which is reached from our diagnostic analysis is that the fit to the voltage drop data seems satisfactory at this point.

The values of the estimated Fourier coefficients $\{c_{6j}, s_{6j}; j=1,...,6\}$ and $\beta_{60}$ are given in Table 3.2. Next to each coefficient we have also listed the value of the coefficient divided by an estimate of its standard error; namely, $\sigma_6(2/n)^{1/2} = .055$ for the $c_{6j}$ and $s_{6j}$ and $\sigma_6/n^{1/2} = .039$ for $\beta_{\lambda 0}$. If $\mu$ actually followed a linear model involving the first six sine and cosine functions and the $\varepsilon$'s were normally distributed, these values would provide tests of the hypotheses that the true Fourier coefficients which are being estimated are actually zero. Of course it is doubtful that $\mu$ follows such a linear model. Nonetheless, this interpretation is still approximately true and we can regard statistics of this type as providing indications of the importance of each of the sine and cosine terms and the intercept in estimating $\mu$ (Exercise 9).

For our purposes it will suffice to regard a term as contributing significantly to the fit if its coefficient exceeds twice its estimated standard deviation. Thus, we might consider deleting the terms involving $\cos(2\pi(4t))$ and $\cos(2\pi(6t))$ from the fitted function. The orthogonality of the sine and cosine functions allows us to do this without reestimating the coefficients for the functions which are retained. The final fit is plotted in Figure 3.2 along with the data. Subsequent diagnostic analysis yields similar conclusions to those for $\mu_6$. The proportion of y variation explained by the fitted curve is an excellent $R^2 = .993$.

As a consequence of our discussions concerning confidence intervals in Section 3.4.4 one might wish to use an estimator based on a larger value of $\lambda$ than the optimal value $\hat{\lambda} = 6$ for construction of interval estimates, e.g., use an estimator based on the first 7 or 8 of the exponential functions. Approximate 95% confidence intervals for the $\mu(t_i)$ based on $\mu_7$ with $\sigma_7$ used to estimate $\sigma$ in (47) are shown in Figure 3. 4. For display purposes the interval end points at each of the $t_i$ have been connected to give two continuous curves which enclose the fitted regression curve. These are not confidence bands, however, and, strictly speaking, the curves are interpretable only at the $t_i$.

**Table 3.2. Coefficient Estimates for the Voltage Drop Data**

| Coefficient | Estimate | Estimate/(Standard Error) |
|---|---|---|
| $\beta_{60}$ | 11.336 | 292.247 |
| $c_{61}$ | -3.222 | -58.736 |
| $s_{61}$ | -1.425 | -25.970 |
| $c_{62}$ | .230 | 4.186 |
| $s_{62}$ | - .348 | -6.350 |
| $c_{63}$ | .156 | 2.849 |
| $s_{63}$ | - .121 | -2.212 |
| $c_{64}$ | .067 | 1.220 |
| $s_{64}$ | - .194 | -3.538 |
| $c_{65}$ | .112 | 2.040 |
| $s_{65}$ | - .162 | -2.959 |
| $c_{66}$ | - .038 | - .701 |
| $s_{66}$ | - .156 | -2.837 |

This example serves to illustrate several points. First it shows how GCV can be used to select $\lambda$ and how our heuristic discussions of interval estimates in Section 3.4.4 can be used to guide the construction of interval estimates in practice. Secondly, it indicates that in some cases the modeling practices familiar from linear regression analysis are also applicable to nonparametric regression. While the distribution theory for statistics used in linear models may not hold precisely in the nonparametric setting, examination of the values of statistics

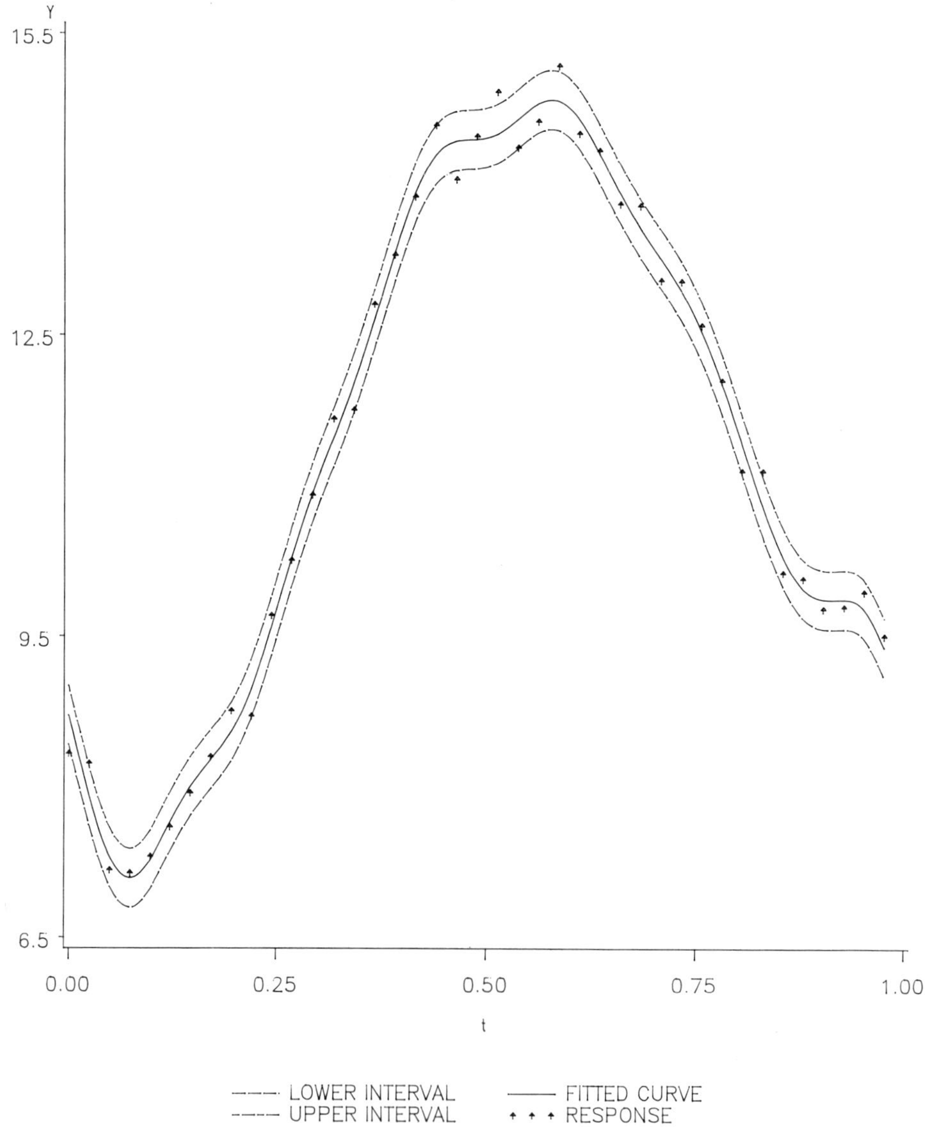


**Figure 3.4** Fit and interval estimates for the voltage drop data.

such as standardized coefficients, studentized residuals, etc. can still provide a valuable aid in the modeling process.

## 3.5 Polynomial Regression

The classical Fourier series estimators are important if for no other reason than the fact that most other nonparametric regression estimators are closely related to them. For the same reason polynomial regression estimators represent an important cornerstone in the theory of nonparametric regression. In this final section we give a brief treatment of polynomial regression from the viewpoint of series estimators.

Suppose now that we have chosen the Legendre polynomials $\{x_j\}$, defined by (10)-(11), for our CONS in $L_2[a,b]$. If the corresponding generalized Fourier series for $\mu$ converges uniformly we can then write model (1) as an (infinite) polynomial regression model. The convergence of such a series can be shown to depend on the convergence of the classical Fourier series for a function derived from $\mu$ (See, for example, Szegö, 1959). This has the consequence that polynomial regression estimators are also closely related to classical Fourier series estimators. To estimate $\mu$ we can then approximate its Legendre Fourier series by the $\lambda$th partial sum, $\Sigma_{j=0}^{\lambda-1}\beta_j x_j$, and estimate the $\beta_j$ through least squares. The resulting estimator is obtained from formula (19) and will be called a *polynomial regression estimator*.

There are a variety of motivations for polynomial regression estimators. One of these is closely related to the frequency motivation for classical Fourier series estimators discussed in Section 3.4. The idea in that case originated from the observation that $\sin(2\pi jt)$ and $\cos(2\pi jt)$ are periodic functions on the interval $[0,1]$. This made them candidates for predictor functions that would allow us to isolate various frequency aspects of the data. Since $\mu$ was smooth it should be aligned with lower frequency components.

The periodicity of $\sin(2\pi jt)$ and $\cos(2\pi jt)$ is reflected by the fact that the functions have $2j - 1$ and $2j$ zeros in $(0,1)$, respectively. A similar property holds for the Legendre polynomials. It can be shown (see Davis, 1975, Chapter X or Szegö, 1959, Chapter III) that the jth Legendre polynomial, $x_j$, $j=0,\ldots,$ has exactly j distinct, real roots in the interior of $[a,b]$. Thus $x_j$ changes sign j times in $(a,b)$. By analogy with the sine and cosine functions, one might therefore view this as meaning that $x_{j+1}$ is of higher frequency than $x_j$ and thereby obtain a frequency interpretation for the Fourier coefficients in the Legendre expansion for $\mu$. Regression analysis using low order Legendre polynomials can then be thought of as an attempt to isolate the low frequency components of the data which should provide the most information about the regression function.

Another motivation for polynomial regression estimators, that is somewhat more standard than the frequency interpretation advanced above, stems from Taylor's Theorem.

**Taylor's Theorem.** Let $\mu \in W_2^m[a,b]$. Then, there exist coefficients $\theta_0,\ldots,\theta_{m-1}$ such that

$$\mu(t) = \sum_{j=0}^{m-1} \theta_j t^j + \int_a^b \frac{(t - u)_+^{m-1}}{(m - 1)!}\mu^{(m)}(u)du , \tag{50}$$

where

$$(x)_+^r = \begin{cases} x^r , & x \geq 0 , \\ 0 , & x < 0 . \end{cases} \tag{51}$$

*Proof.* Write $\mu(t) = \int_a^b (t - u)_+^0 \mu'(u)du - \mu(a)$ and integrate by parts. □

Taylor's Theorem suggests that if, for some nonnegative integer $\lambda$, the remainder term

$$\mathrm{Rem}_\lambda(t) = [(\lambda - 1)!]^{-1}\int_a^b (t - u)_+^{\lambda-1}\mu^{(\lambda)}(u)du \tag{52}$$

is uniformly small then we could write

$$y_k \doteq \sum_{j=0}^{\lambda-1} \theta_j t_k^j + \varepsilon_k \quad , \qquad k=1,\ldots,n.$$

In other words, the data would follow an approximate polynomial regression model. We could then estimate the polynomial coefficients by least squares or some other method. [Actually, polynomial regression does not give estimates of the $\theta_j$ in (50). For a precise formulation see Exercise 17.]

Still another motivation for the use of polynomial regression stems from the Weierstrass approximation theorem (Davis, 1975, Chapter VI) which states that if $\mu \in C[a,b]$ then, given any $\varepsilon > 0$, there exists a polynomial $p_\lambda$ of sufficiently high order $\lambda$ such that $|\mu(t) - p_\lambda(t)| \le \varepsilon$ for all $t \in [a,b]$. Note that in contrast to Taylor's Theorem no differentiability conditions are required for this result to hold. However, the basic implications concerning the possible utility of an approximate polynomial regression model remain unchanged.

Both the Taylor and Weierstrass Theorem arguments for the use of polynomial regression are tantamount to lumping remainder terms (e.g., $\mathrm{Rem}_\lambda(t_1),\ldots,\mathrm{Rem}_\lambda(t_n)$) into the random error component of our model. If $\mu$ is uniformly well appoximated by a polynomial of the order selected and the remainders at the $t_j$ are small relative to the random errors, polynomial regression should work fairly well. If not, problems can arise. Since both the remainder and random errors are unknown there is no way of knowing for certain whether or not the Taylor and Weierstrass Theorem arguments are valid for any given data set.

In view of the uncertainty about the magnitude of the remainder from polynomial approximation of $\mu$ and the random errors it would seem natural to try to modify the polynomial regression estimator to attempt to compensate for the possibility of large remainder terms. This line of reasoning leads to smoothing spline estimators for $\mu$ which are the topic of Chapter 5.

### 3.5.1 Form of the Estimator

There are a number of useful representations for the polynomial regression estimator. In particular, there is no reason to restrict attention to the Legendre polynomial basis when conducting polynomial regression analysis. In fact, this is not the usual approach. Typically, one will use the power basis

$$p_j(t) = t^j \quad , \qquad j = 0,1,\ldots. \tag{53}$$

Let $T_\lambda$ be the $n \times \lambda$ matrix

$$T_\lambda = \{p_j(t_k)\}_{\substack{k=1,n \\ j=0,\lambda-1}} \quad . \tag{54}$$

Then, under the power basis representation

$$\mu_\lambda(t) = (1,t,\ldots,t^{\lambda-1})(T_\lambda'T_\lambda)^{-1}T_\lambda'\underline{y} \ . \tag{55}$$

Note that $\mu_\lambda(t)$ is not altered by changing from the Legendre to the power basis. However, the estimated coefficients for the basis elements,

$$\underline{\theta}_\lambda = (T_\lambda'T_\lambda)^{-1}T_\lambda'\underline{y} \quad , \tag{56}$$

will not coincide with the estimated Fourier coefficients under the Legendre form.

There is another form for the polynomial regression estimator that is instructive. Rather than use the power basis or Legendre basis suppose we define basis functions by

$$x_{rn}(t) = \sum_{j=1}^{\lambda} a_{jr} t^{j-1}$$

with $A_\lambda = \{a_{jr}\}$ a nonsingular $\lambda \times \lambda$ matrix satisfying

$$A_\lambda'T_\lambda'T_\lambda A_\lambda = nI.$$

In other words we choose our basis so that

$$n^{-1}\sum_{r=1}^{n} x_{jn}(t_r)x_{kn}(t_r) = \begin{cases} 1 & , \ j = k, \\ 0 & , \ j \neq k . \end{cases} \tag{57}$$

The $x_{jn}$ then have an orthogonality property on the $t_j$ which mimics that of the Legendre polynomials on $[a,b]$ and will often be quite similar to the Legendre functions (Exercise 13). There is a large literature in Statistics on the construction of polynomials of this type; see, e.g., Narula (1979).

As a result of (57) the polynomial regression estimator now takes on the form

$$\mu_\lambda(t) = \sum_{j=0}^{\lambda-1} \beta_{\lambda j} x_{jn}(t)$$

with

$$\beta_{\lambda j} = n^{-1}\sum_{r=1}^{n} y_r x_{jn}(t_r).$$

This can be rewritten as

$$\mu_\lambda(t) = n^{-1}\sum_{r=1}^{n} y_r K_{\lambda,n}(t,t_r) \tag{58}$$

with

$$K_{\lambda,n}(s,t) = \sum_{j=0}^{\lambda-1} x_{jn}(s)x_{jn}(t) . \tag{59}$$

The function $K_{\lambda,n}(\cdot,\cdot)$ is called a kernel polynomial and plays a parallel role to the Dirichlet kernel for classical Fourier series. Notice, for example, that the kernel sums to one, i.e.,

$$n^{-1}\sum_{r=1}^{n} K_{\lambda,n}(t,t_r) = n^{-1}\underline{t}_\lambda' A_\lambda A_\lambda' T_\lambda' \underline{1} = 1 .$$

This follows from the fact that $A_\lambda$ is nonsingular and, hence, $A_\lambda' T_\lambda' \underline{1}$ ($\underline{1}$ being a vector of all unit elements) is the first column of $A_\lambda^{-1}$.

The preceding discussion shows that the polynomial regression estimator is a weighted average with weights provided by a kernel function. There is, of course, nothing unique about polynomial estimators in this respect and parallels of (58) and (59) hold for any series type estimator. This viewpoint leads us into the study of kernel estimators in the next chapter.

### 3.5.2 Some Asymptotic Theory

In this section results are obtained concerning the asymptotic properties of a polynomial regression type estimator of $\mu$. We will not study $\mu_\lambda$ in (55) directly. For purely technical reasons it will instead be more convenient to focus on a closely related estimator studied by Rafajlowicz (1987). Thus the results in this section are intended to be indicative of what might be expected for the large sample behavior of $\mu_\lambda$ rather than a rigorous derivation of its asymptotic properties.

For specificity assume throughout this section that $[a,b] = [-1,1]$. Then, to estimate $\mu$ we consider a general polynomial estimator of the form

$$\sum_{j=0}^{\lambda-1} b_j x_j \quad ,$$

where the $x_j$ are the Legendre polynomials on $[-1,1]$ defined in (10)-(11) and the $b_j$ are estimates of the Legendre Fourier coefficients

$$\beta_j = \int_{-1}^{1} \mu(t) x_j(t) dt. \qquad (60)$$

The choice of an estimator is therefore equivalent to selecting estimators for the $\beta_j$. For this purpose one might consider a weighted average of the $y_r$ of the form

$$b_j = \sum_{r=1}^{n} W_r y_r x_j(t_r) \quad , \qquad (61)$$

where the $W_r$ are specified weights and the $t_r$ are the design points. We will choose the $t_r$ and $W_r$ simultaneously to regulate the bias of the $b_j$.

The expectation of $b_j$ in (61) is

$$Eb_j = \sum_{r=1}^{n} W_r \mu(t_r) x_j(t_r) \ . \tag{62}$$

We would like $b_j$ to have a small bias in estimating $\beta_j$. Thus, whenever possible, we would like to choose the $W_r$ and $t_r$ in (61) so that the sum in (62) provides a good approximation to the integral in (60), in some sense. However, this is just a numerical quadrature problem; so choosing good values for the weights and design points is tantamount to the choice of a good quadrature rule.

There are many possible quadrature rules available for approximating integrals by sums. For our purposes we will restrict attention to a particular type of Gaussian quadrature rule (see, e.g., Conte and de Boor, 1980) which results from choosing the $t_r$ to be the n distinct roots of $x_n$, the nth degree Legendre polynomial, and by taking

$$W_r = \int_{-1}^{1} \frac{x_n(t)}{x_n'(t_r)(t - t_r)} \, dt, \quad r = 1,\ldots,n. \tag{63}$$

It can be shown that for this choice of the $t_r$ and $W_r$,

$$\sum_{r=1}^{n} W_r p(t_r) = \int_{-1}^{1} p(t)dt \tag{64}$$

if p is any polynomial of order 2n (degree 2n - 1 or less). This means that our choice of the $t_r$ and $W_r$ provides a good quadrature rule in the sense that it gives exact integrals when applied to polynomials of degree 2n -1 or less. This fact can be used to obtain rather precise information about the magnitude of the quadrature error in approximating the integral of a function (see Conte and de Boor, 1980). The actual computation of the $W_r$ and $t_r$ requires numerical techniques to find the roots of $x_n$. Fortunately, there now exists tables of the $t_r$ that can be used for many cases arising in practice (see, for example, Stroud and Secrest, 1966).

The $W_r$ in (63) have a number of other properties that will prove useful subsequently. Using results in Szegö (1959) it can, for example, be shown that

$$W_r \geq 0, \qquad r = 1,\ldots,n \;, \tag{65a}$$

$$\sum_{r=1}^{n} W_r = 1 \tag{65b}$$

and that there exists a positive constant $C_1$ such that

$$\max_{1\leq r\leq n} W_r \leq C_1 n^{-1} \quad . \tag{65c}$$

To summarize, the estimator we propose to use for $\mu(t)$ is

$$m_\lambda(t) = \sum_{j=0}^{\lambda-1} b_j x_j(t) \;, \tag{66}$$

where the $x_j$ are the Legendre polynomials, and the $b_j$ are computed using (61) with the weights defined in (63) and the design points chosen as the n distinct roots of $x_n$. This estimator is closely related to the ordinary polynomial regression estimator of $\mu$ obtained by least-squares estimation. It is, in fact, a type of least-squares estimator since the $b_j$ in (66) are the minimizers of the weighted least-squares criterion

$$\sum_{r=1}^{n} W_r\Big(y_r - \sum_{j=0}^{\lambda-1} a_j x_j(t_r)\Big)^2$$

over all possible choices for $\underline{a}' = (a_0,\ldots,a_{\lambda-1}) \in \mathbb{R}^\lambda$. To see this, let

$$X_\lambda = \{x_j(t_i)\}_{\substack{i=1,n\\ j=0,\lambda-1}}$$

and note that a minimizer must satisfy the normal equations

$$X_\lambda' W X_\lambda \underline{a} = X_\lambda' W \underline{y}$$

with $W = \mathrm{diag}(W_1,\ldots,W_n)$. The matrix $X_\lambda' W X_\lambda$ has typical element $\Sigma_{r=1}^n W_r x_i(t_r) x_j(t_r)$. If $\lambda \le n$, then $x_i x_j$ is a polynomial of degree at most $2n - 2$. Thus, as a result of property (64), $X_\lambda' W X_\lambda = I$ and the minimizer is

$$\underline{b} = X_\lambda' W \underline{y}$$

as claimed. Notice that our choice of weights and design points has had the effect of providing an orthogonality property on the $t_r$ for the Legendre polynomials similar to property (27) for the complex exponentials.

To study the asymptotic behavior of $m_\lambda$ as an estimator of $\mu$ it becomes necessary to restrict attention to regression functions in some appropriate function space. For this purpose we will assume that $\mu$ belongs to the set $\mathrm{Lip}_B^{m,\alpha}[-1,1]$ of all functions on $[-1,1]$ with m continuous derivatives and mth derivatives that satisfy a Lipshitz condition of order $\alpha$ for some $0 < \alpha < 1$. That is, $\mu$ is in the set of all functions $f \in C^m[-1,1]$ for which

$$|f^{(m)}(t) - f^{(m)}(s)| \le B|s - t|^\alpha$$

for some fixed positive constant B and all $s, t \in [-1,1]$. There is a wealth of information available about the approximation powers of polynomials for functions which satisfy these conditions. Let $\mathcal{P}^\lambda$ denote the set of all polynomials of order $\lambda$. Then it can be shown (see Schumaker, 1980) that if $f \in \mathrm{Lip}_B^{m,\alpha}[-1,1]$

$$\inf_{p \in \mathcal{P}^\lambda} \sup_t |f(t) - p(t)| \le C\lambda^{-(m+\alpha)} \tag{67}$$

for C a constant that is independent of f, and that there is a polynomial $\tilde{p} \in \mathcal{P}^\lambda$ such that

$$\inf_{p \in \mathcal{P}^\lambda} \sup_t |f(t) - p(t)| = \sup_t |f(t) - \tilde{p}(t)|. \tag{68}$$

Equation (67) gives a uniform upper bound on the rate of convergence of polynomials over functions in $Lip_B^{m,\alpha}[-1,1]$. Equation (68) states that for any $f \in Lip_B^{m,\alpha}[-1,1]$ there is a corresponding best approximation to f from $\mathcal{P}^\lambda$ in the sense of minimizing the distance $\sup_t |f(t) - p(t)|$ between f and the functions in $\mathcal{P}^\lambda$.

We will now establish a result, due to Rafajlowicz (1987), which gives an upper bound for the convergence rate of the integrated risk corresponding to $m_\lambda$.

**Theorem 3.3 (Rafajlowicz, 1987).** Suppose that $\mu$ admits m continuous derivatives and $\mu^{(m)}$ satisfies a Lipshitz condition of order $\alpha$. Then, provided $\lambda < n$, there exists constants $C_1$ and $C_2$, which are independent of n, such that

$$IR(\lambda) = \int_{-1}^{1} E(\mu(t) - m_\lambda(t))^2 dt \leq \sigma^2 C_1 \lambda/n + C_2 \lambda^{-2(m+\alpha)} . \quad (69)$$

*Proof.* The integrated risk for $m_\lambda$ can be decomposed into a variance and bias component. The variance component is

$$\int_{-1}^{1} \operatorname{Var} m_\lambda(t)dt = \sum_{j=0}^{\lambda-1} \operatorname{Var} b_j ,$$

as a consequence of the orthonormality of the $x_j$ on $[-1,1]$. Now, from (61),

$$\begin{aligned} \operatorname{Var} b_j &= \sigma^2 \sum_{r=1}^{n} W_r^2 x_j(t_r)^2 \\ &\leq \sigma^2 \max_{1 \leq r \leq n} W_r \leq C_1 \sigma^2 n^{-1} , \end{aligned}$$

due to properties (64) and (65) of the $W_r$. Using this bound for the variance of $b_j$ in our expression for the integrated variance gives the first part of the sum in (69).

The integrated squared bias for $m_\lambda$ is

$$IB^2(\lambda) = \int_{-1}^{1} (\mu(t) - Em_\lambda(t))^2 dt.$$

An elementary inequality shows that

$$IB^2(\lambda) \le 2\int_{-1}^{1} (\mu(t) - p(t))^2 dt + 2\int_{-1}^{1} (p(t) - Em_\lambda(t))^2 dt$$

for any polynomial $p \in \mathbb{P}^\lambda$. Since p is arbitrary, we may take p to be $\tilde{p}$, the best approximation to $\mu$ in the sense of the distance $\sup_t|\mu(t) - p(t)|$ (c.f. (68)). Thus, using (67), we find that

$$IB^2(\lambda) \le 2C\lambda^{-(m+\alpha)} + 2\int_{-1}^{1} (\tilde{p}(t) - Em_\lambda(t))^2 dt. \tag{70}$$

Let us now deal with the remaining integral in (70). Since $\tilde{p} \in \mathbb{P}^\lambda$ this means there are coefficients $a_0,\ldots,a_{\lambda-1}$ such that $\tilde{p} = \Sigma_{j=0}^{\lambda-1} a_j x_j$. Thus $\tilde{p} - Em_\lambda = \Sigma_{j=0}^{\lambda-1}(a_j - Eb_j)x_j$ and

$$\int_{-1}^{1} (\tilde{p}(t) - Em_\lambda(t))^2 dt = \sum_{j=0}^{\lambda-1} (a_j - Eb_j)^2 .$$

The $a_j$ are necessarily the Legendre Fourier coefficients for $\tilde{p}$ and, as a result of (64),

$$a_j = \int_{-1}^{1} \tilde{p}(t)x_j(t)dt = \sum_{r=1}^{n} W_r\tilde{p}(t_r)x_j(t_r) . \tag{71}$$

Now set $\delta_k = \tilde{p}(t_k) - \mu(t_k)$ and combine (64), (71) and a simple inequality (Exercise 19) to obtain

$$\sum_{j=0}^{\lambda-1} (a_j - Eb_j)^2 = \sum_{k=1}^{n}\sum_{r=1}^{n} \delta_k\delta_r W_kW_r \sum_{j=0}^{\lambda-1} p_j(t_k)p_j(t_r)$$

$$\le \sum_{r=1}^{n} \delta_r^2 W_r .$$

However, $\delta_r^2 \le C^2\lambda^{-2(m+\alpha)}$ due to (67) and $\Sigma W_r = 1$ from which we conclude that

$$\int_{-1}^{1} (\tilde{p}(t) - Em_\lambda(t))^2 \le C^2\lambda^{-2(m+\alpha)} \quad .$$

The theorem follows after combining appropriate constants. □

An immediate consequence of Theorem 3.3 is that $m_\lambda$ is consistent for $\mu$ provided that $\lambda \to \infty$ in such a way that $\lambda/n \to 0$. If we further suppose the bound (69) is sharp, we can calculate an optimal rate of growth for $\lambda$ and the corresponding optimal rate of decay for $IR(\lambda)$. One finds the asymptotically optimal $\lambda$ in this case is $\lambda_{opt} = O(n^{1/(2(m+\alpha)+1)})$ and $IR(\lambda_{opt}) = O(n^{-2(m+\alpha)/(2(m+\alpha)+1)})$.

### 3.5.3 An Example

To illustrate the use of polynomial regression estimators data was simulated from the model

$$y_j = 40t_j(1 - t_j) + \varepsilon_j \ , \qquad j = 1,\ldots,50,$$

where $t_j = (j - 1)/n$ and the $\varepsilon_j$ are zero mean, uncorrelated, normal errors with common variance $\sigma^2 = .25$. The actual data is given in Exercise 10. The GCV criterion was minimized at $\hat{\lambda} = 3$ (see Table 3.3) and the resulting estimate of the regression function was

$$\mu_3(t) = -.48 + 41.2t - 40.68t^2 \ .$$

The fitted function is therefore in excellent agreement with the true regression curve $\mu(t) = 40t(1 - t)$. It is interesting to note that for these data an estimate of the error variance is $\sigma_3^2 = nMSE(\lambda)/(n - \lambda) = .21$ which is quite close to the true error variance .25.

Table 3.3 gives the values of the GCV criterion and the true value of the loss $L_n(\lambda)$ for several values of $\lambda$. As might be expected $L_n(\lambda)$ and $GCV(\lambda)$ are both minimized for $\hat{\lambda} = 3$. A stochastic analog of the GCV Theorem of Chapter 2 would suggest that the difference between GCV(λ) and $L_n(\lambda)$ should be approximately $n^{-1}\Sigma_{j=1}^{n}\varepsilon_j^2 = .235$ if $L_n(\lambda)$ is small. This is seen to be roughly the case from the values in Table 3.3.

**Table 3.3. GCV(λ) and $L_n(\lambda)$ for Polynomial Regression Example**

| λ | GCV(λ) | $L_n(\lambda)$ | GCV(λ)-$L_n(\lambda)$ |
|---|---|---|---|
| 1 | 9.9132 | 8.9362 | .9768 |
| 2 | 10.1665 | 8.9073 | 1.2592 |
| 3 | 0.2221 | 0.0388 | .1833 |
| 4 | 0.2317 | 0.0389 | .1928 |
| 5 | 0.2415 | 0.0394 | .2021 |
| 6 | 0.2514 | 0.0403 | .2111 |
| 7 | 0.2437 | 0.0548 | .1889 |

## EXERCISES

**1.** Compute the first three Legendre polynomials, using formulas (10) and (11), in the case where [a,b] = [0,1].

**2.** Let $L_2[a,b] \times [a,b]$ denote the collection of all square integrable functions on [a,b] × [a,b], i.e., the set of all functions μ(s,t) such that

$$\int_a^b \int_a^b \mu(s,t)^2 ds dt < \infty.$$

A well known result from functional analysis has the consequence that if $\{x_j(t)\}$ is a CONS for $L_2[a,b]$ then the product sequence

$$\{x_j x_r\}_{j,r=1,\infty}$$

is a CONS for $L_2[a,b] \times [a,b]$. Verify the orthonormality property of the product sequence and give a formula for the generalized Fourier series of a square integrable bivariate regression function. Discuss the implications of this for Fourier series estimation in the case of bivariate dependent variables. [Note: There is nothing special about the bivariate case. Analogous results hold for any number of dependent variables. The variables need not all be defined on the same interval [a,b] and, in fact, a different CONS can be used for each variable.]

**3.** Show that for generalized Fourier series estimators the ordinary cross-validation criterion is

$$CV(\lambda) = n^{-1} \sum_{j=1}^{n} (y_j - \mu_\lambda(t_j))^2/(1 - h_{jj}(\lambda))^2 \quad ,$$

where $h_{jj}(\lambda)$ is the jth diagonal element of $H(\lambda) = X_\lambda(X_\lambda^* X_\lambda)^{-1} X_\lambda^*$. Discuss the relationship between CV and GCV in this case.

**4.** In the discussion after Theorem 3. 1 it was noted that $\inf_{\lambda \le n} L_n(\lambda) \overset{p}{\to} 0$ if $L_n(\lambda_n) \overset{p}{\to} 0$ for any sequence $\lambda_n$. Verify that this is true and then show that $\inf_{\lambda \le n} R_n(\lambda) \to 0$ implies that $\inf_{\lambda \le n} L_n(\lambda) \overset{p}{\to} 0$. [Hint: Use a Chebychev type inequality.]

**5.** Verify identity (16) and thereby verify the sine and cosine form for the classical Fourier series estimator.

**6.** Suppose that $\mu \in W_{2,per}^m[0,1]$, the subset of $W_2^m[0,1]$ whose elements satisfy $\mu^{(j)}(0) = \mu^{(j)}(1)$, j=0,...,m - 1. Assume that (25) holds and that the Fourier coefficients decay algebraically with exponent $(2m + 1 + \delta)$. Show that for m≥1 the asymptotically optimal $\lambda$ for minimizing $R_n(\lambda)$ is $\lambda_{opt} \propto n^{1/(2m+1+\delta)}$ and that $R_n(\lambda_{opt}) = O(n^{-(2m+\delta)/(2m+1+\delta)})$ for classical Fourier series estimators.

**7.** Define the kth partial sum of the Fourier series for $\mu \in C[0,1]$ by

$$S_k(t) = \sum_{j=-k}^{k} \beta_j e^{2\pi ijt} \quad .$$

It can be shown (see, eg., Dym and McKean, (1972) that $\lambda^{-1}\sum_{j=0}^{\lambda-1} S_j(t)$ converges uniformly to $\mu$ if $\mu(0) = \mu(1)$. This suggests that one might estimate $S_k$ by

$$S_{\lambda k}(t) = \sum_{j=-k}^{k} \beta_{\lambda j} e^{2\pi ijt}$$

and $\mu(t)$ by $\mu_\lambda(t) = \lambda^{-1}\sum_{j=0}^{\lambda-1} S_{\lambda k}(t)$. Show that when $t_j = (j - 1)/n$, j=1,...,n,

$$\mu_\lambda(t) = \frac{1}{n}\sum_{j=1}^{n} y_j K_\lambda(t - t_j) \qquad ,$$

where $K_\lambda$ is the *Fejer Kernel*

$$K_\lambda(u) = [\sin(\lambda\pi u)/\sin(\pi u)]^2/\lambda.$$

Graph the function $K_\lambda$ for several values of $\lambda$ and discuss its properties relative to the Dirichlet kernel (30).

**8.** Show that when $t_j = (j - 1)/n$, $j=1,\ldots,n$, CV and GCV coincide for classical Fourier series estimators.

**9.** Show that if, for example, $\mu \in W_{2,per}^2[0,1]$ and the $\varepsilon_j$ satisfy the conditions needed for Theorem 3. 2 then $n^{1/2}(\beta_{\lambda j} - \beta_j)/\sigma$ has an asymptotic standard normal distribution. Discuss the implications of this fact for testing the hypothesis that $\beta_j$ is zero.

**10.** The 50 observations given below were generated from the model

$$y_j = 40t_j(1 - t_j) + \varepsilon_j, \quad j=1,\ldots,50,$$

where $t_j = (j - 1)/n$ and the $\varepsilon_j$ are zero mean, uncorrelated normal random variates with common standard deviation $\sigma = .5$.

a) Fit a classical Fourier series estimator to these data using GCV to select $\lambda$. Through examination of the estimated coefficients, diagnostics, etc. determine any adjustments which should be made to your estimator.

b) Compare the estimated Fourier coefficients to the true Fourier coefficients for $\mu$ in this case.

c) Compute the estimator of $\sigma^2$ in (48) and the least-squares estimator (49) of $\sigma^2$ at the optimal $\lambda$. Compare these estimators to the true value of $\sigma^2$, namely .25.

d) Construct interval estimates of $\mu$ at the $t_i$ using your estimators of $\sigma^2$ in Part c) and discuss their coverage performance.

e) Compute $L_n(\lambda)$ for various values of $\lambda$ and compare the optimal $\lambda$ to the value selected by GCV. Also compare the values of GCV($\lambda$) and $L_n(\lambda)$ and discuss the magnitude of their difference.

| t | y | t | y | t | y |
|---|---|---|---|---|---|
| .00 | .0437 | .34 | 8.9664 | .68 | 8.6527 |
| .02 | .4536 | .36 | 8.5062 | .70 | 8.0524 |
| .04 | .0120 | .38 | 9.1926 | .72 | 8.7368 |
| .06 | 2.2703 | .40 | 9.2407 | .74 | 7.9800 |
| .08 | 2.5558 | .42 | 10.5346 | .76 | 7.3618 |
| .10 | 3.0928 | .44 | 9.3915 | .78 | 6.9728 |
| .12 | 4.1561 | .46 | 9.3649 | .80 | 6.7658 |
| .14 | 4.2316 | .48 | 10.3919 | .82 | 5.4360 |
| .16 | 4.8360 | .50 | 10.3574 | .84 | 5.2303 |
| .18 | 5.9017 | .52 | 10.0387 | .86 | 5.2017 |
| .20 | 5.6193 | .54 | 9.7512 | .88 | 4.7096 |
| .22 | 7.2262 | .56 | 8.8896 | .90 | 3.3387 |
| .24 | 7.3737 | .58 | 9.6229 | .92 | 2.0831 |
| .26 | 7.6464 | .60 | 10.1354 | .94 | 2.2804 |
| .28 | 7.7062 | .62 | 9.3907 | .96 | 1.2281 |
| .30 | 8.4472 | .64 | 9.9147 | .98 | 1.6561 |
| .32 | 8.1836 | .66 | 8.6358 | | |

**11.** Let f be a square integrable continuous density on [0,1] with f(0) = f(1). This has the implication that

$$f(t) = 1 + \sum_{j\neq 0} \beta_j e^{2\pi i j t}$$

with

$$\beta_j = \int_0^1 f(x)e^{-2\pi i j x}\,dx, \quad j=0, \pm 1, \ldots .$$

Given a random sample $X_1, \ldots, X_n$ with density f define the jth sample Fourier coefficient by

$$\tilde{\beta}_j = n^{-1} \sum_{r=1}^{n} e^{-2\pi i j X_r}$$

An estimator of f is then provided by

$$f_\lambda(t) = \sum_{j=-\lambda}^{\lambda} \tilde{\beta}_j e^{2\pi i j t} \quad .$$

a) Discuss the properties of $f_\lambda$ as a density estimator. Point out any possible shortcomings.

b) Show that $E\tilde{\beta}_j = \beta_j$ and that $\mathrm{Cov}(\tilde{\beta}_j, \tilde{\beta}_k) = E(\tilde{\beta}_j - \beta_j)(\tilde{\beta}_k^* - \beta_k^*) = n^{-1}(\beta_{j-k} - \beta_j\beta_k^*)$. In particular establish that $\mathrm{Var}\,\tilde{\beta}_j = n^{-1}(1 - |\beta_j|^2)$.

c) Using your results in Part a) obtain a simple expression for the integrated risk

$$IR(\lambda) = E\int_0^1 (f(t) - f_\lambda(t))^2 dt$$

in terms of the Fourier coefficients for f.

d) Obtain an unbiased estimate of the difference between $IR(\lambda)$ and $IR(\lambda + 1)$ and discuss how this might be used to guide the selection of a "good" value for the number of terms included in the Fourier series of f.

[Kronmal and Tarter, 1968, Tarter and Kronmal 1976, Hart, 1985]

**12.** One of the difficulties with the sample Fourier coefficients defined in Exercise 11 is that, for fixed n, the variance of $\tilde{\beta}_j$ will tend to grow with j (Why?). One approach which attempts to compensate for this is to damp the $\tilde{\beta}_j$ for larger $|j|$ when they are used in an estimator. An estimator of the density f in Exercise 11 which does this is, for n even,

$$f_\theta(t) = 1 + \sum_{\substack{j=-n/2 \\ j\neq 0}}^{n/2} \frac{\theta_j}{\theta_j + n^{-1}} \tilde{\beta}_j e^{2\pi ijt} \quad ,$$

where $\theta_{-j} = \theta_j$ and $\theta_1 > \theta_2 > \ldots > \theta_{n/2} > 0$.

Suppose that the $\beta_j$ are independent, zero mean, complex normal random variables with Var $\beta_j = E|\beta_j|^2 = \theta_j$, $j=0, \pm 1, \ldots$. If one approximates the distribution of $\tilde{\beta}_j$ by a complex normal with mean $\beta_j$ and variance $1/n$, show that $f_\theta$ can then be viewed as a Bayesian estimate of f. What difficulties do you see with this prior distribution for the $\beta_j$? [Hint: What characteristics should a density estimator have?]

[See Wahba, 1981c, and references therein.]

**13.** Compute the first three polynomials which are orthonormal on $t_j = (j - 1)/n$, $j=1,\ldots,n$, i.e., construct functions $x_{0n}$, $x_{1n}$, $x_{2n}$ from the functions, 1, t, $t^2$ which satisfy (57). Compare these to the first three Legendre polynomials you computed in Exercise 1. [It may help to recall that

$$\sum_{j=0}^{n-1} j = n(n-1)/2, \quad \sum_{j=0}^{n-1} j^2 = (2n-1)n(n-1)/6,$$

$$\sum_{j=0}^{n-1} j^3 = n^2(n-1)^2/4$$

and

$$\sum_{j=0}^{n-1} j^4 = (3n^2 - 3n - 1)(2n-1)n(n-1)/30.]$$

**14.** The data given below represents the voltage drop in the battery of a guided missile motor during its time of flight.

a) Fit a polynomial regression estimator to this data. Through examination of estimated coefficients, diagnostics, etc. determine any adjustments which should be made to the estimator.

b) Obtain approximate interval estimates for the true regression function at the $t_i$ using the interval estimates discussed in Section 3.3.

| Time (seconds) | Voltage Drop | Time (seconds) | Voltage Drop |
|---|---|---|---|
| .0 | 8.33 | 10.5 | 14.92 |
| .5 | 8.23 | 11.0 | 14.37 |
| 1.0 | 7.17 | 11.5 | 14.63 |
| 1.5 | 7.14 | 12.0 | 15.18 |
| 2.0 | 7.31 | 12.5 | 14.51 |
| 2.5 | 7.60 | 13.0 | 14.34 |
| 3.0 | 7.94 | 13.5 | 13.81 |
| 3.5 | 8.30 | 14.0 | 13.79 |
| 4.0 | 8.76 | 14.5 | 13.05 |
| 4.5 | 8.71 | 15.0 | 13.04 |
| 5.0 | 9.71 | 15.5 | 12.06 |
| 5.5 | 10.26 | 16.0 | 12.05 |
| 6.0 | 10.91 | 16.5 | 11.15 |
| 6.5 | 11.67 | 17.0 | 11.15 |
| 7.0 | 11.76 | 17.5 | 10.14 |
| 7.5 | 12.81 | 18.0 | 10.08 |
| 8.0 | 13.30 | 18.5 | 9.78 |
| 8.5 | 13.88 | 19.0 | 9.80 |
| 9.0 | 14.59 | 19.5 | 9.95 |
| 9.5 | 14.05 | 20.0 | 9.51 |
| 10.0 | 14.48 | | |

**Source: Montgomery and Peck (1982). Reprinted with permission from John Wiley and Sons, Inc.**

**15.** Suppose that $\mu \in C[0,1]$ and the $\varepsilon_j$ are independent and identically distributed. Show that $\hat{\sigma}^2$ in (48) is a consistent estimator of $\sigma^2$.

**16.** Let $\{x_j\}_{j=0}^{\infty}$ be a CONS with $x_0(t) \equiv 1$. Show that a generalized Fourier series estimator based on $x_j$ can be written as

$$\mu_\lambda(t) = \sum_{j=1}^{n} K_{\lambda,n}(t,t_j)y_j \Big/ \sum_{j=1}^{n} K_{\lambda,n}(t,t_j)$$

for some kernel function $K_{\lambda,n}$ depending on $\lambda$ and n.

**17.** The $\lambda$th partial sum for the Legendre series for $\mu$ is

$$\sum_{j=0}^{\lambda-1} \beta_j x_j \quad ,$$

where

$$\beta_j = \int_a^b \mu(t)x_j(t)dt$$

and the $x_j$ are defined by (10)-(11). Thus an expression for $\mu$ obtained from Proposition 3.1 is

$$\mu(t) = \sum_{j=0}^{\lambda-1} \beta_j x_j(t) + E_\lambda(t),$$

with $E_\lambda(t) = \mu(t) - \Sigma_{j=0}^{\lambda-1}\beta_j x_j(t)$. In contrast, Taylor's Theorem gives

$$\mu(t) = \sum_{j=0}^{\lambda-1} \alpha_j x_j(t) + \text{Rem}_\lambda(t)$$

for coefficients $\alpha_0,\ldots,\alpha_{\lambda-1}$ and $\text{Rem}_\lambda$ defined by (52). Relate the coefficients and remainders in these two expressions and draw conclusions about the utility of the least-squares estimators $\beta_{\lambda 0},\ldots,\beta_{\lambda,\lambda-1}$ of the $\beta_j$ as estimators of the $\alpha_j$.

**18.** The estimated Fourier coefficients obtained from least squares are optimal in the sense that they minimize the sum of squared errors. Show that the true Fourier coefficients have a similar optimality property. Specifically, show that $\beta_1,\ldots,\beta_\lambda$

with

$$\beta_j = \int_a^b \mu(t)x_j(t)dt$$

minimize

$$||\mu - \sum_{j=1}^{\lambda} b_j x_j||$$

over all possible values of $b_1, \ldots, b_\lambda$.

**19.** Let the $W_r$ and $t_r$ be defined as in Section 3.5.2 and let $\underline{x}(t_r) = (x_0(t_r), \ldots, x_q(t_r))'$ where the $x_j$ are the Legendre polynomials. Show that for any set of real numbers $\delta_1, \ldots, \delta_n$

$$\sum_{k=1}^{n} \sum_{r=1}^{n} \delta_k \delta_r W_k W_r \underline{x}(t_k)'\underline{x}(t_r) \leq \sum_{r=1}^{n} \delta_r^2 W_r .$$

[Hint: Minimize $\Sigma_{r=1}^{n}(\delta_r - \underline{c}'\underline{x}(t_r))^2$ with respect to $\underline{c} \in \mathbb{R}^n$.]

# 4
# Kernel Estimators

## 4.1 Introduction

Suppose, as before, that we have n observations $(t_1,y_1),\ldots,(t_n,y_n)$ which follow the model

$$y_j = \mu(t_j) + \varepsilon_j , \qquad j = 1,\ldots,n. \tag{1}$$

The $\varepsilon_j$ are zero mean, uncorrelated random variables and the $t_j$ are nonstochastic design points assumed to satisfy

$$a \le t_1 < t_2 < \ldots < t_n \le b \tag{2}$$

for finite constants a and b. The assumption of distinct and scalar valued $t_j$ is for ease of presentation and can be relaxed with most of the results which follow held intact (see, eg., Exercise 23). However we will not pursue this in any great detail.

In this chapter we discuss what are known as kernel estimators of $\mu$ in model (1). Like series estimators, kernel estimators are linear estimators in the sense described in Chapter 1. They arise from a particularly simple method of choosing the weights in (1.3).

Our study of kernel estimators begins with a discussion of the motivations and various forms for these estimators that is presented in the next section. Sections 4.3 and 4.6 contain a treatment of the asymptotic theory for kernel estimators while Sections 4.4 and 4.5 detail methods for choosing the kernel (or weight) function and bandwidth (or smoothing) parameter. Extensions of the basic kernel smoothing methodology are described in Sections 4.7, 4.8 and 4.11. Some applications of kernel regression estimates are presented in Section 4.9 while Section 4.10 summarizes some of what is known about the behavior of kernel estimators when the independent variable, t, is also random.

## 4.2 Kernel Estimators

Kernel estimators are estimators of the form (1.3) with weights deriving from a single function that is independent of the design. The idea behind such estimators can be seen by referring back to our work with classical Fourier series estimators in Chapter 3.

Let 2p + 1 be the number of terms included in a Fourier series estimator of $\mu$, i.e., the estimator is $\Sigma_{j=-p}^{p} \beta_{pj} e^{2\pi ijt}$ with $\beta_{pj} = n^{-1}\Sigma_{r=1}^{n} y_r e^{-2\pi ijt_r}$. Define $\lambda = 1/(2p + 1)$ (Note: This is a departure from our previous notation where $2\lambda + 1$ was the number of terms). Then, for equally spaced $t_j$ the estimator can be expressed as

$$\sum_{j=1}^{n} y_j K_\lambda(t - t_j) \tag{3a}$$

with $K_\lambda$ the (reparametrized) Dirichlet kernel now defined to be

$$K_\lambda(u) = \sin(2\pi u/\lambda)/\sin(\pi u). \tag{3b}$$

A plot of $K_\lambda$ was given in Figure 3.1 for several values of p and is seen from this to be symmetric about and have a maximum at zero. The function tends to diminish in magnitude away from zero, although the behavior of the Dirichlet kernel leaves something to be desired in this regard.

Now let us note that estimator (3) is perfectly well defined for any nonnegative value of p, whether it is integral or not, and could be used even if the $t_j$ were not equally spaced. Of course, if the $t_j$ are not uniform and/or p is not integral, $\mu_\lambda$ would, strictly speaking, no longer be a classical Fourier series estimator. Nonetheless, it would still represent an estimator of $\mu$ since, due to the properties of $K_\lambda$, $\mu_\lambda(t)$ will tend to be a type of weighted average of observations near t.

Using a kernel function such as (3b) to generate the weights in (1.3) is a particularly simple approach. In this case the weights involve the design only through their dependence on $t_j$ and do not require the solution of equation systems as do Fourier series estimators (e.g., (3.19a)) and other estimators to be studied later.

The previous discussion suggests that a convenient way to generate weights is through the use of a function such as (3b). Such functions should have finite support, e.g., [-1,1], with a maximum at zero; the latter property insures that $y_j$ contributes most to its own prediction. Also we would like our function K to satisfy the moment conditions

$$\int_{-1}^{1} K(u)du = 1 , \tag{4a}$$

$$\int_{-1}^{1} uK(u)du = 0 , \tag{4b}$$

$$\int_{-1}^{1} u^2K(u)du = \alpha \neq 0 \tag{4c}$$

and

$$\int_{-1}^{1} K(u)^2du < \infty. \tag{4d}$$

Condition (4a) is roughly equivalent to having the weights sum to one while (4b) is a type of symmetry condition that is automatically satisfied if K is symmetric about zero. The reasons for condition (4c)-(4d) will be seen shortly. A function K with these properties will hereafter be referred to as a *kernel of order 2* or simply a *kernel* if the context is clear. In subsequent discussions we will have occasion to consider kernels of higher order than two which arise from imposing additional moment conditions on the function K. We will also use the term kernel for these functions when the specification of order is immaterial to the discussion.

To generate the weights in (1.3) we could now use

$$K(t,t_j;\lambda) = \frac{1}{\lambda} K(\lambda^{-1}(t - t_j)) , \quad \lambda > 0 . \tag{5}$$

This is almost, but not quite, what happens with classical Fourier series estimators. In that case $K_\lambda(u)$ cannot be written as a function involving the argument $u/\lambda$ due to the fact that its denominator $\sin(\pi u)$ does not involve $\lambda$. However, if we approximate the sine function using the first term in its Taylor series expansion we find that

$$K_\lambda(u) = \frac{\sin(\pi u/\lambda)}{\sin(\pi u)} \doteq \frac{1}{\lambda} \frac{\sin(\pi u/\lambda)}{(\pi u)/\lambda}$$

which does have the form of (5).

The parameter $\lambda$ in (5) is called the *bandwidth* or smoothing parameter. Note that if $K(u)$ has support on $[-1,1]$ then $K(u/\lambda)$ has support on $[-\lambda,\lambda]$, so the bandwidth determines, among other things, how far away observations are allowed to be from t and still contribute to the estimation of $\mu(t)$. The bandwidth also governs the peakedness of the kernel and, hence, the degree of dependence of the estimator on information near t. Small values of $\lambda$ will result in rougher (wigglier) estimators that rely heavily on the data near t. In contrast, larger $\lambda$'s allow more averaging to occur and thereby give smoother estimators. If K is nonnegative then, as a result of (4a), it is a probability density on $[-1,1]$ and, in that case, $\lambda$ is a scale parameter in the statistical sense. We will discuss the selection of $\lambda$ in more detail in Section 4.5.

Note that there is no real loss in assuming that K has support on $[-1,1]$. This is because any kernel with finite support can be rescaled to have support on $[-1,1]$. It should be noted, however, that kernels which do not have finite support are sometimes used. As seen from Exercise 7 the use of kernels with support on the entire line results in estimators with global bias difficulties which are localized to the boundaries when K has finite support. Thus kernels with finite support seem preferable.

Table 4.1 provides several examples of kernels that are often used in practice. The uniform, triangular and quadratic kernels all satisfy (4). However, the kernel titled "minimum variance" violates (4c). The properties of this latter kernel will be discussed in more detail in Section 4.7. Issues related to the choice of a good kernel function are discussed in Sections 4.4 and 4.7. Notice, however, that the first three kernels in Table 4.1 do not exhibit the oscillation properties away from their peaks that were present in the Dirichlet kernel. This seems preferable, although not mandatory, for reasons dicussed in Section 4.1.

In keeping with our analogy with the Fourier series estimator (3), the estimator we obtain from (5) is

$$\mu_\lambda(t) = (n\lambda)^{-1} \sum_{j=1}^{n} K(\lambda^{-1}(t - t_j))y_j . \tag{6a}$$

There are several variants on this theme which include

$$\mu_\lambda(t) = \sum_{j=1}^{n} (t_j - t_{j-1})\frac{1}{\lambda}K(\lambda^{-1}(t - t_j))y_j, \tag{6b}$$

$$\mu_\lambda(t) = \sum_{j=1}^{n} K(\lambda^{-1}(t - t_j))y_j / \sum_{j=1}^{n} K(\lambda^{-1}(t - t_j)) , \tag{6c}$$

$$\mu_\lambda(t) = \sum_{j=1}^{n} (\frac{1}{\lambda}\int_{s_{j-1}}^{s_j} K(\lambda^{-1}(t - s))ds)y_j , \tag{6d}$$

where

$$s_0 = a, \; s_{j-1} \le t_j \le s_j, \quad j=1,\dots,n - 1, \quad s_n = b,$$

and

$$\mu_\lambda(t) = \int_a^b \frac{1}{\lambda}K(\lambda^{-1}(t - s))h_y(s)ds , \tag{6e}$$

where

$$h_y(s) = \begin{cases} y_1 & , \quad s \le t_1, \\ y_j\left(\dfrac{t_{j+1} - s}{t_{j+1} - t_j}\right) + y_{j+1}\left(\dfrac{s - t_j}{t_{j+1} - t_j}\right) & , \quad t_j \le s \le t_{j+1}, \\ y_n & , \quad s \ge t_n \end{cases} .$$

**Table 4.1 Examples of Kernels**

| Kernel | K(u) | $V = \int K(u)^2 du$ | $M = \int u^m K(u) du$ | $(V^m M)^{2/(2m+1)}$ |
|---|---|---|---|---|
| **Uniform** | $\begin{cases} \frac{1}{2}, & \lvert u \rvert \le 1, \\ 0, & \lvert u \rvert > 1, \end{cases}$ | $\frac{1}{2}$ | $\frac{1}{3}$ | .3701 |
| **Triangular** | $\begin{cases} 1-\lvert u \rvert, & \lvert u \rvert \le 1, \\ 0, & \lvert u \rvert > 1 \end{cases}$ | $\frac{2}{3}$ | $\frac{1}{6}$ | .3531 |
| **Quadratic (optimal) (m=2)** | $\begin{cases} \frac{3}{4}(1-u^2), & \lvert u \rvert \le 1, \\ 0, & \lvert u \rvert > 1 \end{cases}$ | .6 | .2 | .3491 |
| **Minimum Variance (m=4)** | $\begin{cases} \frac{3}{8}(3-5u^2), & \lvert u \rvert \le 1, \\ 0, & \lvert u \rvert > 1 \end{cases}$ | $\frac{9}{8}$ | -.08571 | .6432 |

We will refer to all five estimators (6a)-(6e) as kernel estimators.

Let us pause a moment to discuss the relationships between and the history behind estimators (6a)-(6e). First note that estimator (6b) is essentially a generalization of estimator (6a) that replaces the factor $n^{-1}$ by $t_j - t_{j-1}$, which is more appropriate for unequally spaced data. Estimator (6c) is a modification of (6a) which ensures that the weights sum to one. Note that, out of estimators (6a)-(6c), only (6c) will, strictly speaking, ever be a weighted average of the Y values.

To understand the motivation for estimators (6d) and (6e) first observe that, e.g., estimator (6a) has expectation $(n\lambda)^{-1}\Sigma_{j=1}^{n}K(\lambda^{-1}(t - t_j))\mu(t_j)$. For equally spaced design points and large n this is approximately equal to

$$\lambda^{-1}\int_0^1 \mu(s)K(\lambda^{-1}(t - s))ds .$$

Thus, estimator (6d) is in the spirit of (6a) - (6c) but uses $\lambda^{-1}y_j\int_{s_{j-1}}^{s_j} K(\lambda^{-1}(t - s))ds$ to estimate $\lambda^{-1}\int_{s_{j-1}}^{s_j} \mu(s)K(\lambda^{-1}(t - s))ds$ rather than $(n\lambda)^{-1}y_jK(\lambda^{-1}(t-t_j))$ or $\lambda^{-1}y_j(t_j-t_{j-1})K(\lambda^{-1}(t - t_j))$. A typical choice for $s_j$ is $s_j = (t_{j+1} + t_j)/2$. Provided $t \in [\lambda,1 - \lambda]$ estimator (6d) is also a weighted average of the $y_j$.

Estimator (6e) is similar to (6d). The difference is that a piecewise linear approximation, $h_y(s)$, is used to estimate $\mu(s)$ for $t_j \leq s \leq t_{j+1}$ rather than the piecewise constant approximation of $y_j$, for $t_j \leq s < t_{j+1}$, used by estimator (6d). Other properties of this estimator are explored in Exercises 1, 3, 19 and 20.

Estimators (6a) and (6b) stem from the work of Priestley and Chao (1972). The use of estimator (6c) with nonstochastic $t_j$ was initially proposed by Benedetti (1974, 1975). Estimator (6c) is often used in the case where the $t_j$ are also random. In this case its use was suggested by Nadaraya (1964) and Watson (1964). For this reason it is often called the Nadaraya-Watson estimator. We will consider the case of random t's in Section 4.10.

Estimator (6d) was originally studied by Gasser and Müller (1979) and Cheng and Lin (1981). The closely related estimator (6e) was proposed by Clark (1977) and has also been studied by Cheng and Lin (1981).

All five estimators (6a)-(6e) have very similar properties for uniformly spaced $t_j$. It appears that estimator (6d) has received more attention in the literature and that it tends to have a bias of smaller order than that of estimators (6a) - (6c) (see, Gasser and Müller, 1979). It is also more appropriate for unequally spaced design points than (6a) and (6c).For these reasons we will focus on (6d) in Sections 4.3-4.9. However, it should be emphasized that similar developments are also possible for the other estimators.

Each of estimators (6a)-(6e) can be extended to include the case where t is vector valued. If $\underline{t}$ is a p-vector, a simple way this can be accomplished is to let

$$K(u_1,\ldots,u_p) = \prod_{j=1}^{p} K_j(u_j) ,$$

where $K_j$ satisfies (4). Then, an estimator of $\mu$ is, e.g.,

$$\mu_\lambda(\underline{t}) = \frac{1}{n} \sum_{j=1}^{n} y_j \prod_{r=1}^{p} \lambda_r^{-1} K_r(\lambda_r^{-1}(t_r - t_{jr})) ,$$

where $\underline{t}_j = (t_{j1},\ldots,t_{jp})'$ is the vector of values for the p independent variables corresponding to $y_j$. Analogous results to those for the case of univariate t can be obtained for this estimator as well as more general estimators based on a more general p-variate kernel function (see Exercises 21). We will not pursue this further here and will concentrate on univariate t in all subsequent work.

In the next section we will study the asymptotic properties of estimator (6d). Before doing so, however, it will be useful to complete the circle and relate kernel estimators back to the

Fourier series estimators which were used to motivate their study. By doing this we will gain some insight into the way that kernel estimators actually smooth the data.

First let us note that, for general $\mu$, problems arise for estimation of $\mu(t)$ when $t \in [a,\lambda)$ or $t \in (b - \lambda, b]$. To see what happens at the boundaries, suppose $t \in [a,\lambda)$, i.e., we are within a bandwidth of the left boundary. This is depicted in Figure 4.1 for $[a,b] = [0,1]$, equally spaced $t_j$ with $t = 1/n$ and $\lambda$ slightly less than $3/n$. For the situation in Figure 4.1 we have a full bandwidth's worth of data on the right hand side of $t$ (namely, observations at $2/n$ and $3/n$) but have only the observation at 0 on the left hand side of $t$. If we had observed a Y value at $-1/n$ there would be no problem and the estimator could be constructed from a full complement of data. However, the observed $t_j$ are all within $[0,1]$. As a result, there is an imbalance in the amount of information available from the data about $\mu$ on each side of $t$. The illustration in Figure 4.1 is indicative of what happens in general. There is nothing special about the lower boundary and similar problems arise for $t \in (b - \lambda, b]$. The lack of data at the boundaries causes a boundary bias problem for kernel estimators unless they are modified in some fashion. We will discuss this further in Section 4.3.

When $\mu$ is periodic ($\mu(a) = \mu(b)$) the kernel estimator can be modified to avoid boundary difficulties such as those discussed above. The basic idea can again be seen from Figure 4.1. We noted that if we had data at $-1/n$, in that case, we would have balanced information about $\mu$ on both sides of $t$. However, if $\mu$ is periodic with period 1 this means that $\mu(-1/n) = \mu(1 - 1/n) = \mu((n - 1)/n)$. Thus $y_n$ in Figure 4.1 can be viewed as an observation at $-1/n$ and we can borrow from the upper end of the data to complete our estimator. This example extends immediately to the general problem of estimation in the periodic case. The resulting modified kernel estimator has close ties to classical Fourier series.

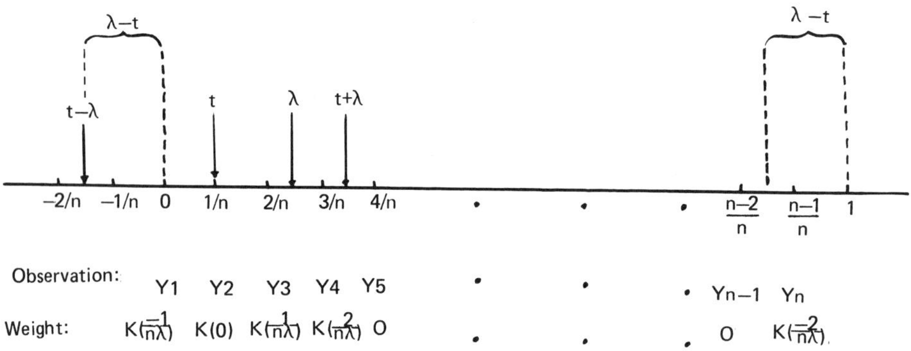


**Figure 4.1** Treatment of data in the periodic case.

Thus let us assume that $\mu$ is periodic and, for simplicity, that $t_j = (j - 1)/n$, $j = 1,\ldots,n$. Let n be odd and suppose we wish to estimate $\mu(t_r)$. The periodic nature of the regression function means that we can always formulate the problem as if half of the observations are below $t_r$ and half are above. This is accomplished by simply rearranging the data values. For example, if $r \geq (n + 1)/2$ we defined new observations by

$$\begin{aligned}
\tilde{y}_{1,r} &= y_{r+1-(n+1)/2}, & \tilde{t}_{1,r} &= (r - (n + 1)/2)/n, \\
&\vdots \\
\tilde{y}_{(n+1)/2,r} &= y_r, & \tilde{t}_{r,r} &= (r - 1)/n, \\
&\vdots \\
\tilde{y}_{n-r+(n+1)/2,r} &= y_n, & \tilde{t}_{n-r+(n+1)/2,r} &= (n - 1)/n, \\
\tilde{y}_{n-r+1+(n+1)/2,r} &= y_1, & \tilde{t}_{n-r+1+(n+1)/2,r} &= 1, \\
&\vdots \\
\tilde{y}_{n,r} &= y_{r-(n+1)/2}, & \tilde{t}_{n,r} &= 1 + (r - 1 - (n + 1)/2)/n .
\end{aligned}$$

A similar arrangement holds when $r \leq (n + 1)/2$. In either case one finds that the index of the original Y response corresponding to a $\tilde{y}_{j,r}$ is $j + r - (n + 1)/2$ modulo the possible addition or subtraction of a factor of n. An analogous development can be pursued when n is even.

Assuming K to be symmetric the above discussion suggests estimation of $\mu(t_r)$ by

$$\begin{aligned}
\tilde{\mu}_\lambda(t_r) &= n^{-1} \sum_{j=1}^{n} \frac{1}{\lambda} K(\lambda^{-1}(\tilde{t}_{r,r} - \tilde{t}_{j,r}))\tilde{y}_{j,r} \\
&= n^{-1} \sum_{j=1}^{n} \frac{1}{\lambda} K((n\lambda)^{-1}(\frac{n + 1}{2} - j))\tilde{y}_{j,r} . \qquad (7)
\end{aligned}$$

Now consider $\tilde{\underline{\mu}}_\lambda = (\tilde{\mu}_\lambda(t_1),\ldots,\tilde{\mu}_\lambda(t_n))'$ as an n-vector and observe that the n vectors

$$\underline{x}_j = n^{-1/2}(1, e^{2\pi ij/n}, \ldots, e^{2\pi ij(n-1)/n})',$$

$$j = -(n - 1)/2, \ldots, (n - 1)/2,$$

form an orthonormal basis for $\mathbb{R}^n$. Thus,

$$\tilde{\underline{\mu}}_\lambda = \sum_j \beta_{\lambda j} \underline{x}_j \tag{8a}$$

with

$$\beta_{\lambda j} = \underline{x}_j^* \tilde{\underline{\mu}}_\lambda \tag{8b}$$

(recall, '*' indicates complex conjugate transpose).

To explore the consequences of (8) we must explicitly compute $\beta_{\lambda j}$. It is found that

$$\beta_{\lambda j} = n^{-1/2} \sum_{r=1}^{n} \tilde{\mu}_\lambda(t_r) e^{-2\pi i j(r-1)/n}$$

$$= n^{-3/2} \sum_{r=1}^{n} \sum_{k=1}^{n} \frac{1}{\lambda} K((n\lambda)^{-1}(\frac{n+1}{2} - k)) \tilde{y}_{k,r} e^{-2\pi i j(r-1)/n}$$

$$= n^{-3/2} \sum_{k=1}^{n} \left( \frac{1}{\lambda} K((n\lambda)^{-1}(\frac{n+1}{2} - k)) e^{-2\pi i j(n + 1 - 2k)/2n} \right.$$

$$\left. \times \sum_{r=1}^{n} \tilde{y}_{k,r}\, e^{-2\pi i j(r + k - (n+3)/2)/n} \right) .$$

A simple computation, using the relationship between the indices of the $\tilde{y}$'s and the original observations, along with the identity $e^{2\pi i j} = 1$ reveals that

$$n^{-1} \sum_{r=1}^{n} \tilde{y}_{k,r}\, e^{-2\pi i j(r+k-(n+3)/2)/n} = n^{-1} \sum_{r=1}^{n} y_r e^{-2\pi i j(r-1)/n}$$

$$= \tilde{\beta}_j ,$$

where $\tilde{\beta}_j$ is the jth estimated Fourier coefficient for the Y data. Since this is independent of k we now see that

$$\beta_{\lambda j} = n^{1/2} \tilde{\beta}_j K_{\lambda j} , \tag{9}$$

where

$$K_{\lambda j} = n^{-1} \sum_{k=1}^{n} \frac{1}{\lambda} K((n\lambda)^{-1}(\frac{n+1}{2} - k))e^{-2\pi ij(n+1-2k)/2n}$$

is a discrete Fourier coefficient for the kernel. Thus

$$\tilde{\mu}_\lambda(r/n) = n^{-1/2} \sum_{j=-(n-1)/2}^{(n-1)/2} \beta_{\lambda j} e^{2\pi ijr/n}$$

$$= \sum_{j=-(n-1)/2}^{(n-1)/2} \tilde{\beta}_j K_{\lambda j} e^{2\pi ijr/n} \quad . \tag{10}$$

Equation (10) has the implication that $\tilde{\mu}_\lambda$ is essentially a Fourier series estimator. The estimator uses all n of the possible exponential functions which can be fitted to the data and its jth estimated Fourier coefficient is $\tilde{\beta}_j K_{\lambda j}$. Now $\tilde{\beta}_j$ is just the ordinary Fourier coefficient obtained from the Y data. Thus, in the periodic case, kernel estimation corresponds to Fourier series estimation using all n exponential functions with the jth sample Fourier coefficient multiplied by the jth Fourier coefficient of the kernel, $K_{\lambda j}$.

Multiplication of $\tilde{\beta}_j$ by $K_{\lambda j}$ tends to work as a damping factor. To see this note that

$$K_{\lambda j} \doteq \int_{-1/2}^{1/2} \frac{1}{\lambda} K(\frac{x}{\lambda}) e^{-2\pi ijx} dx$$

$$= \int_{-1/2\lambda}^{1/2\lambda} K(u) e^{-2\pi ij\lambda u} du$$

$$\doteq \int_{-\infty}^{\infty} K(u) e^{-2\pi ij\lambda u} du$$

for $\lambda$ sufficiently small and n large. Thus, $K_{\lambda j}$ is approximately the Fourier transform of K evaluated at $\tau = 2\pi j\lambda$.

Fourier transforms for some of the kernels in Table 1 are well known. For example, the Fourier transform for the rectangular kernel on [-1, 1] is

$$f(u) = \frac{\sin u}{u} \quad .$$

Similarly, for the triangular kernel $K(u) = (1 - |u|)$, $|u| < 1$,

$$f(u) = \left(\frac{\sin(u/2)}{u/2}\right)^2$$

and for the quadratic kernel

$$f(u) = 3(\sin u - u\cos u)/u^3 \quad .$$

Graphs of these three functions are shown in Figure 4.2. If we now think about evaluating them at points $2\pi j\lambda$ the effect, for any fixed $\lambda$, will be that $f(2\pi j\lambda)$ tends to grow smaller as j increases in magnitude. Thus, $K_{\lambda j}$ tends to decrease as $|j|$ increases meaning that we are damping down the contribution of the Fourier coefficients corresponding to higher frequencies.

As a result of the above analysis it now becomes apparent that the choice of a kernel corresponds to a particular way of down weighting Fourier coefficients in the Fourier representation of the estimator (10). The converse is also true. Thus, for example, we recognize that the kernel estimator based on the uniform kernel and the ordinary (classical) Fourier series estimator studied in the previous chapter exhibit an interesting dual nature. A kernel estimator that utilizes the uniform kernel will uniformly weight those observations within a bandwidth of the point of estimation and give zero weights to all other data values. However, the weights assigned to the Fourier coefficients are, up to an approximation, dictated by a Dirichlet kernel. In contrast, the Fourier series estimator (3) assigns weights to the observations using the Dirichlet kernel (3.b). If 2p + 1 is the number of terms retained in the series, a Fourier series estimator can also be expressed as

$$\sum_{|j| \le p} \tilde{\beta}_j e^{2\pi ijt} = 2 \sum_{|j| \le (n-1)/2} \tilde{\beta}_j K(j/p) e^{2\pi ijt}$$

with K the uniform kernel. Thus the weights assigned to the Fourier coefficients in this case correspond to the uniform

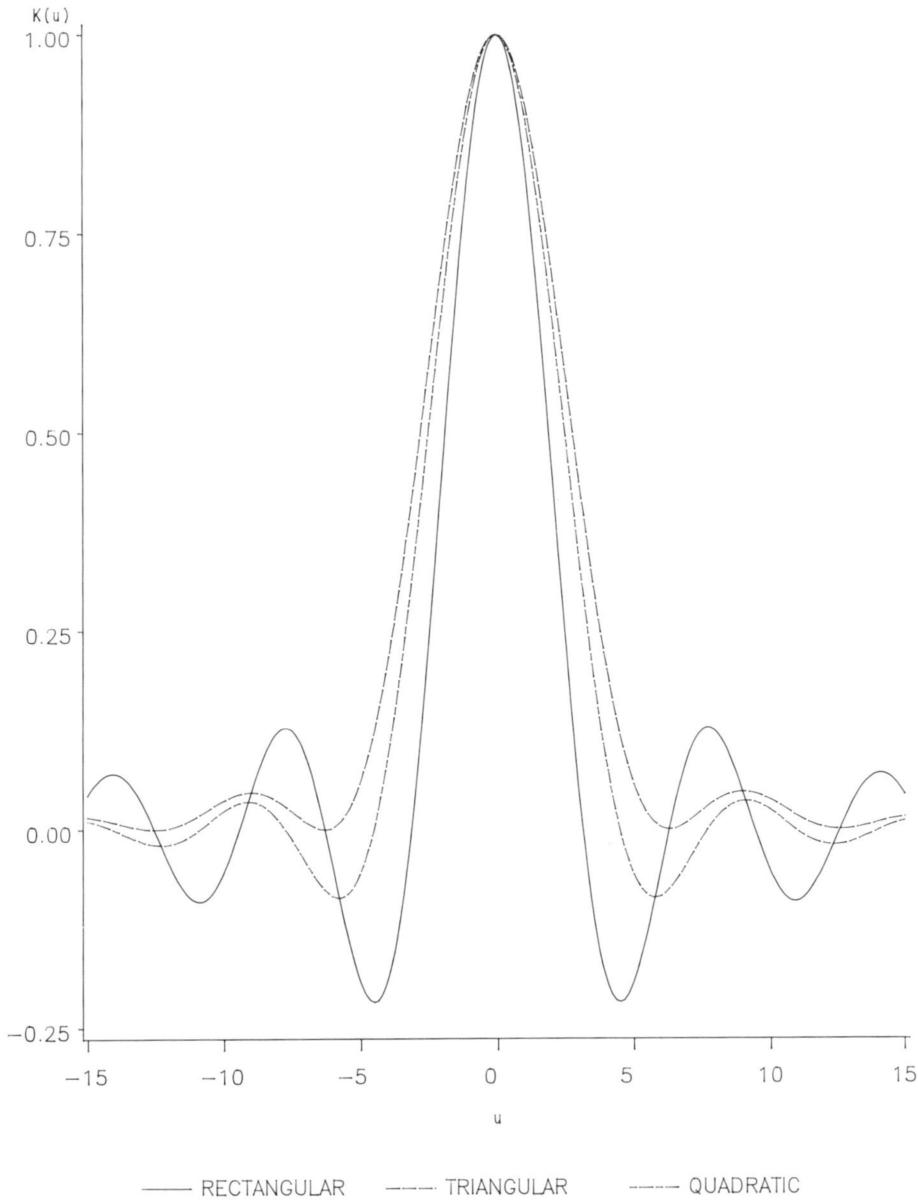


**Figure 4.2** Fourier transforms of certain kernels.

kernel. One might conclude from this that a kernel estimator which uses the uniform kernel does the same thing in the t or "time" domain that a Fourier series estimator does in the frequency domain. A similar comparison can be made between a kernel estimator constructed from the triangular kernel and the Fejer form of Fourier series estimator (Exercise 3.7).

As noted in Chapter 3 the fact that $\mu$ is smooth makes us think that most of the information about the regression function should be captured in the coefficients corresponding to lower frequencies while the coefficients for higher frequencies will contain mostly information about the random noise. When looked at from this viewpoint the kernel estimator seems to smooth in a rather natural way. It uses all the Fourier coefficients, but, for larger $\lambda$, relies primarily on the information from low frequencies by down weighting the contributions of higher frequencies to the estimator. When $\lambda$ is small the estimator will use more information from the high frequency portion of the data than when $\lambda$ is large. Apart from boundary effects this same conclusion carries over to kernel estimators in general.

## 4.3 Asymptotic Theory

In the next seven sections we focus on estimator (6d). Thus $\mu_\lambda$ is used exclusively to mean the estimator whose value at t is

$$\mu_\lambda(t) = \sum_{j=1}^{n} y_j \int_{s_{j-1}}^{s_j} \frac{1}{\lambda} K(\lambda^{-1}(t - s))ds \tag{11a}$$

with

$$s_0 = a \quad , s_{j-1} \le t_j \le s_j, \quad j=1,\ldots,n-1, \quad s_n = b. \tag{11b}$$

In the present section our objective is to study the asymptotic behavior of the risk corresponding to $\mu_\lambda$ as $n\to\infty$ and $\lambda\to 0$. The development follows that of Gasser and Müller (1979).

For simplicity let us assume that $[a,b] = [0,1]$ and

$$t_j = j/n, \qquad j=1,\ldots,n. \tag{12}$$

Also set,

$$s_j = (t_{j+1} + t_j)/2 \quad , \quad j=1,\ldots,n. \tag{13}$$

The kernel used in computing (11) is assumed to satisfy conditions (4a)-(4d). Kernels of this type are called second order kernels. Higher order kernels are discussed in Section 4.7. It will also be assumed that K is in $C^1[-1,1]$.

Eventually we want to obtain an asymptotic expression for

$$R_n(\lambda) = n^{-1} \sum_{r=1}^{n} E(\mu(r/n) - \mu_\lambda(r/n))^2 \quad . \tag{14}$$

It is difficult to do so directly. Consequently, we begin with the somewhat more modest undertaking of obtaining an asymptotic form for the risk at a point $t \varepsilon (0,1)$, defined by

$$R_n(t,\lambda) = E(\mu(t) - \mu_\lambda(t))^2 \quad . \tag{15}$$

Under certain conditions it will be possible to sum the asymptotic expression for $R_n(t;\lambda)$ over the $t_j$ to obtain the desired form for $R_n(\lambda)$.

To study $R_n(t;\lambda)$ let us first note that

$$R_n(t;\lambda) = (\mu(t) - E\mu_\lambda(t))^2 + \text{Var } \mu_\lambda(t) \ .$$

Thus, we can study the asymptotic behavior of the bias, $\mu(t) - E\mu_\lambda(t)$, and the variance of $\mu_\lambda(t)$ separately and then combine our results to obtain the large sample behavior of $R_n(t;\lambda)$. An initial step in this regard is provided by the following lemma.

**Lemma 4.1.**

$$\text{Var } \mu_\lambda(t) = \frac{\sigma^2}{n\lambda^2} \int_0^1 K(\lambda^{-1}(t - u))^2 du + O((n\lambda)^{-2}) \ . \tag{16}$$

*Proof.* First note that the exact variance for $\mu_\lambda(t)$ is

$$\text{Var } \mu_\lambda(t) = \frac{\sigma^2}{\lambda^2} \sum_{j=1}^{n} [ \int_{s_{j-1}}^{s_j} K(\lambda^{-1}(t - s))ds]^2 \quad . \tag{17}$$

So, consider the difference between (17) and the first term in (16).

Using the mean value theorem for integrals we find that

$$\left| \sum_{j=1}^{n} \left( \int_{s_{j-1}}^{s_j} K(\lambda^{-1}(t - s))ds \right)^2 - n^{-1} \int_0^1 K(\lambda^{-1}(t - s))^2\, ds \right|$$

$$= \left| \sum_{j=1}^{n} \left\{ K(\lambda^{-1}(t - \theta_j))^2 (s_j - s_{j-1})^2 - n^{-1} K(\lambda^{-1}(t - \xi_j))^2 (s_j - s_{j-1}) \right\} \right|$$

for $s_{j-1} \le \theta_j, \xi_j \le s_j$. This last expression can be written, using (13), as

$$\left| \sum_{j=1}^{n} n^{-2} (K(\lambda^{-1}(t - \theta_j))^2 - K(\lambda^{-1}(t - \xi_j))^2) \right|$$

$$\le n^{-2} \sum_{j=1}^{n} \left| K(u_j)^2 - K(v_j)^2 \right| \tag{18}$$

with $u_j = (t - \theta_j)/\lambda$ and $v_j = (t - \xi_j)/\lambda$. Since K is in $C^1[-1,1]$ so is $K^2$ which implies that

$$|K(u_j)^2 - K(v_j)^2| \le C|\theta_j - \xi_j|/\lambda$$

for some positive constant C. Hence (18) is bounded above by

$$n^{-2} C \sum_{u_j, v_j \in [-1,1]} |u_j - v_j|$$

$$\le n^{-2} C(n\lambda)^{-1} O(n\lambda)$$

$$= O(n^{-2}). \tag{19}$$

The last inequality makes use of the fact that the cardinality of the set of $u_j, v_j$ which fall in $[-1,1]$ is $O(n\lambda)$. As a result of (19) the disparity between (17) and the integral in (16) is $\lambda^{-2} O(n^{-2}) = O(1/(\lambda n)^2)$. □

Lemma 4.1 gives the expression we need for the variance of $\mu_\lambda(t)$. By making a change of variable we see that

$$\text{Var } \mu_\lambda(t) = \frac{\sigma^2}{n\lambda} \int_{(t-1)/\lambda}^{t/\lambda} K(u)^2 du + O((n\lambda)^{-2}). \tag{20}$$

Thus, for $R_n(t;\lambda)$ to decay to zero we not only need that $\lambda \to 0$ as $n \to \infty$ but also that $n\lambda \to \infty$. Recall that the same type of condition was needed for Fourier series estimators. In that case we required $n/(2p + 1) \to \infty$ with p the number of terms included in the estimators.

The next lemma provides a companion result to Lemma 4.1 for the expectation of $\mu_\lambda(t)$.

**Lemma 4.2.** If $\mu$ is continuously differentiable then

$$E\mu_\lambda(t) = \frac{1}{\lambda} \int_0^1 K(\lambda^{-1}(t - s))\mu(s)ds + O(n^{-1}) . \tag{21}$$

*Proof.* The proof rests on the following string of equalities and inequalities:

$$\left| \sum_{j=1}^{n} \mu(t_j) \int_{s_{j-1}}^{s_j} K(\lambda^{-1}(t - s))ds - \int_0^1 K(\lambda^{-1}(t - s))\mu(s)ds \right|$$

$$= \left| \sum_{j=1}^{n} \int_{s_{j-1}}^{s_j} (\mu(t_j) - \mu(t))K(\lambda^{-1}(t - s))ds \right|$$

$$= \left| \sum_{j=1}^{n} (\mu(t_j) - \mu(\xi_j)) \int_{s_{j-1}}^{s_j} K(\lambda^{-1}(t - s))ds \right| \quad (\text{for } s_{j-1} \le \xi_j \le s_j)$$

$$\le \lambda n^{-1} \sup_t |\mu'(t)| \sum_{j=1}^{n} \left| \int_{u_{j-1}}^{u_j} K(u)du \right| \qquad (u_j = (t - s_j)/\lambda)$$

$$\le \lambda n^{-1} \sup_t |\mu'(t)| \sup_u |K(u)| \sum_{u_j \varepsilon [-1,1]} |u_j - u_{j-1}|$$

$$= O(\lambda/n) .$$

□

Now suppose that $\mu \varepsilon C^2[0,1]$. A change of variable and Taylor expansion about t allows us to write the integral in (21) as

$$\frac{1}{\lambda}\int_0^1 K(\lambda^{-1}(t - s))\mu(s)ds = \int_{(t-1)/\lambda}^{t/\lambda} K(u)\mu(t - \lambda u)du$$

$$= \mu(t)\int_{(t-1)/\lambda}^{t/\lambda} K(u)du - \lambda\mu'(t)\int_{(t-1)/\lambda}^{t/\lambda} uK(u)du$$

$$+ \frac{\lambda^2}{2}\int_{(t-1)/\lambda}^{t/\lambda} \mu''(t + \theta_u\lambda u)u^2K(u)du$$

for $\theta_u$ a number depending on u that lies in [0,1]. For $\lambda$ sufficiently small, $[(t - 1)/\lambda, t/\lambda] \subset [-1,1]$. Thus, this expression eventually reduces to

$$\mu(t) + \frac{\lambda^2}{2}\int_{-1}^1 \mu''(t + \theta_u\lambda u)u^2K(u)du$$

as a result of properties (4a) and (4b). Now note that since $\mu''$ is continuous

$$\int_{-1}^1 [\mu''(t + \theta_u\lambda u) - \mu''(t)]u^2K(u)du \to 0$$

as $\lambda \to 0$. We have just proved the following corollary.

**Corollary 4.1.** If $\mu \varepsilon C^2[0,1]$ then

$$E\mu_\lambda(t) - \mu(t) = \frac{\lambda^2}{2}\mu''(t)\int_{-1}^1 u^2K(u)du + o(\lambda^2) + O(n^{-1}). \quad (22)$$

It is now possible to see the import of conditions (4a) -(4d) required for K. Condition (4a) assures us that the lead term in $E\mu_\lambda(t)$ is $\mu(t)$. Condition (4b) is a restriction which eliminates the term of order $\lambda$ in the bias thereby assuring that the bias is $O(\lambda^2) + O(n^{-1})$. Requirement (4b) and (4c) have the implication that K is matched to a $C^2[0,1]$ function in the sense that the bias is made to decay as rapidly as possible for a regression function that is known to have only two continuous derivatives. If $\mu$ has more than two derivatives other moments of K could be required to vanish and a bias of still smaller order could be obtained. This is discussed in Section 4.7. Finally, requirement (4d) is needed to insure that $\mu_\lambda(t)$ has finite asymptotic variance.

Corollary 4.1 provides an indication of where the bias of $\mu_\lambda$ will tend to be largest: namely, where $\mu''(t)$ is largest in magnitude. So $\mu_\lambda$ will tend to be more biased in areas where the derivative (or slope) of $\mu$ is changing rapidly. This means, as one might expect, that the estimator does not perform as well in areas where the regression function is very nonlinear.

If

$$\text{Bias}(t) = E\mu_\lambda(t) - \mu(t)$$

is positive, this indicates that $\mu_\lambda(t)$ tends to overestimate $\mu(t)$ while if it is negative $\mu_\lambda(t)$ tends to underestimate $\mu(t)$. Assuming that $\int u^2 K(u)du > 0$ we see that $\mu_\lambda$ will tend to overestimate where $\mu''$ is positive and underestimate when $\mu''$ is negative. When $\mu''$ is positive this means that $\mu$ is concave upwards and if $\mu'' < 0$ this implies that $\mu$ is concave downwards. The implication of this is that $\mu_\lambda$ will tend to overestimate the valleys (or troughs) of $\mu$ and underestimate the peaks of the function. Since peaks and valleys can usually be located by visual inspection of the data, we will frequently have some idea of whether $\mu_\lambda$ will tend to under or overestimate $\mu$. This is not as much help as it might seem since the actual bias and its magnitude will be unknown. However, the important point is that inspection of the data may suggest areas where the estimator may not behave as well as others and also the direction in which it will miss $\mu$ in such regions.

An illustration of the type of bias behavior discussed above is provided by Figure 4.3. There we have plotted $\mu(t) = (\sin 2\pi t)^2$ along with $\mu(t) + \lambda^2\mu''(t)/6$, for $\lambda = .1$, which is an approximation to $E\mu_\lambda(t)$ for this case when a rectangular kernel on $[-1,1]$ is used. We see, for example, that $\mu_\lambda$ will tend to overestimate $\mu$ on $[0,1/8]$ where $\mu$ is concave upwards and underestimate on $[1/8,3/8]$ where $\mu$ is concave downwards.

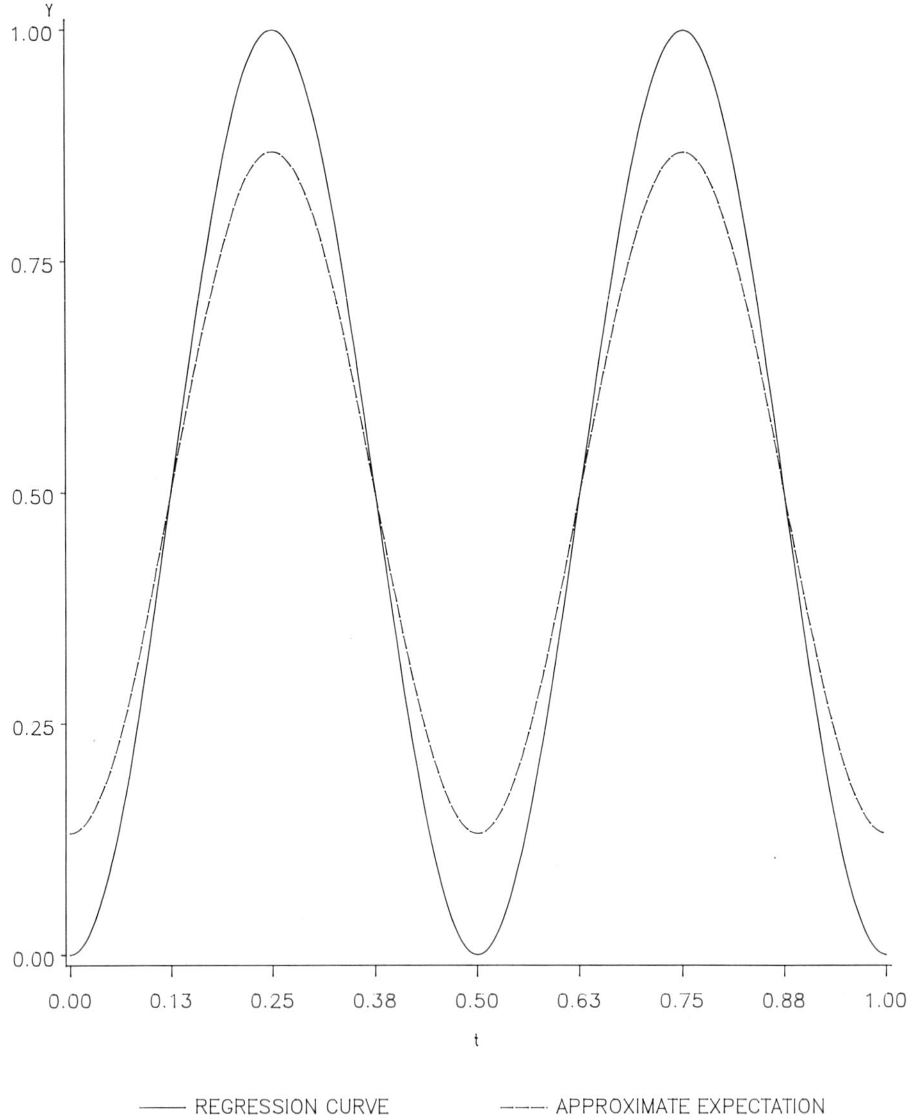


**Figure 4.3** Comparison of a function and the expectation of its kernel estimate.

By combining (20) and (22) it is possible to establish the asymptotic properties of $R_n(t;\lambda)$. These are stated formally in the next theorem.

**Theorem 4.1.** Suppose that $\mu \in C^2[0,1]$ and (12)-(13) hold. Then, if $n\to\infty$, $\lambda \to 0$ in such a way that $n\lambda \to \infty$

$$R_n(t;\lambda) \sim \frac{\sigma^2}{n\lambda} \int_{-1}^{1} K(u)^2 du + \lambda^4 \frac{\mu''(t)^2}{4} \left[ \int_{-1}^{1} u^2 K(u) du \right]^2 . \quad (23)$$

The asymptotically optimal bandwidth is

$$\lambda_{opt} = \left\{ \sigma^2 n^{-1} \int_{-1}^{1} K(u)^2 du / \mu''(t)^2 \left[ \int_{-1}^{1} u^2 K(u) du \right]^2 \right\}^{1/5} , \quad (24)$$

where $\mu''(t) \neq 0$, and

$$R_n(t;\lambda_{opt}) \sim \frac{1.25}{n^{4/5}} \left\{ \left| \mu''(t) \right| \sigma^4 \left[ \int_{-1}^{1} u^2 K(u) du \right] \left[ \int_{-1}^{1} K(u)^2 du \right]^2 \right\}^{2/5} . \quad (25)$$

*Proof.* Equation (23) is immediate from (20) and (22). (24) follows from differentiation of (23), and (25) is obtained by substituting (24) into (23). □

Theorem 4.1 has the consequence that the optimal risk is larger where $|\mu''(t)|$ is larger and the optimal bandwidth is larger where $|\mu''(t)|$ is smaller. The first result is a natural consequence of the fact that $\mu(t)$ is more difficult to estimate if $\mu$ is very wiggly near t. To understand the second result note, for example, that if $\mu''(t) = 0$ this means that $\mu$ is nearly linear around t. But if $\mu$ is locally linear then all the observations near t will provide information about the behavior of $\mu$. Consequently the bandwidth should be widened to allow more averaging and thereby reduce the estimator's variance. Conversely, if $|\mu''(t)|$ is large this means that $\mu$ is changing rapidly near t so that only the data obtained at points very close to t will provide useful information about $\mu(t)$. The bandwidth should then be narrowed to reduce bias.

While Theorem 4.1 tells us the optimal bandwidth and risk for any particular point it does not tell us how to choose $\lambda$ to obtain optimal global behavior in terms of minimizing (14). Since

$$R_n(\lambda) = n^{-1} \sum_{j=1}^{n} R_n(j/n;\lambda) \tag{26}$$

it is tempting to use (23) to obtain approximations for the $R_n(j/n;\lambda)$ and use the result as an asymptotic formula for $R_n(\lambda)$. This approach fails because if $t_j$ is within a bandwidth of the boundary (23) is no longer valid. Even if we renormalize (11) to where the weights sum to one for $t \in [0,\lambda)$ or $t \in (1 - \lambda,1]$, it is easy to check that the bias at the boundary will be $O(\lambda) + O(n^{-1})$. The variance will remain $O((n\lambda)^{-1})$. If we were then to compute $R_n(\lambda)$ using (26) it would be found that

$$R_n(\lambda) = O(\lambda^3) + O((n\lambda)^{-1}) + O(\lambda^4)$$

as $n\to\infty$, $\lambda \to 0$ and $n\lambda \to \infty$. The $O(\lambda^3)$ portion of $R_n(\lambda)$ comes from summing over the $O(n\lambda)$ squared bias terms in the boundary regions. The $O(\lambda^3)$ term dominates $O(\lambda^4)$ and the best rate of decay is only $n^{-3/4}$ rather than the $n^{-4/5}$ we would have anticipated.

If $\mu$ is smoothly periodic a modified version of $\mu_\lambda$ can be constructed along the lines of discussions in Section 4.2. For this estimator the approximation (23) remains valid for all t. It follows immediately that the optimal $\lambda$ in this case is $O(n^{-1/5})$ and the optimal risk is $O(n^{-4/5})$. We leave the details behind this for the reader. The important implication is that unless $\mu$ is periodic the behavior of the regression function at the boundary will globally affect the estimation risk if the estimator is not modified in some fashion. Note that we have encountered this phenomenon before in our study of Fourier series estimators.

Two methods for altering kernel estimators to remove boundary bias have been studied by Gasser and Müller (1979) and Rice (1984a). We will discuss the first of these here. The second is examined in the exercises.

It suffices to show how to alter the estimator for one of the boundaries, so we choose the lower region $[0,\lambda)$. If $t \in [0,\lambda)$, then it can always be written as

$$t = q\lambda$$

for some $0 \le q < 1$. For any such point assume it is possible to construct a modification of the original kernel, $K_q$, which satisfies

$$\int_{-1}^{q} K_q(u)\,du = 1 \quad , \tag{27a}$$

$$\int_{-1}^{q} uK_q(u)\,du = 0 \ , \tag{27b}$$

$$\int_{-1}^{q} u^2 K_q(u)\,du = \alpha_q \ne 0 \ , \tag{27c}$$

and

$$\int_{-1}^{q} K_q(u)^2\,du < \infty. \tag{27d}$$

Also assume that $K_q$ is differentiable in u and continuous in q with

$$K_q \to K \qquad \text{as } q \to 1. \tag{27e}$$

A kernel satisfying (27) is called a *boundary kernel*. Many kernels of interest can be modified to obtain boundary kernels. For example, a boundary kernel corresponding to the rectangular kernel is found to be

$$K_q(u) = 4\,\frac{q^3 + 1}{(q + 1)^4} + 6\,\frac{(1 - q^2)}{(q + 1)^4}\,u \ . \tag{28}$$

The derivation of this kernel is discussed in Exercise 10. Other examples can be bound in Gasser and Müller (1979) and Gasser, Müller and Mammitzsch (1985).

Corresponding to a t in the boundary region we will use a modified estimator $\tilde{\mu}_\lambda(t)$ obtained by replacing K with a boundary kernel $K_q$ whenever t is within a bandwidth of the boundary. Tracing back through the arguments which led to (20) and (22) we find, for example, that if $t = q\lambda$, $0 \le q < 1$,

$$\mathrm{Var}(\tilde{\mu}_\lambda(q\lambda)) \sim \frac{\sigma^2}{n\lambda}\int_{-1}^{q} K_q^2(u)du$$

$$\le \frac{\sigma^2}{n\lambda} \max_{0\le q\le 1} \int_{-1}^{q} K_q^2(u)du = O((n\lambda)^{-1})$$

and that

$$E\tilde{\mu}_\lambda(q\lambda) - \mu(q\lambda) = \mu''(0)\frac{\lambda^2}{2}\int_{-1}^{q} u^2 K_q(u)du + o(\lambda^2) + O(n^{-1}) \quad .$$

The latter expression has the implication that

$$(E\tilde{\mu}_\lambda(q\lambda) - \mu(q\lambda))^2 \le \mu''(0)^2 \frac{\lambda^4}{4}\max_{0\le q\le 1}\left[\int_{-1}^{q} u^2 K_q(u)du\right]^2 + o(\lambda^4) + O(\lambda^2 n^{-1})$$

$$= O(\lambda^4) + O(\lambda^2 n^{-1}).$$

Let $B = \{t_j : t_j < \lambda \text{ or } t_j > 1 - \lambda\}$. Then, as the upper boundary can be handled the same way as the lower, it follows that for $n\to\infty$, $\lambda\to 0$ and $n\lambda \to \infty$,

$$R_n(\lambda) = \frac{1}{n}\sum_{t_j\in B} E(\tilde{\mu}_\lambda(t_j) - \mu(t_j))^2 + \frac{1}{n}\sum_{t_j\notin B} E(\mu_\lambda(t_j) - \mu(t_j))^2$$

$$\sim O(\lambda)[O((n\lambda)^{-1}) + O(\lambda^4) + O(\lambda^2 n^{-1})] + \frac{\sigma^2}{n\lambda}\int_{-1}^{1} K(u)^2 du$$

$$+ \lambda^4\left(\frac{1}{4}\frac{1}{n}\sum_{t_j\notin B}\mu''(t_j)^2\right)\left(\int_{-1}^{1} u^2 K(u)du\right)^2 \qquad (29)$$

$$\sim \frac{\sigma^2}{n\lambda}\int_{-1}^{1} K(u)^2 du + \lambda^4\left(\frac{1}{4}\int_{0}^{1}\mu''(t)^2 dt\right)\left(\int_{-1}^{1} u^2 K(u)du\right)^2 .$$

In obtaining this we have employed the continuity of $\mu''$ to replace $n^{-1}\Sigma\mu''(t_j)^2$ by an integral. The conclusions drawn from (29) are summarized in Theorem 4.2.

**Theorem 4.2.** If $\mu \in C^2[0,1]$ and (12) and (13) hold, the asymptotic minimizer of $R_n(\lambda)$ is

$$\lambda_{opt} = \left\{ \sigma^2 n^{-1} \int_{-1}^{1} K(u)^2 du / [\int_0^1 \mu''(t)^2 dt][\int_{-1}^{1} u^2 K(u) du]^2 \right\}^{1/5} \tag{30}$$

and the asymptotically optimal risk is

$$R_n(\lambda_{opt}) \sim \frac{1.25}{n^{4/5}} \left\{ \int_0^1 \mu''(t)^2 dt \right\}^{1/5} \left\{ \sigma^4 \left[ \int_{-1}^{1} K(u)^2 du \right]^2 \left[ \int_{-1}^{1} u^2 K(u) du \right] \right\}^{2/5}. \tag{31}$$

Theorem 4.2 has the interpretation that if $\mu_\lambda$ is altered appropriately at the boundary it is possible to obtain a global rate of $n^{-4/5}$ for $R_n(\lambda)$. Notice that this is the rate one would have anticipated from our discussion in Chapter 1, as $n^{-4/5}$ is the best uniform rate of convergence for any linear estimator of a function in the set $W_2^2[0,1]$ that contains $C^2[0,1]$.

Theorem 4.2 is, unfortunately, not very useful for practical selection of $\lambda$ since $\lambda_{opt}$ involves the unknown function and variance. It is possible to estimate $\mu''$ using, for example, the approach in Section 4.8, and thereby obtain an estimate of $\int_0^1 \mu''(t)^2 dt$ in (30). A consistent estimator for $\sigma^2$ can be constructed along the lines of discussions in Section 4.6 (see Exercise 22). Thus, in theory it is possible to obtain an estimator of $\lambda_{opt}$. Härdle, Hall and Marron (1985) give convincing arguments against this practice, however. In certain cases they have investigated it can be shown that, even under ideal conditions where $\sigma^2$ and $\int_0^1 \mu''(t)^2 dt$ are known, this plug-in approach provides estimators with no better asymptotic risk behavior than some data driven methods for selecting $\lambda$. The increased variability that results when $\sigma^2$ and $\int_0^1 \mu''(t)^2 dt$ must be estimated makes the utility of the plug-in approach suspect. Also, it should be noted that by estimating $\lambda_{opt}$ we are only estimating an approximation to the best bandwidth and this approximation may not perform well for small samples.

In concluding this section let us point out a few probabilistic implications of Theorems 4.1 and 4.2. First, a consequence of Theorem 4.2 is that $\mu_\lambda(t) \to \mu(t)$ in probability. This can actually be strengthened to $\mu_\lambda(t) \to \mu(t)$ almost surely (see Exercise 8). By examining the proof of Theorem 4.2 it is seen that what has also been shown is integrated risk consistency, i.e.,

$$\int_0^1 E(\mu_\lambda(t) - \mu(t))^2 dt \to 0$$

as $n \to \infty$, $\lambda \to 0$ and $n\lambda \to \infty$. The conclusions to be drawn are that, under certain restrictions, $\mu_\lambda$ is both a locally and globally consistent estimator of $\mu$. For discussion of the consistency properties of the other estimators in Section 4.2 see Priestley and Chao (1972), Benedetti (1974, 1977) and Cheng and Lin (1981).

## 4.4 Selecting a Kernel

In order to use a kernel estimator it is necessary to select a kernel on which to base the estimator. The natural question to ask here is whether there is a best kernel to choose. This section gives a partial answer to that question.

To determine a 'best' choice for a kernel we must first decide what 'best' means. In the spirit of Chapter 2, we might take this to mean minimization, with respect to K, of the risk once an optimal choice has been made for $\lambda$. As a result of Theorem 4.2 we see this is asymptotically equivalent to minimizing

$$R(K) = \left[ \left( \int_{-1}^1 u^2 K(u) du \right) \left( \int_{-1}^1 K^2(u) du \right)^2 \right]^{2/5} .$$

Note that R(K) also appears in (25) so a K which is optimal for $R_n(\lambda)$ will also be optimal for $R_n(t;\lambda)$.

The problem, therefore, is to find a kernel K which minimizes R(K) subject to restrictions (4a)-(4c) for some fixed $\alpha$ (unless $\alpha$ is fixed R(K) can be made as small as one likes). Results in Gasser and Müller (1979) reveal that the solution is a quadratic kernel with two sign changes in [-1,1]. To avoid such oscillations, Gasser and Müller (1979) impose the restriction that the kernel be positive on [-1,1]. The result is the quadratic kernel in Table 4.1. This kernel is well known in the density estimation literature due to the work of Epanechnikov (1969). Benedetti (1974, 1977) also pointed out the optimality of the quadratic kernel for nonparametric regression.

Alternatively, we can think of smoothing as merely a way of obtaining an estimator of $\mu(t_i)$ with smaller variance than $y_i$. From this viewpoint smoothing is really a variance reduction technique and bias becomes something that must simply be tolerated. By taking this perspective one is instead led to select K to minimize the variance of $\mu_\lambda$. Asymptotically (see Lemma 4.1) this suggests the selection of a minimum variance kernel by the minimization of

$$V(K) = \int_{-1}^{1} K(u)^2 du$$

subject to (4a) and (4b). To obtain the minimizer first note that (4a) and (4b) determine a second order (i.e., linear) polynomial, $\theta_0 + \theta_1 u$, as the solution to

$$\int_{-1}^{1} (\theta_0 + \theta_1 u) du = 1$$

and

$$\int_{-1}^{1} u(\theta_0 + \theta_1 u) du = 0.$$

We see that $\theta_0 = 1/2$, $\theta_1 = 0$ is the solution which gives the rectangular kernel in Table 4.1.

Let $K_r(u)$ be the rectangular kernel. Then any other kernel K on [-1,1] can be written as $K_r + \delta K$ with $\delta K$ a function satisfying

$$\int_{-1}^{1} \delta K(u)du = \int_{-1}^{1} \delta K(u)udu = 0.$$

Since

$$\int_{-1}^{1} K(u)^2 du = \int_{-1}^{1} K_r(u)^2 du + 2\int_{-1}^{1} K_r(u)\delta K(u)du + \int_{-1}^{1} \delta K(u)^2 du$$

$$= \int_{-1}^{1} K_r(u)^2 du + \int_{-1}^{1} \delta K(u)^2 du > V(K_r) ,$$

the rectangular kernel must be the asymptotically minimum variance kernel.

Table 4. 1 gives a comparison of the values of R(K) and V(K) for several kernels, including the quadratic and rectangular kernels. We see that not much is lost, in terms of R(K), by use of the rectangular over the optimal quadratic kernel. On the other hand, both the quadratic and rectangular kernels have very similar values for V(K). One might conclude from all this that the actual choice of a kernel, once $\lambda$ has been optimized, is not very important. This appears to be essentially true. The choice of $\lambda$ is what seems to be crucial rather than which kernel is chosen. We should point out that this conclusion pertains to kernels of the same order and does not extend across orders; see Section 4.7.

It is also possible to consider the use of optimal boundary kernels. For example, a minimum variance (lower) boundary kernel $K_q$ must minimize

$$V_q(K) = \int_{-1}^{q} K(u)^2 du$$

subject to

$$\int_{-1}^{q} K(u)du = 1$$

and

$$\int_{-1}^{q} uK(u)du = 0.$$

The solution is the boundary kernel (28) that was mentioned in the previous section. For further discussion of optimal boundary kernels see Gasser and Müller (1979) and Gasser, Müller and Mammitzsch (1985).

## 4.5 Selection of λ

Perhaps the most important problem associated with the use of a kernel estimator is the selection of a good value of $\lambda$. In this section we discuss the use of data driven techniques for smoothing parameter selection.

For the kernel estimator (6d) the vector of fitted values can be written as $\underline{\mu}_\lambda = H(\lambda)\underline{y}$, where $H(\lambda)$ has typical element

$$h_{jk}(\lambda) = \frac{1}{\lambda} \int_{s_{k-1}}^{s_k} K(\lambda^{-1}(t_j - s))ds.$$

Thus $\mathrm{tr}[H(\lambda)] = \lambda^{-1} \Sigma_{j=1}^{n} \int_{s_{j-1}}^{s_j} K(\lambda^{-1}(t_j - s))ds$ and the GCV criterion of Chapter 2 becomes

$$GCV(\lambda) = n^{-1} \sum_{j=1}^{n} (y_j - \mu_\lambda(t_j))^2 \Big/ \left[(1 - (n\lambda)^{-1} \sum_{j=1}^{n} \int_{s_{j-1}}^{s_j} K(\lambda^{-1}(t_j - s))ds\right]^2.$$

This can be minimized to effect a choice for $\lambda$. For nonperiodic regression functions this criterion should be modified to allow for the use of a modified estimator at the boundary. Since the modified estimator will still be linear so that $\underline{\mu}_\lambda = H(\lambda)\underline{y}$, for some matrix $H(\lambda)$, this causes no real conceptual difficulties.

Similar considerations obtain for selecting $\lambda$ for the other estimators in Section 4.2. Also, other criteria can be used to select $\lambda$ such as ordinary cross-validation.

Figure 4.4 gives a plot of data generated from the model

$$y_j = [\sin(2\pi j/100)]^2 + \varepsilon_j, \quad j = 0,\dots,100,$$

with the $\varepsilon_j$ independent normal variates having mean zero and

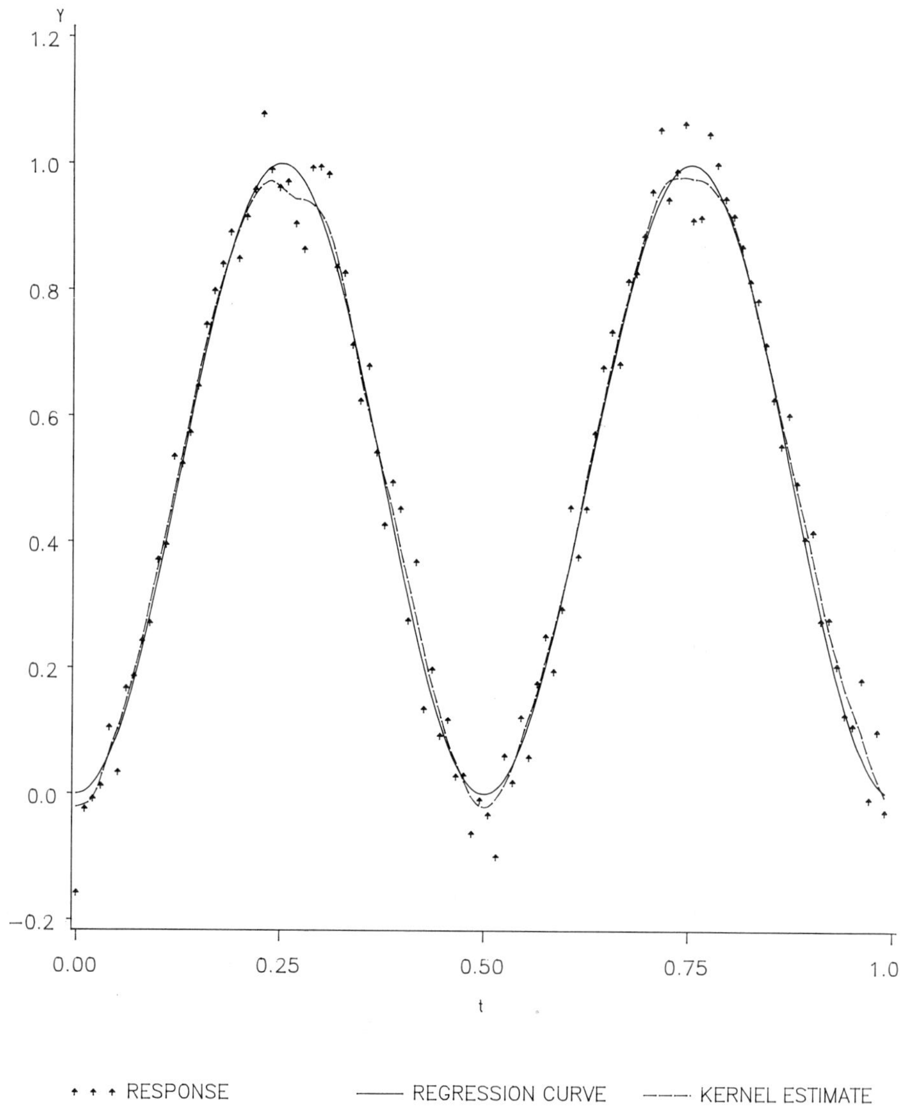


**Figure 4.4** Regression curve and kernel estimate for simulated data.

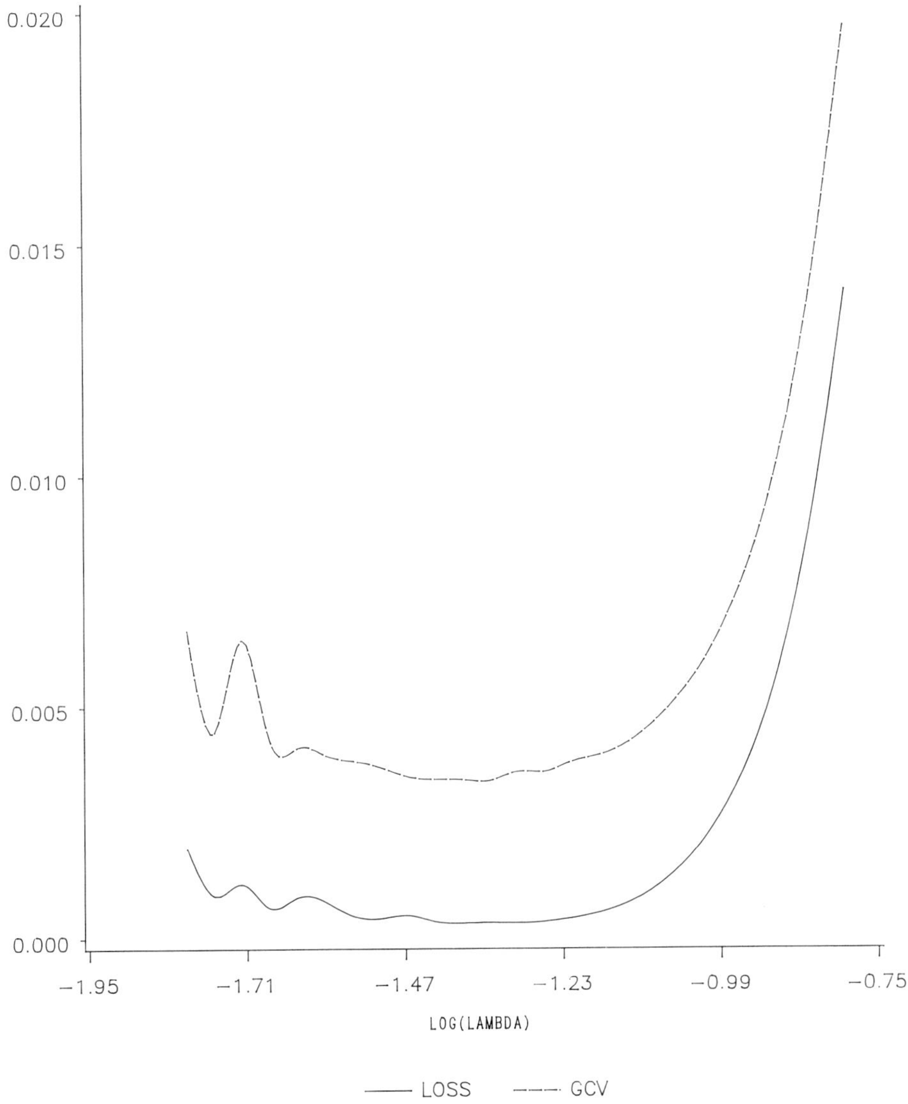


**Figure 4.5** Loss and GVC criteria for simulated data.

standard deviation $\sigma = .05$. A periodic form of the estimator was used with a quadratic kernel. The value of $\lambda$ was selected by GCV and the resulting fit is seen in Figure 4.4 to be quite good. The optimal value of $\lambda$ was found to be approximately $\hat{\lambda} = .0413$. This should be compared to the asymptotically optimal value of $\lambda$ which can be calculated from (30) as $\lambda_{opt} \doteq .0412$.

Since the true regression function is known we can actually compute the loss for this example. We have done this for various values of $\lambda$ and plotted the results in Figure 4.5 as a function of the logarithm (base 10) of $\lambda$. Also plotted are the corresponding values of the GCV criterion. Notice how GCV($\lambda$) parallels the loss, $L_n(\lambda)$. The disparity between the two is roughly $\sigma^2 = .0025$ near their minima and they are both minimized by essentially the same value of $\lambda$. This type of result is what would be anticipated from the GCV Theorem in Chapter 2.

There are now available a number of results concerning the consistency and optimality properties of various data driven bandwidth selection procedures. We devote the remainder of this section to summarizing some of the key results in this area.

Wong (1983) has shown that if the $t_j$ are uniformly spaced the selection of $\lambda$ by the CV criterion results in a consistency property for estimator (6c). More specifically, let $\mu_{\hat{\lambda}}$ denote estimator (6c) with $\lambda$ selected by CV. Then, Wong demonstrates that $n^{-1}\Sigma_{j=1}^{n}(\mu_{\hat{\lambda}}(t_j) - \mu(t_j))^2 \xrightarrow{P} 0$, i.e., the loss converges in probability to zero. There is undoubtably nothing unique about estimator (6c) in this regard and similar results should also hold for the other estimators in Section 4.2.

Rice (1984b) has considered the use of the unbiased risk estimator (2.14) for selection of $\lambda$. He deals with estimator (6a) in the case of equally spaced data and a smoothly periodic regression function. Thus, he uses an estimator of the form (10) except the $K_{\lambda j}$ are replaced by the true (rather than discrete) Fourier coefficients of the kernel. Rice is able to establish that the minimizer of the unbiased risk estimate, $\hat{\lambda}$, is consistent for the minimizer of $R_n(\lambda)$ in the sense that $\hat{\lambda}/\lambda_{opt} \xrightarrow{P}$

1 and even proves asymptotic normality for a suitably normalized version of $\hat{\lambda}$. He then shows that selection of $\lambda$ by several other criteria, which include CV and GCV, is asymptotically equivalent to the unbiased risk approach so that all will have asymptotically efficient minimizers. Some of Rice's results carry over to estimation of the derivative of the regression function as well (see Rice, 1986).

Perhaps the strongest results to date concerning the behavior of bandwidth estimators are due to Härdle, Hall and Marron (1985). Assuming that $\mu \in C^2[0,1]$, $\mu^{(j)}(0) = \mu^{(j)}(1)$, $j = 0,1,2$, and the $t_j$ are equally spaced, they are able to relate various bandwidth estimates to the minimizers of both the loss and the risk for estimator (6a). Let $\hat{\lambda}_{opt}$ denote the minimizer of $L_n(\lambda) = n^{-1}\Sigma_{j=1}^n(\mu(t_j) - \mu_\lambda(t_j))^2$ and let $\hat{\lambda}$ be an estimator of the optimal bandwidth obtained via minimization of any one of several criteria which include CV, GCV, and Rice's unbiased risk estimate (see Härdle, Hall and Marron, 1985, for further details). Then, Härdle, Hall and Marron show that

$$n^{3/10}(\hat{\lambda} - \hat{\lambda}_{opt})$$

and

$$n^{3/10}(\hat{\lambda}_{opt} - \lambda_{opt})$$

both have zero mean, normal limiting distributions. They also establish that

$$n(L_n(\hat{\lambda}) - L_n(\hat{\lambda}_{opt}))$$

and

$$n(L_n(\hat{\lambda}_{opt}) - L_n(\lambda_{opt}))$$

have limiting chi-squared distributions, apart from a constant of proportionality.

One of the surprising aspects of the work by Härdle, Hall and Marron (1985) is the very slow rate $n^{-3/10}$ for convergence of $\hat{\lambda}$ to the minimizer of the loss. The essential conclusion to be drawn from this is that no bandwidth selector can be expected to give a very good estimate of $\hat{\lambda}_{opt}$. On the other hand, the same can be said for $\lambda_{opt}$. Thus data driven methods can be expected to perform no worse than $\lambda_{opt}$ in terms of estimating the optimal bandwidth $\hat{\lambda}_{opt}$.

The poor performance for bandwidth estimators is somewhat discouraging. However, from the standpoint of the loss corresponding to an estimator with an estimated bandwidth, things are not so bleak. The difference $L_n(\hat{\lambda}) - L_n(\hat{\lambda}_{opt})$ is $O_p(n^{-1})$. Consequently, $\mu_{\hat{\lambda}}$ will tend to behave quite similarly to the optimal estimator in terms of its loss.

It may at first seem puzzling that although $\hat{\lambda}$ does not perform well as an estimator of $\hat{\lambda}_{opt}$, $L_n(\hat{\lambda})$ is still close to $L_n(\hat{\lambda}_{opt})$. There is a reasonable explanation for this, however. It turns out that (Marron, 1986) the loss tends to be quite flat near its minimum. This means that it will take very large samples to effectively estimate $\hat{\lambda}_{opt}$ but that values reasonably close to $\hat{\lambda}_{opt}$ will provide estimators with something close to the optimal loss.

## 4.6 Asymptotic Distribution Theory

Since kernel estimators are essentially weighted averages it should be expected that they would be asymptotically normal under certain restrictions. This type of result can be established through the use of the Liapunov Central Limit Theorem of Section 3.4.4. Again, for simplicity we will restrict attention to estimator (6d) under the assumption that (12)-(13) hold. The same type of results can, of course, be shown to hold for the other estimators in Section 4.2 (see, e.g., Benedetti, 1974, 1977) and more general designs.

For the application of the Liapunov Theorem define

$$V_{jn} = \lambda^{-1} y_j \int_{s_{j-1}}^{s_j} K(\lambda^{-1}(t - s))ds \quad .$$

Then, for any fixed $t \in (0,1)$

$$\sigma_n^2 = \sum_{j=1}^{n} E(V_{jn} - EV_{jn})^2 = \frac{\sigma^2}{\lambda^2} \sum_{j=1}^{n} \left(\int_{s_{j-1}}^{s_j} K(\lambda^{-1}(t - s))ds\right)^2$$

$$\sim \frac{\sigma^2}{n\lambda} \int_{-1}^{1} K(u)^2 du,$$

as $n \to \infty$, $\lambda \to 0$ and $n\lambda \to \infty$ due to Lemma 4.1.

Suppose we choose $\theta = 2 + \alpha$ for some $\alpha > 0$. Then we must work with

$$\sum_{j=1}^{n} E|V_{jn} - EV_{jn}|^\theta = \sum_{j=1}^{n} E|\varepsilon_j \lambda^{-1} \int_{s_{j-1}}^{s_j} K(\lambda^{-1}(t - s))ds|^\theta .$$

To avoid some needless technicalities let us assume that

i) the $\varepsilon_j$ are independent and identically distributed with $E|\varepsilon_1|^\theta < \infty$ and

ii) $K$ is differentiable, nonnegative and $\int_{-1}^{1} K(u)^\theta du < \infty$.

Then,

$$|\sum_{j=1}^{n} \left( \int_{s_{j-1}}^{s_j} K(\lambda^{-1}(t - s))ds\right)^\theta - n^{-(1+\alpha)} \int_0^1 K(\lambda^{-1}(t - s))^\theta ds|$$

$$\le \sum_{j=1}^{n} |\left(\int_{s_{j-1}}^{s_j} K(\lambda^{-1}(t - s))ds\right)^\theta - n^{-(1+\alpha)} \int_{s_{j-1}}^{s_j} K(\lambda^{-1}(t - s))^\theta ds|$$

$$= n^{-\theta} \sum_{j=1}^{n} |K(\lambda^{-1}(t - \xi_j))^\theta - K(\lambda^{-1}(t - \theta_j))^\theta| \quad (s_{j-1} \le \xi_j,\ \theta_j \le s_j)$$

$$= n^{-\theta} \sum_{u_j, v_j \in [-1,1]} |K(u_j)^\theta - K(v_j)^\theta| \quad (u_j = (t - \xi_j)/\lambda, v_j = (t - \theta_j)/\lambda)$$

$$\le \lambda^{-1} n^{-(1+\theta)} O(n\lambda) = O(n^{-\theta}) .$$

Thus, as $n \to \infty$ and $\lambda \to 0$

$$\sum_{j=1}^{n} \left(\frac{1}{\lambda} \int_{s_{j-1}}^{s_j} K(\lambda^{-1}(t - s))ds\right)^{\theta} = (n\lambda)^{-(1+\alpha)} \int_{-1}^{1} K(u)^{\theta} du + O((n\lambda)^{-\theta})$$

$$= O((n\lambda)^{-(1+\alpha)}) .$$

The above relations allow us to see that as $n \to \infty$, $\lambda \to 0$ and $n\lambda \to \infty$,

$$\frac{\sum_{j=1}^{n} E|V_{jn} - EV_{jn}|^{\theta}}{(\sigma_n)^{\theta}} = \frac{O((n\lambda)^{-(1+\alpha)})}{O((n\lambda)^{-(1+\alpha/2)})} = O((n\lambda)^{-\alpha/2}) .$$

So $\mu_\lambda(t) - E\mu_\lambda(t)$ will be asymptotically normal if $n\lambda \to \infty$. The implications of this fact are summarized in Theorem 4.3.

**Theorem 4.3.** Assume that (12)-(13) hold, the $\varepsilon_j$ are independent and identically distributed with $E|\varepsilon_1|^{2+\alpha} < \infty$ for some $\alpha > 0$ and that K is nonnegative and differentiable on $[-1,1]$. Then, for any fixed $t \in (0,1)$

$$\frac{(n\lambda)^{1/2}(\mu_\lambda(t) - E\mu_\lambda(t))}{\sigma[\int_{-1}^{1} K(u)^2 du]^{1/2}}$$

converges in distribution to a standard normal random variable provided $n \to \infty$, $\lambda \to 0$ in such a way that $n\lambda \to \infty$.

Notice that no smoothness restrictions are required on $\mu$ for this to hold. This is because $V_{jn} - EV_{jn}$ is independent of $\mu$ and what has actually been proved is a theorem about a weighted average of independent identically distributed random variables. Smoothness of $\mu$ is needed when we try to obtain a somewhat more useful result concerning asymptotic normality of $\mu_\lambda(t) - \mu(t)$.

To establish asymptotic normality for $\mu_\lambda(t) - \mu(t)$ we can proceed as in Section 3.4.4 and write

$$\frac{\mu_\lambda(t) - \mu(t)}{V^{1/2}} = \frac{\mu_\lambda(t) - E\mu_\lambda(t)}{V^{1/2}} + \frac{E\mu_\lambda(t) - \mu(t)}{V^{1/2}}$$

with $V = (n\lambda)^{-1}\sigma^2\int_{-1}^{1}K(u)^2du$. The first term on the right hand side of this expression is asymptotically normal as a result of Theorem 4.3. All that is needed are conditions under which

$$\frac{E\mu_\lambda(t) - \mu(t)}{V^{1/2}} \to 0 . \tag{32}$$

From Corollary 4.1 if $\mu \in C^2[0,1]$ the numerator in (32) is $O(\lambda^2)$ while the denominator is $O((n\lambda)^{-1/2})$. Thus, we need $n^{1/2}\lambda^{5/2}\to 0$, i.e., $n\lambda^5 \to 0$. This proves the following corollary.

**Corollary 4.2.** Suppose that the conditions of Theorem 4.3 are met and that $\mu \in C^2[0,1]$. Then,

$$(n\lambda)^{1/2}\frac{\mu_\lambda(t) - \mu(t)}{\sigma[\int_{-1}^{1}K(u)^2du]^{1/2}}$$

has an asymptotic standard normal distribution provided $n\to\infty$, $\lambda\to 0$ in such a way that $n\lambda \to \infty$ and $n\lambda^5 \to 0$.

Under the conditions of Corollary 4.2 an approximate 100(1 - $\alpha$)% confidence interval for $\mu(t)$ is provided by

$$\mu_\lambda(t) \pm Z_{\alpha/2}\sigma[\frac{1}{n\lambda}\int_{-1}^{1}K(u)^2du]^{1/2} . \tag{33}$$

Unfortunately, the optimal value of $\lambda$ in Theorem 4.2 does not satisfy the conditions of Corollary 4.2 since $\lambda^5_{opt}n = O(n^{-1/5})^5 n = O(1)$. This is a consequence of the fact that the optimal $\lambda$ attempts to balance variance and squared bias with the result that both have the same asymptotic rate of decay.

Since Corollary 4.2 does not hold for the optimal $\lambda$ this may suggest that (33) will not perform well with $\lambda$ selected by GCV or some other criterion which attempts to estimate the minimizer of $R_n(\lambda)$. An ad hoc way around this problem is to choose $\lambda$ by GCV, say, and then use a smaller value of $\lambda$ for interval estimation to ensure that the bias is small relative to the variance. The difficulties with this approach have already been extolled in Section 3.4.4 in the analogous case of Fourier series estimators. Another possibility is examined in Exercises 17 and 18.

As a rule, $\sigma^2$ will be unknown in (33). When that is the case it can be estimated by any consistent estimator of $\sigma^2$ without affecting the asymptotic character of (33). One possible consistent estimator is

$$\sigma_\lambda^2 = \sum_{j=1}^{n} \left[\lambda^{-1} \int_{s_{j-1}}^{s_j} K(\lambda^{-1}(t_j - s))ds\right](y_j - \mu_\lambda(t_j))^2$$

(Exercise 22).

Gasser, Sroka and Steinmetz (1986) propose another possible estimator of $\sigma^2$. This estimator is based on what are called pseudo-residuals. The ith pseudo-residual is defined by

$$\tilde{\varepsilon}_i = c_i y_{i-1} + d_i y_{i+1} - y_i \quad , \quad i = 2,\ldots,n - 1 \ ,$$

where

$$c_i = (t_{i+1} - t_i)/(t_{i+1} - t_{i-1})$$

and

$$d_i = (t_i - t_{i-1})/(t_{i+1} - t_{i-1}) \ .$$

One finds that $E\tilde{\varepsilon}_i^2 \doteq (c_i^2 + d_i^2 + 1)\sigma^2$, which suggests estimating $\sigma^2$ by

$$\hat{\sigma}^2 = (n - 2)^{-1} \sum_{j=2}^{n-1} \varepsilon_i^2/(c_i^2 + d_i^2 + 1) \ .$$

This estimator is similar in spirit to estimator (3.48) proposed by Rice (1984b). Under very mild conditions $\hat{\sigma}^2$ is found to be a strongly consistent estimator of $\sigma^2$. The pseudo-residuals have other uses. Specifically, they can be used to aid in the detection of outliers and heteroskedasticity. We refer the reader to Gasser, Sroka and Steinmetz (1986) for further discussion of these topics.

An example of the use of the confidence intervals in (33), obtained from the data in Figure 4.4, is shown in Figure 4.6. The figure shows the true function along with the 95% interval estimates at each of the $t_j$ obtained using the estimator from Section 4.5 and the Gasser, Sroka and Steinmetz estimator of $\sigma^2$. Again, for presentation purposes, we have connected the interval bounds. However, the intervals are only to be interpreted point wise and not as confidence bands. It is of interest to note that for this data set our estimate of $\sigma^2$ is .00289 (the true value being .0025) and 98 out of the 101 function values were actually covered by their respective intervals.

## 4.7 Higher Order Kernels, Variable Bandwidths and Optimal Design

In this section we broaden the definition of kernel estimators to take advantage of additional information about the smootheness of $\mu$. Also, we see what transpires when the assumption of uniformly spaced design points used in Section 4.3 is relaxed.

Suppose it is known that $\mu \in C^m[0,1]$ for $m>2$. Then, referring back to the arguments which led to Corollary 4.1, one might expect this knowledge could be exploited to obtain a similar expression to (22) except with a better order for the bias. The idea is that the bias of $\mu_\lambda(t)$ can now be written as

$$E\mu_\lambda(t) - \mu(t) \doteq \sum_{j=1}^{m-1} \frac{(-\lambda)^j}{j!} \mu^{(j)}(t) \int_{-1}^{1} u^j K(u)du$$

$$+ (-1)^m \frac{\lambda^m}{m!} \mu^{(m)}(t) \int_{-1}^{1} u^m K(u)du \quad . \tag{34}$$

So, the leading term in the bias will be of order $\lambda^m$ if the first $m - 1$ terms in the sum vanish. This means that the kernel should satisfy (4d) and

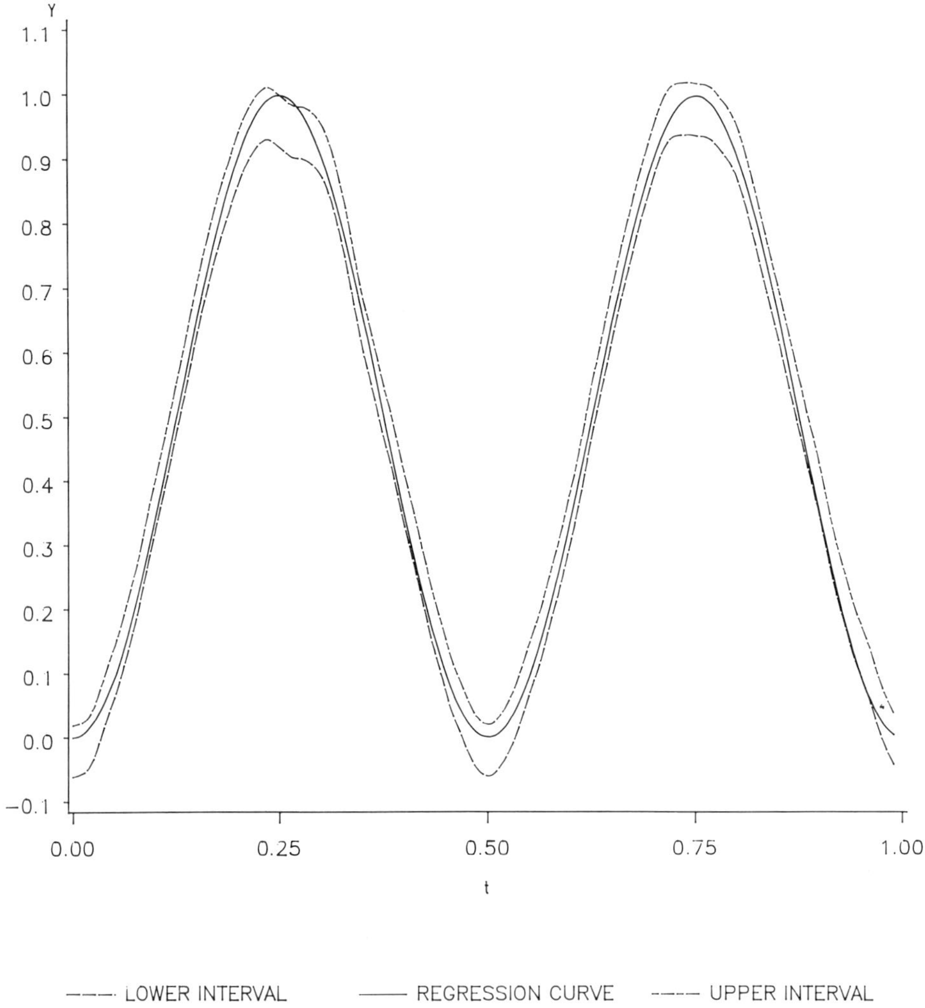


**Figure 4.6** Regression curve and interval estimates for simulated data.

i) $\int_{-1}^{1} K(u)du = 1,$

ii) $\int_{-1}^{1} u^j K(u)du = 0, \quad j=1,\ldots,m-1,$

and

iii) $\int_{-1}^{1} u^m K(u)du = \alpha \neq 0 .$

A kernel of this type is called a *kernel of order m*. Kernels satisfying (4a)-(4c) are therefore kernels of order 2.

The use of higher order kernels provides one generalization of the estimators in Section 4.2. It would now be straightforward to extend the results in Section 4.3 and obtain rates of convergence for the risk of these estimators. This is done, for example, in Gasser and Müller (1979). Rather than do so here we will pursue a slightly more general development and also treat the case where the design points are not uniformly spaced.

Assume that design points have been spaced according to a positive, differentiable density p on [0,1]. The jth design point $t_j$ is defined by the relation

$$\int_0^{t_j} p(t)dt = j/n, \quad j=1,\ldots,n, \tag{35}$$

i.e., the $t_j$ are the n-tiles of p. When $p(t) \equiv 1$ the result is uniformly spaced points on [0,1]. A sequence of the form (35) is often called a regular sequence (c.f., Sacks and Ylvisaker, 1968).

Consider now estimator (6d) with the $s_j$ defined by (13). If the $t_j$ are generated by p using relation (35) and $n\to\infty$, $\lambda \to 0$ in such a way that $n\lambda \to \infty$ it can then be shown (see Müller, 1984b) that

$$\operatorname{Var} \mu_\lambda(t) \sim \frac{\sigma^2}{n\lambda} \frac{1}{p(t)} \int_{-1}^{1} K(u)^2 du \equiv \frac{\sigma^2}{n\lambda} \frac{1}{p(t)} V$$

and

$$E\mu_\lambda(t) - \mu(t) \sim (-1)^m \frac{\lambda^m}{m!} \mu^{(m)}(t) \int_{-1}^{1} u^m K(u)du \equiv (-1)^m \frac{\lambda^m}{m!} \mu^{(m)}(t)M.$$

Thus, an asymptotic expression for the risk at t is

$$R_n(t;\lambda) \sim \frac{\sigma^2}{n\lambda}\frac{1}{p(t)}V + \frac{\lambda^{2m}}{m!^2}\mu^{(m)}(t)^2M^2 \quad . \tag{36}$$

Differentiating (36) with respect to $\lambda$ one finds the asymptotically optimal bandwidth at t to be

$$\lambda_{opt}(t) = \left[\frac{\sigma^2}{n}\frac{m!^2}{2m}\frac{V/M^2}{p(t)\mu^{(m)}(t)^2}\right]^{1/(2m+1)} \tag{37}$$

and

$$R_n(t;\lambda_{opt}(t)) \sim \text{Const}(m,\sigma^2)n^{-\frac{2m}{2m+1}}\left[V/p(t)\right]^{\frac{2m}{2m+1}}$$

$$\times \left[M^2\mu(t)^2\right]^{\frac{1}{2m+1}} \quad , \tag{38}$$

where $\text{Const}(m,\sigma^2)$ is a constant involving m and $\sigma^2$. The main conclusions to be drawn from this are that i) increased smoothness of $\mu$ translates into faster rates of convergence and ii) the bandwidth will be wider where p is smaller. Both of these could be anticipated. For example, in a region where p is small there will tend to be fewer $t_j$ and, hence, the bandwidth must be broadened to obtain more information about $\mu$.

If (mth order) boundary kernels are used, one can integrate (36) over [0,1] to obtain a global measure of the performance of $\mu_\lambda$. The result is an asymptotic expression for the integrated risk: namely,

$$IR_n(\lambda) = \int_0^1 E(\mu_\lambda(t) - \mu(t))^2dt$$

$$\sim \frac{\sigma^2}{n\lambda}V\int_0^1\frac{1}{p(t)}dt + \frac{\lambda^{2m}}{m!^2}M^2\int_0^1\mu^{(m)}(t)^2dt \quad . \tag{39}$$

From (39) the asymptotically optimal global bandwidth and risk are found to be

$$\lambda_{opt} = \left[ \frac{m!^2 (\int_0^1 \frac{1}{p(t)} dt) V/M^2}{n(2m) \int_0^1 \mu^{(m)}(t)^2 dt} \right]^{\frac{1}{2m+1}} \tag{40}$$

and

$$IR_n(\lambda_{opt}) = \text{Const}(m,\sigma^2) n^{-\frac{2m}{2m+1}} \left[ V \int_0^1 \frac{1}{p(t)} dt \right]^{\frac{2m}{2m+1}}$$

$$\times \left[ M^2 \int_0^1 \mu^{(m)}(t)^2 dt \right]^{\frac{1}{2m+1}} . \tag{41}$$

Again we see that if $\mu \in C^m[0,1]$ the rate of convergence can be improved to $n^{-2m/(2m+1)}$.

When the design points are unequally spaced the use of a global bandwidth, optimal or otherwise, may give poor results. This is because the variability of the design can cause the estimator to be computed from much fewer observations in some regions than in others. Thus, in this case it makes more sense to have the bandwidth depend on the density of the design in some fashion. Estimators for which $\lambda=\lambda(t)$ varies with t are called *variable bandwidth kernel estimators*.

The best bandwidth to choose for any given t is $\lambda_{opt}(t)$ in (37) which would result in the best possible variable bandwidth estimator, at least from an asymptotic viewpoint. Of course we cannot actually compute $\lambda_{opt}(t)$ in practice. However, consideration of its use with $\mu_\lambda$ leads to some interesting conclusions regarding selection of the design.

If boundary kernels are used and $\inf|\mu^{(m)}(t)|>0$, Müller (1984b) shows it is possible to compute the integrated risk for the variable bandwidth estimator $\mu_\lambda$ with $\lambda = \lambda_{opt}(t)$. The result is obtained from (38) and found to be

$$\tilde{IR}_n = \text{Const}(m,\sigma^2)\, n^{-\frac{2m}{2m+1}}\, V^{\frac{2m}{2m+1}}\, M^{\frac{2}{2m+1}} \left\{ \int_0^1 \frac{\mu^{(m)}(t)^2}{p(t)^{2m}} dt \right\}^{1/(2m+1)} \tag{42}$$

See Müller (1984b) for more details concerning the derivation of (42).

Both (41) and (42) involve the design density p. The question now arises of how p should be chosen to minimize these expressions. Müller (1984b) shows that (41) is minimized by

$$p_{opt}(t) \equiv 1 \tag{43}$$

and that the minimizer of (42) is

$$\tilde{p}_{opt}(t) = |\mu^{(m)}(t)|^{2/(4m+1)} / \int_0^1 |\mu^{(m)}(s)|^{2/(4m+1)} ds. \tag{44}$$

Thus we see that, if a globally optimal bandwidth is to be used, the local behavior of $\mu$ does not influence design selection and, when it is possible to do so, one is best off by uniformly spacing the design points. In contrast, if the optimal variable bandwidth kernel estimator is used, design points should be concentrated where $|\mu^{(m)}|$ is large. This is a reflection of the fact that the function is changing more rapidly where $|\mu^{(m)}|$ is large, and hence, more data is required in such regions to adequately estimate $\mu$. From a qualitative viewpoint the last conclusion suggests that when any type of variable bandwidth kernel estimator is being used design points should be concentrated where $|\mu^{(m)}|$ is believed to be large.

It is possible to generalize the results of this section still further and include cases where the $\varepsilon_j$ have nonconstant variances and a weighted version of the integrated risk is used. We refer the reader to Müller (1984b) and Hart and Wehrly (1986) for further discussions of these topics. Staniswallis (1986) has developed a data driven method of estimating locally optimum bandwidths.

Examination of (42) reveals that the asymptotically optimal risk for a kernel estimator based on an mth order kernel involves the kernel only through the factors

$$V(K) = \int_{-1}^{1} K(u)^2 du$$

and

$$M(K) = \int_{-1}^{1} u^m K(u) du.$$

Thus, it is possible to parallel the developments in Section 4.4 and consider the construction of optimal and minimum variance kernels for kernel estimators of order m. For example, a minimum variance kernel must minimize V(K) subject to

$$\int_{-1}^{1} K(u) du = 1$$

and

$$\int_{-1}^{1} u^j K(u) du = 0 , \quad j = 1, \ldots, m - 1.$$

Exercise 9 concerns derivation of a general expression for such minimum variance kernels from which it follows that $3(3 - 5u^2)/8$ is the minimum variance kernel of order 4. Further discussions of optimal and minimum variance kernels can be found in Gasser and Müller (1979), Müller (1984a) and Gasser, Müller and Mammitzsch (1985).

## 4.8 Estimation of Derivatives

Frequently, it is necessary to estimate some functional of $\mu$ such as its derivative at some point t or, perhaps, its integral over some interval. Although other functionals can be considered, derivative estimation seems to be one of the most important problems so we will concentrate on that topic in this section.

To understand the idea behind derivative estimation suppose that $\mu$ admits m derivatives and we wish to estimate the dth derivative for some $d < m$. Then, by arguments which are now standard, the expectation of estimator (6d) based on some general kernel, that need not satisfy (4), is

$$E\mu_\lambda(t) \doteq \sum_{j=0}^{m} (-1)^j \mu^{(j)}(t) \frac{\lambda^j}{j!} \int_{-1}^{1} u^j K(u) du .$$

It is immediately clear from this expression how one should choose K and alter $\mu_\lambda$ to obtain an estimator of $\mu^{(d)}$. Specifically, we require that

$$\begin{aligned} \int_{-1}^{1} u^j K(u)du &= 0, \qquad j = 0,\ldots,m-1,\ j \neq d, \\ &= (-1)^d d!\ , \quad j = d, \\ &= \alpha \neq 0\ , \quad j = m\ , \end{aligned}$$

and then define the estimator as

$$\mu_{\lambda d}(t) = \lambda^{-(d+1)} \sum_{j=1}^{n} y_j \int_{s_{j-1}}^{s_j} K(\lambda^{-1}(t - s))ds. \tag{45}$$

It can be shown that $\mathrm{Var}(\mu_{\lambda d}(t)) = O(1/n\lambda^{2d+1})$, $E\mu_{\lambda d}(t) - \mu^{(d)}(t) = O(\lambda^{m-d})$ and the optimal rate of decay for $R_{nd}(t;\lambda) = E(\mu_{\lambda d}(t) - \mu^{(d)}(t))^2$ is $n^{-2(m-d)/(2m+1)}$ (see Exercise 11).

There exist parallels of the work in Sections 4.3-4.7 for the estimator $\mu_{\lambda d}$. Some of these are explored in Georgiev (1984), Müller (1984a,b), Gasser, et. al. (1984), Gasser and Müller (1984) and Gasser, Müller and Mammitzsch (1985).

An illustration of the use of derivative estimation that will also be of interest for the next section is provided by growth data on a boy and a girl from a growth study undertaken at the University of Zürich. The data was kindly provided by Dr. L. Molinari.

Figure 4.7 shows data on the growth in stature of a boy and girl taken from the above mentioned Zürich study. In growth studies it is often of interest to estimate the velocity curve or derivative of the growth curve. Comparisons are then made between sexes concerning the timing of growth spurts, peak velocity, etc. Thus we now consider estimation of the velocity curves corresponding to the data in Figure 4.7.

To estimate the velocity curves estimator (45) was used with d = 1 and

$$K(u) = 105(-9u^5 + 14u^3 - 5u)/32\ ,$$

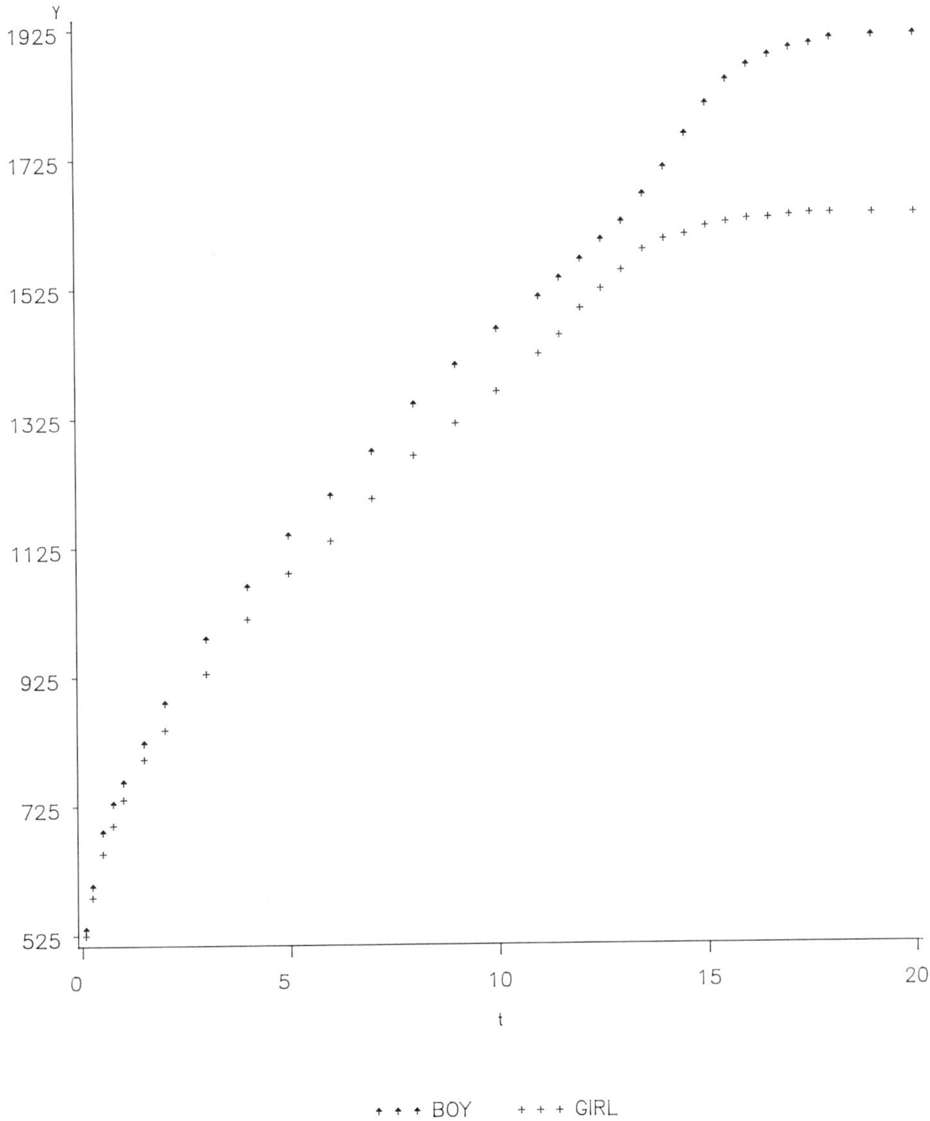


**Figure 4.7** Growth in stature of a boy and a girl.

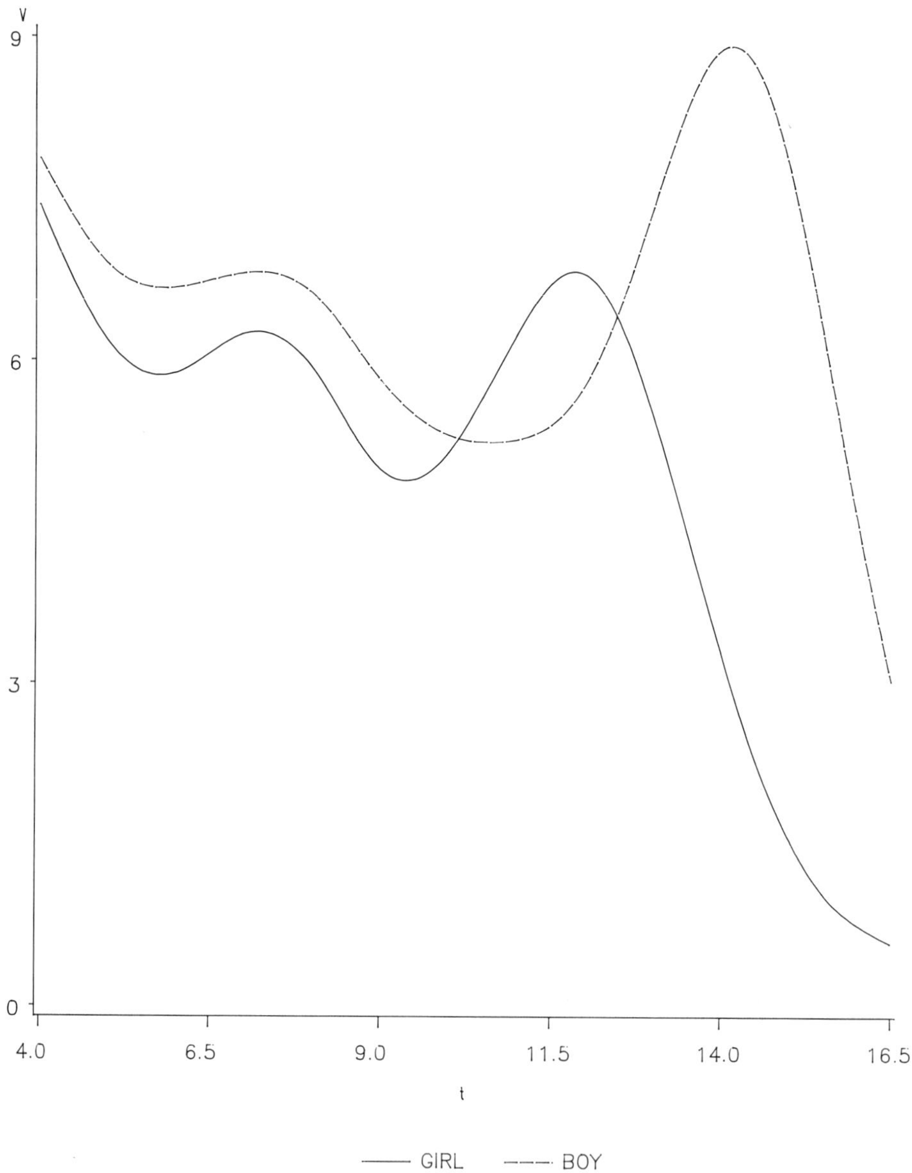


**Figure 4.8** Estimated velocity curves for a girl and a boy.

this being the optimal kernel for derivative estimation when m = 5 and d = 1 (c.f., Gasser, Müller and Mammitzsch, 1985). A bandwidth of $\lambda$ = 3.4 was used for both subjects. This value was chosen to provide similar estimates to those obtained by Gasser, et. al. (1984) who have undertaken an extensive study of the Zürich growth data. Their study will be described in more detail in the next section.

A plot of the estimated velocity curves for the data in Figure 4.7 is shown in Figure 4.8. (To avoid the need for boundary corrections the boundary regions have been excluded from the plot.) There are a number of features of the estimated curves which merit comments. Both curves seem to contain two distinct growth spurts. One of these is seen to begin at about ten years of age for the girl and twelve years for the boy. This is the so called pubertal spurt that accompanies the advent of puberty. The maximum velocity of the pubertal spurt occurs at around twelve years of age for the girl and fourteen years for the boy. It appears from Figure 4.8 that the boy has a later but more intense pubertal spurt than the girl. These findings are typical of results from many growth studies.

Both curves in Figure 4.8 also have distinct peaks at roughly eight years of age. This is due to the mid-growth spurt described, for example, in Gasser, et. al. (1984). The presence of this mid-growth spurt has some interesting implications for current parametric models used in growth curve analysis. These will be discussed in the next section.

## 4.9 Applications

In this section we give a brief account of the work by Gasser, et al. (1984) on growth curve estimation. It provides an excellent example of the use of nonparametric regression methodology. Moreoever, it illustrates the usefulness of asymptotic theory to guide finite sample decisions as well as providing examples of the use of derivative estimates and higher order kernels.

The way humans grow is of interest to a variety of fields and industries. These include pediatrics and the clothing industry. Due to this and other reasons there is a long history of the study of human growth in both statistics and biology. Several longitudinal growth studies were undertaken in 1954/1955, one at the University of Zürich. The results described here were obtained from that study.

Let $\mu(t)$ denote the height of an individual at time t. $\mu$ is often called a growth curve. Growth curves are known to exhibit several traits. Perhaps the most pronounced of these is a growth spurt at puberty whose velocity peaks at around 12 years of age for girls and 14 years for boys. Two classical models used to describe this pubertal spurt (PS) are the logistic model $\alpha + \beta/(1 + \exp(-\gamma(t - \delta)))$ and the Gompertz function $\alpha + \beta\exp(-\exp(-\gamma(t - \delta)))$.

Parametric models for global fits have also been proposed. Among the best of these are two due to Preece and Baines (1978):

$$\mu(t) = \alpha - 2(\alpha - \mu(\beta))/(\exp(\gamma(t - \beta)) + \exp(\delta(t - \beta)))$$

and

$$\mu(t) = \alpha - 4(\alpha - \mu(\beta))/[(\exp(\gamma(t - \beta)) + \exp(\delta(t - \beta)))(1 + \exp(\nu(t - \beta)))].$$

Nonlinear regression techniques are usually used to estimate the parameters in these models.

In growth studies the focus tends to be on the derivative of the growth curve rather than $\mu$ itself. Thus the interest is in the velocity of growth. Once the parameters have been estimated in one of the Preece and Baines models it is an easy matter to estimate velocity by differentiating the fitted curve.

Recently, Gasser et. al. (1984) have considered the use of nonparametric regression methods for estimation of growth curves. As we have already noted, nonparametric estimates are not tied to a parametric model, and, as a result, should be more flexible and let the data have a stronger voice in the fit. We shall see that this property leads to some interesting conclusions for growth curve analysis both in terms of new results and criticisms of the standard parametric models.

Gasser, et. al. (1984) decided to estimate the growth curves for 45 boys and 45 girls from the Zürich growth study. The reader is referred to their paper for more details on the data. Let $y_{kj}$ be the measured height of the kth individual at age $t_j$. They postulated the model

$$y_{kj} = \mu_k(t_j) + \varepsilon_{kj}, \qquad j = 1,\ldots,n,$$

with the $\varepsilon_{kj}$ being random errors, due to measuring errors and other factors. It was assumed that the $\varepsilon_{kj}$ had means zero and variances $\sigma_k^2$. There were n=34 readings for each subject ranging from age 4 weeks to 20 years.

Kernel estimators were used to estimate the $\mu_k$. Estimates of the velocity curves, $\mu_k'$, were obtained using the kernel estimators discussed in Section 4.8 with d=1. In addition a new curve was estimated: the second derivative or acceleration curve $\mu_k''$. The acceleration curves were also estimated using kernel estimators this time with d=2. Boundary kernels were used for all three curve estimates to avoid boundary bias problems.

A problem arose in estimation of the bandwidths for the different curves. If individually optimized bandwidths were used then asymptotic considerations suggest that the fitted curve for the kth individual, $\mu_{\lambda(k)}$, should approximately satisfy

$$\int_a^b \operatorname{Var}(\mu_{\lambda(k)}(t))dt \propto \int_a^b \operatorname{Bias}(\mu_{\lambda(k)}(t))^2 dt,$$

where $\operatorname{Var}(\mu_{\lambda(k)}(t))$ is the variance of the fit at t for the kth individual and $\operatorname{Bias}(\mu_{\lambda(k)}(t)) = E\mu_{\lambda(k)}(t) - \mu_k(t)$. This is because if a kernel of order m is used with an optimal bandwidth for estimating $\mu$

$$\frac{\sigma_k^2}{n\lambda}\int_{-1}^1 K(u)^2 du \propto \lambda^{2m}\left(\int_a^b \mu_k^{(m)}(t)^2 dt\right)\left(\int_{-1}^1 u^m K(u)du\right)^2$$

$$\propto n^{-\frac{2m}{2m+1}} \left[\sigma_k^2 \int_{-1}^{1} K(u)^2 du\right]^{\frac{2m}{2m+1}} \left[\left(\int_a^b \mu_k^{(m)}(t)^2 dt\right)\left(\int_{-1}^{1} u^m K(u) du\right)^2\right]^{\frac{1}{2m+1}}$$

(see, e.g., Theorem 4.2). Similar formulae hold for estimation of $\mu_k'$ and $\mu_k''$. An examination of this last expression shows that there is a confounding of residual variance $\sigma_k^2$ and the factor

$$\left[\int_a^b \mu_k^{(m)}(t)^2 dt\right]^{1/(2m+1)} .$$

This has the consequence that identical curves corresponding to individuals with different residual variances would be estimated with different bandwidths. Thus, the estimated curves would show a systematic interindividual variation due to differences in the estimator biases. Since comparison and analysis of curves across individuals is what is desired this argues against the use of bandwidth selection methods such as GCV for each individual.

The above discussion shows the usefulness of asymptotic analysis. It provides us with closed form expressions from which conclusions can be drawn with minimal labor. In this particular case it is not clear that the interrelationship between bias and residual error could have been quantified in any other fashion. While asymptotic theory does not tell us how to select the bandwidths it has led us away from the potential pitfall of doing what seems natural and optimizing the bandwidths for each subject. Gasser, et. al. (1984) eventually developed a method for bandwidth selection that utilized all the children to select a bandwidth that, with some modifications, was deemed appropriate for all subjects.

Given a value for $\sigma_k^2$ and a true growth curve $\mu_k$ it would be possible to actually compute the finite sample risks for the function estimate and its derivatives and to minimize each with respect to $\lambda$. To analyze the influence of the choice of a kernel on the estimators, Gasser et. al. (1984) did precisely this under the assumption that $\mu_k$ had the form of the second Preece and

Baines function with values for the parameters taken from Preece and Baines (1978). Several values were selected for $\sigma_k^2$. The results indicated that optimal higher order kernels were far superior to lower order and/or minimum variance kernels both for estimation of the growth curve and its derivatives. The surprising part of this result is that it is not asymptotic but instead applies to n=34 data points. This has the implication that kernel choice may be important to an estimator's performance even for small sample sizes. A further indication of this is provided by a simulation study in Gasser, Müller and Mammitzsch (1985).

After deciding on the use of higher order ($m = d + 4$) optimal kernels and a unified bandwidth selection method the growth data from the ninety subjects were analyzed with somewhat surprising results. First, through use of the estimated velocity and acceleration curves, Gasser, et. al. were able to quantify and establish the existence of a mid-growth spurt (MS) at about age seven (c.f. Figure 4.8). This characteristic of growth had been reported in some of the early human growth literature but had disappeared from the literature upon the advent of modern parametric models. This was because the modern models made no allowance for its presence.

The Preece and Baines models did not take the MS into account and their residual analysis did not pinpoint the discrepancy in their models concerning this fact. One possible explanation for this is that the object of interest is the velocity curve but the observed data and what is actually fitted is in terms of the growth curve. This makes diagnostic analysis somewhat difficult.

The fact that the Preece and Baines models did not incorporate the MS was shown by Gasser, et. al. to result in fitted models where the PS started too early in some cases, this being dependent on sex. Thus, the failure to include the MS led to the creation of an artificial difference between the growth of boys and girls.

The velocity peak for the PS was found to be asymmetric using nonparametric methods (c.f. Figure 4.8). It rises slower and falls off faster. This is in contrast to the logistic curve which has a symmetric peak. Although the peak for the Gompertz function is asymmetric it turns out to be asymmetric in the wrong direction.

Another interesting result of the Gasser, et. al. study was also related to their acceleration curve estimates. They were able to see from these that the phase of the PS from onset to maximal acceleration was more or less independent of that from maximal acceleration to maximal deceleration. The differences between boys and girls were primarily due to the first phase. This represents a refinement in what was known: namely, that boys tend to have a later and more intense PS.

The Gasser, et. al. study illustrates how nonparametric techniques can

i) lead to the discovery of new traits or characteristics of a real world phenomenon and

ii) provide a useful tool for the critical examination of parametric models.

While these are not their only uses, either would suffice to motivate the use of nonparametric estimators. For further details on the use of nonparametric estimators in growth curve analysis the reader is referred to the Gasser, et. al. paper and references therein.

Another interesting example of the practical use of kernel regression estimators is provided by the work of Clark (1979, 1980). He uses kernel estimators as a tool for the calibration of radiocarbon dating methods.

The problem can be described as follows. Radiocarbon or carbon-14 is absorbed by all living organisms until they die. At that time the radiocarbon level begins to decay exponentially. More specifically, if z denotes the level of radiocarbon in a organisms that died at time x, then

$$z = A(x)e^{-\theta x}$$

where $\theta$ = log2/(half life of carbon-14) and $A(x)$ is the concentration of radiocarbon in the organism at its time of death.

If it can be assumed that radiocarbon levels in living organisms are constant over time, i.e., $A(x) \equiv A_0$, then

$$z = A_0 e^{-\theta y}$$

and

$$y = \theta^{-1}(\log A_0/z)$$

is the time of death of the organism. The number y is called the radiocarbon age of the organism and is known to be only an approximation to its true age. This is because $A(x)$ is not constant, being affected, for example, by the burning of fossil fuels, nuclear testing, etc.

There is a simple relationship between the radiocarbon age y and true age x of an organism. Specifically

$$y = \theta^{-1}\log(A_0/A(x)e^{-\theta x}) = x - \theta^{-1}\log(A(x)/A_0)$$

$$= \mu(x) \ .$$

The function $\mu$ is called the calibration function. If it were known, then, given an exact value for radiocarbon age, it would be a simple matter to solve the equation $y = \mu(x)$ for x and thereby find the true age of an organism. Unfortunately, $\mu$ is unknown and there has been considerable interest in determining its form.

As noted by Clark (1979), the form of $\mu$ is of interest to geophysicists and archaeologists as well as others. For example, revised radiocarbon dates have been used to cast doubts on the "diffusion theory" of archaeology that the ultimate source of European civilization was the Near East.

Clark (1979, 1980) considers the problem of estimating $\mu$ from data on radiocarbon dates. He works with radiocarbon dates from bristlecone pine trees. These trees have extremely long lifespans and their age can be determined directly, and without error, by counting tree rings. Thus, given radiocarbon dates $y_1,\ldots,y_n$ on samples from n trees one can postulate the model (Clark, 1979).

$$y_i = \mu(t_i) + \varepsilon_i, \quad i = 1,\ldots,n ,$$

where $t_i$ is the true age of the tree (determined by tree ring information) and the $\varepsilon_i$ are random measurement errors.

Using the above model Clark (1979) fitted a set of data on radiocarbon dates for bristlecone pine trees using a kernel estimator of the form (6e). The bandwidth was selected through a cross-validation type procedure. The resulting estimate $\mu_\lambda$ of the calibration function $\mu$ was found to be essentially linear. This is in marked contrast to other models which had been postulated for $\mu$. For example, Suess (1970) proposed a model of the form

$$\beta_1 + \sin(\beta_2 t + \beta_3) + \beta_4 \sin(\beta_5 t)$$

for $\mu$. This function is quite wiggly, particularly when compared to Clark's "linear" fit.

The actual problem of interest is to use the estimator $\mu_\lambda$ of $\mu$ obtained using tree ring data to determine the age of some new observation given only its corresponding radiocarbon date. Thus we consider obtaining a new observation

$$Y_0 = \mu(T_0) + \varepsilon_0.$$

We will observe $Y_0$, the radiocarbon date, but not the object of interest $T_0$, the true age. The problem of estimating $X_0$ can actually be solved by solving the parallel prediction problem for $Y_0$. This is because if we have a prediction interval

$$I(t) = (\mu_\lambda(t) - C_\alpha, \ \mu_\lambda(t) + C_\alpha)$$

such that

$$\text{Prob}(Y_0 \ \varepsilon \ I(t) | T_0 = t) = 1 - \alpha$$

then the inverse image of I(t),

$$J(Y_0) = \{t: Y_0 \in I(t)\} ,$$

satisfies

$$\mathrm{Prob}(T_0 \in J(Y_0)|T_0 = t) = \mathrm{Prob}(Y_0 \in I(t)|T_0 = t)$$

$$= 1 - \alpha .$$

Thus, $P(T_0 \in J(Y_0)) = 1 - \alpha$ and the problem of estimating $T_0$ reduces to the problem of selecting the constant $C_\alpha$ in the prediction interval.

In order to construct prediction intervals Clark (1980) notes that one approach would be to take $C_\alpha$ as $Z_{\alpha/2}(\mathrm{Var}(Y_0 - \mu_\lambda(T_0))^{1/2}$. However, this ignores the fact that $\mu_\lambda$ is biased. Thus he also considers an alternative method with $C_\alpha = Z_{\alpha/2}(\mathrm{Var}(Y_0 - \mu_\lambda(T_0)) + \hat{B}^2)^{1/2}$ and $\hat{B}$ an estimate of the bias of $\mu_\lambda(T_0)$. The estimate $\hat{B}$ that he uses is obtained by a type of jackknife procedure and is actually an estimate of the average bias of $\mu_\lambda$ over the observed tree ring determined ages $t_1,\ldots,t_n$. Thus, $C_\alpha$ is actually of the form $Z_{\alpha/2}\tilde{P}(\lambda)^{1/2}$, with $\tilde{P}(\lambda)$ an estimate of the predictive risk corresponding to the kernel estimator with bandwidth $\lambda$. The latter type of interval was found to be far superior to those based on the variance alone (see Clark, 1980, for a detailed comparison). We have already discussed the possibility of replacing the variance by the risk of a biased estimator for the construction of ordinary confidence intervals. Clark's approach is therefore the predictive parallel of this procedure.

## 4.10 Random t's

Let us now alter our original assumption about the probability structure of the $(t_j,y_j)$ and allow the $t_j$ to also be random. Specifically, assume that $(t_1,y_1),\ldots,(t_n,y_n)$ are independent and identically distributed as a continuous bivariate random variable $(T,Y)$. In this case it is still possible to write down our model as

$$y_j = \mu(t_j) + \varepsilon_j \ , \quad j = 1,\ldots,n.$$

However, now $\mu$ is the conditional mean,

$$\mu(t) = E[Y|T = t],$$

and the $\varepsilon_j$, while independent, will not be identically distributed. In particular, the assumption of a common variance for the $\varepsilon_j$ will not be satisfied. Thus, a somewhat different approach is needed from that of previous sections to derive estimators of $\mu$ in this setting.

An understanding of how to estimate $\mu$ can be obtained by going back to the actual definition of the conditional mean. Let (T,Y) have an associated density f(t,y) with

$$f_T(t) = \int_{-\infty}^{\infty} f(t,y)dy \tag{46}$$

the marginal density for T. Then, the conditional density for Y, given T, is

$$f_{Y|T}(y|t) = \frac{f(t,y)}{f_T(t)} \tag{47}$$

and $\mu$ is defined by

$$\mu(t) = \int_{-\infty}^{\infty} y f_{Y|T}(y|t)dy. \tag{48}$$

We assume that both (47) and (48) are well defined. Thus, if we can estimate $f_{Y|T}$ we can estimate $\mu$ by plugging the estimator into (48). This leads us quite naturally into the subject of density estimation.

Estimation of multivariate density functions is a topic with a vast literature. It traces its origin back to work by Rosenblatt (1956) and Parzen (1962b). A bibliography of this literature can be found in Wertz and Schneider (1979) and a review in Wegman (1972).

Typically it is assumed that n independent identically distributed p-vector observations $\underline{x}_1,\ldots,\underline{x}_n$ have been obtained from a distribution with density $f(x_1,\ldots,x_p)$. The estimator we will consider for f is

$$f_\lambda(x_1,\ldots,x_p) = \frac{1}{n\lambda^p} \sum_{r=1}^{n} \prod_{j=1}^{p} K_j(\lambda^{-1}(x_j - x_{rj})) , \quad (49)$$

where $x_{rj}$ is the element of $\underline{x}_r$ for the jth variable and each of the kernel functions is assumed to have support on [-1,1] and satisfy (4a)-(4d). Classically, the $K_j$ have not been restricted to have finite support (although optimal kernels do have finite support: see Epanechnikov, 1969, and Müller, 1984a). Also, different bandwidths can be used for each variable. We will find the somewhat more restrictive form (49) best suited to our purposes however. Some of the properties of kernel density estimates are explored in Exercises 5-6.

Returning to estimation of (48) we see that what is actually needed is an estimator of the joint density f. As a result of (49) one possible estimator is

$$f_\lambda(t,y) = \frac{1}{n\lambda^2} \sum_{r=1}^{n} K_1(\lambda^{-1}(t - t_r))\, K_2(\lambda^{-1}(y - y_r)). \quad (50)$$

This provides an estimator of $f_T$: namely,

$$\begin{aligned} f_{\lambda T}(t) &= \int_{-\infty}^{\infty} f_\lambda(t,y)dy \\ &= \frac{1}{n\lambda} \sum_{r=1}^{n} K_1(\lambda^{-1}(t - t_r)) . \quad (51) \end{aligned}$$

Putting (50) and (51) together gives

$$f_\lambda(y|t) = f_\lambda(t,y)/f_{\lambda T}(t)$$

as an estimator of $f_{Y|T}$. To estimate $\mu$ use this estimator in (48) which gives, for $\lambda$ sufficiently small,

$$\mu_\lambda(t) = \frac{1}{f_{\lambda T}(t)} \int_{-\infty}^{\infty} y f_\lambda(t,y)dy$$

$$= \frac{1}{f_{\lambda T}(t)} \frac{1}{n\lambda^2} \sum_{r=1}^{n} K_1(\lambda^{-1}(t - t_r)) \int_{-\infty}^{\infty} y K_2(\lambda^{-1}(y - y_r))dy$$

$$= \frac{1}{f_{\lambda T}(t)} \frac{1}{n\lambda} \sum_{r=1}^{n} K_1(\lambda^{-1}(t - t_r)) \int_{-1}^{1} (y_r + \lambda u) K_2(u)du$$

$$= \sum_{r=1}^{n} K_1(\lambda^{-1}(t - t_r)) y_r / \sum_{r=1}^{n} K_1(\lambda^{-1}(t - t_r)) . \qquad (52)$$

This is just estimator (6c) which was defined in the introduction.

Estimator (52) has a long history which stretches back to the work of Nadaraya (1964) and Watson (1964). In fact it is sometimes referred to as the Nadaraya-Watson estimator. This is actually the viewpoint from which estimator (6c) was first derived. Benedetti (1974, 1975) later considered its use when t is not a random variable.

There is quite a bit known about estimator (52). Under certain restrictions it is known to have a variety of consistency properties. Some of these are explored by Nadaraya (1964, 1965, 1970), Noda (1976), Rosenblatt (1969), Schuster (1972), Schuster and Yakowitz (1979), Devroye (1978a, 1981), Devroye and Wagner (1980), Johnston (1982), Mack and Silverman (1982), Greblicki, Krzyzak and Pawlak (1984) and Härdle (1984). For example it follows from Devroye (1978a) that

$$\sup_{a \le t \le b} |\mu_\lambda(t) - \mu(t)| \overset{a.s.}{\to} 0$$

provided $n \to \infty$, $\lambda \to 0$, $n\lambda \to \infty$ and certain other mild conditions are satisfied. The behavior of integrated risk type measures computed from $\mu_\lambda$ have been studied by Konakov (1977) and Hall (1984).

Schuster (1972) has shown that $\mu_\lambda$ is asymptotically normal. If $K_1$ is continuous and $n \to \infty$, $\lambda \to 0$ in such a way that $n\lambda^3 \to \infty$ and $n\lambda^5 \to 0$ then $(n\lambda)^{1/2}(\mu_\lambda(t) - \mu(t))$ has asymptotically a zero mean normal distribution with variance

$$\mathrm{Var}(Y|T=t) \int_{-1}^{1} K^2(u)du / f_T(t),$$

where Var(Y|T=t) is the conditional variance of Y when T=t. While not directly comparable, one may wish to contrast this with results in Section 4.6. Notice, for example, the presence of the condition $n\lambda^5 \to 0$ in both cases.

Härdle and Marron (1985a) consider the problem of bandwidth selection for a variant of estimator (6c) in the case of random t's where $f_T$ is known. The estimator they study is

$$\mu_\lambda(t) = (n\lambda)^{-1} \sum_{j=1}^{n} K(\lambda^{-1}(t - t_j))y_j/f_T(t).$$

They assess the performance of a particular bandwidth by its corresponding value for the (weighted) loss

$$L_n(\lambda) = n^{-1} \sum_{j=1}^{n} (\mu_\lambda(t_j) - \mu(t_j))^2 w(t_j) \quad ,$$

where w is some nonnegative weight function. A rule for selecting an estimator $\hat{\lambda}$ of the minimizer of $L_n(\lambda)$ is said to be asymptotically optimal if

$$L_n(\hat{\lambda}) / \inf_{\lambda \varepsilon \Lambda} L_n(\lambda) \xrightarrow{a.s.} 1 ,$$

where $\Lambda$ is an appropriately restricted subset of the positive real line.

Let

$$\mathrm{MSEW}(\lambda) = n^{-1} \sum_{j=1}^{n} (y_j - \mu_\lambda(t_j))^2 w(t_j) .$$

Then Härdle and Marron consider a general criterion of the form

$$G(\lambda) = \mathrm{MSEW}(\lambda) A(h(\lambda)) \quad ,$$

where A is some general function and

$$h(\lambda) = (n\lambda)^{-1} K(0) n^{-1} \sum_{j=1}^{n} f_T(t_j)^{-1}.$$

The particular choice $A(x) = (1 - x)^{-2}$ gives a weighted GCV criterion while others give bandwidth selection rules stemming from work by Akaike (1970, 1974), Shibata (1981) and Rice (1984b). It is shown that such rules are asymptotically optimal if and only if

$$\int_a^b \operatorname{Var}(Y|T = x)w(x)dx = (b - a)\int_a^b \operatorname{Var}(Y|T= x)f_T(x)w(x)dx ,$$

where $f_T$ is assumed to be supported and bounded above zero on some interval [a,b] which is the support of w. This will be true, for example, when $f_T$ is uniform on [a,b]. However, this condition will not be satisfied in general. Consequently,a criterion such as GCV must be modified to perform satisfactorily when the t's are random. One way to do this is to use $w(t) = [\operatorname{Var}(Y|T = t)]^{-1}$ in the definition of MSEW($\lambda$). Since $\operatorname{Var}(Y|T = t)$ will generally be unknown this will require estimation of the conditional variance function in practice.

Let $\mu_{\lambda(i)}$ denote the estimator of $\mu$ for the case when $(t_i, y_i)$ has been deleted from the data. More specifically

$$\mu_{\lambda(i)}(t) = ((n - 1)\lambda)^{-1} \sum_{j\neq i} K(\lambda^{-1}(t - t_j))y_j/f_T(t).$$

The cross-validation criterion for selecting $\lambda$ is then given by

$$\operatorname{CVW}(\lambda) = n^{-1} \sum_{j=1}^{n} (y_j - \mu_{\lambda(j)}(t_j))^2 w(t_j) .$$

Härdle and Marron (1985a) establish that CVW($\lambda$) always gives an asymptotically optimal selection rule and, consequently, has advantages over GCVW($\lambda$) and other related rules in that it does not require knowledge or estimation of $\operatorname{Var}(Y|T = t)$.

The bandwidth selection results of Härdle and Marron also apply to estimator (6c), other performance criteria and multivariate t. See Härdle and Marron (1985a,b) for further details.

Estimator (6d) has also been studied in the case of random t's. It has similar properties to estimator (52) in this setting.

For example, it has the same asymptotic distribution as estimator (52). See Härdle (1984b) and Härdle and Luckhaus (1984).

There are cases where one may wish to estimate a conditional mean but the data is not a random sample as assumed above. For example, suppose that $Y_1,\ldots,Y_n$ are observations from a Markov process. It may then be of interest to estimate $E[Y_{n+1}|Y_n=y_n]$. Estimator (52) can be used for this purpose as well. A discussion of how this is done is given by Yakowitz (1985) (see also, Rosenblatt, 1970, and Roussas, 1969, for related work). Consistency of estimator (52) for $\mu$ when the $(t_j,y_j)$ are a $\phi$-mixing process has been shown by Bierens (1983).

The reader interested in further discussion and references concerning estimation of the regression function with random t's may wish to refer to the review paper by Collomb (1981) on this subject.

### 4.11 Robust Smoothers

Sacks and Ylvisaker (1978) identified two kinds of robustness for an estimator: model and distributional robustness. Model robustness implies that an estimator can adapt well to departures from an assumed model. In practice, distributional robustness often translates to insensitivity of the estimator to a few bad data values or outliers. Kernel estimators can be viewed as being model robust in the sense that they are capable of fitting rather general regression curves. However, they are not resistant to the effects of outliers.

Since estimator (6d) solves problem(1.4) with $K(t,t_j;\lambda) = \lambda^{-1}\int_{s_{j-1}}^{s_j} K(\lambda^{-1}(t - s))ds$, provided $t \in [\lambda, 1 - \lambda]$, it is clear that the estimator will be distorted by an outlier near t. This is due to the well known tendency of least-squares estimators to be dominated by outliers. The same kind of qualitative conclusions apply to all the estimators discussed in Section 4.2. Thus, kernel estimators are not robust in the distributional sense.

A class of estimators which includes some robust alternatives to the estimators in Section 4.2 is provided by M-type kernel estimators. These estimators owe their origin to Stuetzle and Mittal (1979) and are generalization of the M-estimators for location parameters proposed by Huber (1964, 1981). To define these estimators, assume that $\rho$ is some nonnegative symmetric function. Then, $\mu_\lambda(t)$ is defined as any minimizer of

$$\sum_{j=1}^{n} \left(\int_{s_{j-1}}^{s_j} \frac{1}{\lambda} K(\lambda^{-1}(t - s))ds\right)\rho(y_j - \theta) \tag{53}$$

with respect to $\theta$. If $\rho$ is differentiable with derivative $\rho' = \psi$ this is equivalent to looking for solutions to

$$\sum_{j=1}^{n} \left(\int_{s_{j-1}}^{s_j} \frac{1}{\lambda} K(\lambda^{-1}(t - s))ds\right) \psi(y_j - \theta) = 0 . \tag{54}$$

The estimator of $\mu(t)$ derived from (53) is called an M-type kernel estimator. There is, of course, a parallel of (53) for estimator (6c) as well.

Notice than when $\rho(u) = u^2$ in (53) this gives estimator (6d) again. Thus, M-type kernel estimators are generalizations of the kernel estimator concept.

Choices for $\rho$ that are more robust than $\rho(u) = u^2$ include $\rho(u) = |u|$ and

$$\rho(u) = \begin{cases} u^2/2 , & |u| \le c , \\ c|u| - c^2/2 , & |u| > c, \end{cases} \tag{55}$$

for some $c > 0$. While the absolute value function is not differentiable at zero, (55) is with

$$\psi(u) = \begin{cases} -c, & u < -c , \\ u, & -c \le u \le c , \\ c, & u > c . \end{cases} \tag{56}$$

The use of $\rho(u) = |u|$ leads to an estimator that is similar in nature to a median. This is because the median of n values

$x_1, \ldots, x_n$ is a minimizer of

$$\sum_{i=1}^{n} |x_i - \theta| .$$

The function (55) can be viewed as a combination of $\rho(u) = u^2$ and $\rho(u) = |u|$ where the influence of an extreme observation is ameliorated by the use of absolute values for $|u|$ greater than some point c.

The case of $\rho(u) = |u|$ is indicative of the types of problems posed by the use of M-type kernel estimators. For $\psi$ in this case we can use

$$\psi(u) = \begin{cases} -1 & , \ u < 0 \ , \\ 0 & , \ u = 0 \ , \\ 1 & , \ u > 0 \ . \end{cases} \tag{57}$$

Then, since (54) depends only on the signs of $y_j - \theta$ there will typically be an infinite number of solutions to (54). This is the same problem that arises with the sample median. For other $\psi$ functions there may be only one solution to (54). However, it will generally not be expressable in a closed form and iterative techniques will be required for its computation. The reader may wish to refer to Huber (1981) for discussions concerning the computation of M-estimators which will also have applications to computation of M-type kernel estimators.

Despite the fact that M-type kernel estimators cannot be defined explicitly, many of their properties are known. They are (pointwise) consistent estimators of $\mu$ whether the t's are fixed or random provided the $\psi$ function is monotone and bounded and certain other conditions are met. They are also asymptotically normal. For further details see Härdle (1984b), Härdle and Gasser (1984, 1985) and Härdle and Luckhaus (1984). The extension of M-type kernel estimators to the problem of derivative estimation is considered in Härdle and Gasser (1985).

An example of the types of gains that can be realized from the use of robustified kernel estimators is provided by Härdle and Gasser (1984). They consider the use of an M-type kernel estimator for analysis of Raman spectra in physical chemistry. An estimator is needed to aid in identifying the location and size of peaks and troughs of spectral bands. The difficulty with conventional estimators in this setting is that huge spikes can be added to the relatively small instrumental noise by vibrations of the experimental installations or simply switch-on operations. M-type estimators provide a means of protecting against the distortion of an estimator by these spikes. An explicit example of the efficacy of robust procedures for this type of data can be found in Härdle and Gasser (1984).

## Exercises

1. Show that Clark's estimator (estimator 6e) can be written in the form (1.3) with weights $K_j(t) = K(t,t_j;\lambda)$ defined by

$$K_1(t) = \int_a^{t_1} \frac{1}{\lambda} K(\lambda^{-1}(t - z))dz + \int_{t_1}^{t_2} (\frac{t_2 - z}{t_2 - t_1}) \frac{1}{\lambda} K(\lambda^{-1}(t - z))dz,$$

$$K_i(t) = \int_{t_{i-1}}^{t_i} (\frac{z - t_{i-1}}{t_i - t_{i-1}}) \frac{1}{\lambda} K(\lambda^{-1}(t - z))$$

$$+ \int_{t_i}^{t_{i+1}} \left(\frac{t_{i+1} - z}{t_{i+1} - t_i}\right) \frac{1}{\lambda} K(\lambda^{-1}(t - z))dz \ ,$$

$$i = 2,\ldots,n - 1 \ ,$$

$$K_n(t) = \int_{t_{n-1}}^{t_n} (\frac{z - t_{n-1}}{t_n - t_{n-1}}) \frac{1}{\lambda} K(\lambda^{-1}(t - z))dz$$

$$+ \int_{t_n}^{b} \frac{1}{\lambda} K(\lambda^{-1}(t - z))dz \ .$$

Use this to verify that $\Sigma K_j(t) = 1$ provided $[(t - t_n)/\lambda, (t - a)/\lambda] \subset [-1,1]$.

2. A large literature has evolved in statistics concerning parameter estimates which can be written as statistical functionals. These are estimators which have the form $G(F_n)$, where $F_n$ is the empirical distribution function of the data and $G$ is a function mapping the set of distribution functions onto the line.

For example, given a bivariate random sample $(t_1,y_1),\ldots,(t_n,y_n)$ the empirical distribution function is $F_n(t,y) = n^{-1}$(no. of points $(t_j,y_j)$ with $t_j \le t$ and $y_j \le y$). One functional of $F_n$ in this case is the sample correlation coefficient

$$r = G(F_n) = \frac{1}{n}\sum_{j=1}^{n}(t_j - \bar{t})(y_j - \bar{y})\Big/\left[\frac{1}{n}\sum_{j=1}^{n}(t_j - \bar{t})^2\,\frac{1}{n}\sum_{j=1}^{n}(y_j - \bar{y})^2\right]^{1/2}$$

$$= \iint(t - \bar{t})(y - \bar{y})dF_n(t,y)\Big/ \left[\iint(t - \bar{t})^2 dF_n(t,y)\iint(y - \bar{y})^2 dF_n(t,y)\right]^{1/2},$$

where $\bar{t}_n = \iint t\,dF_n(t,y)$ and $\bar{y}_n = \iint y\,dF_n(t,y)$. For a general bivariate d.f. $F$ the functional is defined by

$$\rho = G(F) = \iint(t - \mu_T)(y - \mu_Y)dF(t,y)\Big/ \left(\iint(t - \mu_T)^2 dF(t,y)\iint(y - \mu_Y)^2 dF(t,y)\right)^{1/2},$$

with $\mu_T = \iint t\,dF(t,y)$ and $\mu_Y = \iint y\,dF(t,y)$, and is well defined for any distribution having finite second moments.

The influence curve (IC) for a functional is useful for both robustness and asymptotic distribution theory considerations. The IC is defined by

$$IC_F(t_0,y_0) = \frac{d}{d\alpha}G((1-\alpha)F + \alpha\delta_{(t_0,y_0)})\Big|_{\alpha=0},$$

where $\delta_{(t_0,y_0)}$ is the distribution function which places point mass one at $(t_0,y_0)$. For robustness purposes it is desirable to have the IC both bounded and continuous as a function of $(t_0,y_0)$. If it is unbounded this essentially means outliers can have an unbounded influence on the estimator. A discontinuous IC indicates the estimator can be adversely influenced by rounding and grouping errors.

Let $\mu_\lambda(x,\psi)$ be an M-type kernel estimator of $\mu$ at x for some particular $\psi$ function. The functional corresponding to this estimator can be defined in general as the solution to

$$\iint K(\lambda^{-1}(x - t))\psi(y - G_F(x))dF(t,y) = 0 .$$

a) Verify that when $F=F_n$, $G_{F_n}(x)$ is the M-type kernel estimator corresponding to estimator (6c).

b) Derive the IC for $G_F(x)$ and discuss its properties. How should $\psi$ be chosen to insure boundedness and continuity? Discuss the properties of the IC for $\psi$ functions $\psi(u) = u$, and (56)-(57). In particular, what conclusions can be drawn concerning the non-robustified estimator (6c)?

[See, e.g., Huber (1981) or Randles and Wolfe (1979) for discussions of M-estimators and influence curves.]

**3.** Suppose that $\mu(t) = \theta_0 + \theta_1 t$. Show that, for example, estimator (6d) is nearly unbiased in this case. Under what conditions is the estimator actually unbiased for a linear regression function with finite n? Further show that Clark's estimator (6e) will actually be unbiased for a linear regression function.

**4.** Suppose that $\mu(t) = \Sigma_{j=1}^m \theta_j t^{j-1}$. What conditions are needed for estimator (6d) to be nearly unbiased for $\mu(t)$ in this case?

**5.** Let $x_1,\ldots,x_n$ be a random sample from a distribution with twice continuously differentiable density f. Define

$$f_\lambda(t) = \frac{1}{n\lambda}\sum_{j=1}^{n} K(\lambda^{-1}(t - x_j)) ,$$

where K is a symmetric probability density on [-1,1] which satisfies

$$\int_{-1}^{1} K(u)^2 du < \infty$$

and

$$\int_{-1}^{1} u^2 K(u) du = \alpha \neq 0 \ .$$

Show that if $n \to \infty$, $\lambda \to 0$ and $n\lambda \to \infty$ then

$$f(t) - Ef_\lambda(t) \sim \lambda^2 f''(t)\alpha/2$$

and

$$E(f_\lambda(t) - Ef_\lambda(t))^2 \sim \frac{f(t)}{n\lambda} \int_{-1}^{1} K(u)^2 du \quad .$$

Use these expressions to obtain an asymptotic formula for the risk $R_n(t;\lambda) = E(f_\lambda(t) - f(t))^2$. Derive the asymptotically optimal bandwidth and risk.

[Rosenblatt, 1956, Parzen, 1962b. For discussion of the integrated risk see, e.g., Rosenblatt, 1971, or Prakasa Rao, 1983.]

**6.** Give conditions under which the kernel density estimator $f_\lambda$ in Exercise 5 will have an asymptotic normal distribution.

[Parzen, 1962b].

**7.** Consider the case where K is a symmetric kernel with support on the entire line, e.g., $K(u) = e^{-u^2/2}/(2\pi)^{1/2}$, $-\infty < u < \infty$. Show that there exists no $\lambda > 0$ such that, for estimator (6d), $E\mu_\lambda(t) - \mu(t) = \lambda^2 C_t + o(\lambda^2)$, where $C_t$ is a constant depending on $\mu$,t and K. What does this suggest about the difference between kernels with finite and nonfinite support?

**8.** Consider estimator (6d) under the conditions of Section 4.3. Assume that $\mu$ and K are continuously differentiable, the $\varepsilon_j$ are independent and identically distributed with finite fourth moments and that $n \to \infty$, $\lambda = \lambda(n) \to 0$ in such a way that

$$\sum_{n=1}^{\infty} \frac{1}{(n\lambda(n))^2} < \infty \quad .$$

Show that $\mu_\lambda(t) \to \mu(t)$ almost surely. [Hint: First show that if

$\{Z_j\}_{j=1}^n$ is a collection of zero mean identically distributed random variables and $\{a_j\}_{j=1}^n$ is a set of constants then

$$E\left(\sum_{j=1}^n Z_j a_j\right)^4 < EZ_1^4\left(\sum_{j=1}^n a_j^2\right)^2 .$$

(Benedetti, 1977). Then employ the Borel-Cantelli Lemma.]

**9.** A minimum variance kernel of even order m must minimize $\int_{-1}^{1} K(u)^2 du$ subject to

$$\int_{-1}^{1} K(u)du = 1 ,$$

$$\int_{-1}^{1} u^j K(u)du = 0, \quad j=1,\ldots,m-1 ,$$

$$\int_{-1}^{1} u^m K(u)du \neq 0 .$$

Show that such minimum variance kernels are symmetric polynomials of order m − 1 with nonzero coefficients uniquely determined by

$$\begin{bmatrix} 1 & \frac{1}{3} & \cdots & \frac{1}{m-1} \\ \frac{1}{3} & \frac{1}{5} & \cdots & \frac{1}{m+1} \\ \vdots & \vdots & & \vdots \\ \frac{1}{m-1} & \frac{1}{m+1} & \cdots & \frac{1}{2m+3} \end{bmatrix} \begin{bmatrix} \theta_0 \\ \theta_2 \\ \vdots \\ \theta_{m-2} \end{bmatrix} = \begin{bmatrix} 1 \\ 0 \\ \vdots \\ 0 \end{bmatrix} .$$

Verify that $3(3 - 5u^2)/8$ is the minimum variance kernel of order 4.

[Gasser and Müller, 1979, Müller, 1984a].

**10.** A minimum variance (lower) boundary kernel of order m must minimize $\int_{-1}^{q} K(u)^2 du$ subject to

$$\int_{-1}^{q} K(u)du = 1 ,$$

and

$$\int_{-1}^{q} u^j K(u)du = 0 , \quad j = 1,\ldots,m-1 .$$

Show that such kernels are polynomials of order m with coefficients uniquely determined by

$$\begin{bmatrix} 1+q & \frac{1}{2}(1-q^2) & \cdots & \frac{1}{m}(1-(-q)^m) \\ \frac{1}{2}(1-q^2) & \frac{1}{3}(1+q^3) & \cdots & \frac{1}{m+1}(1-(-q)^{m+1}) \\ \vdots & \vdots & & \vdots \\ \frac{1}{m}(1-(-q)^m) & \frac{1}{m+1}(1-(-q)^{m+1}) & \cdots & \frac{1}{2m-1}(1-(-q)^{2m-1}) \end{bmatrix} \begin{bmatrix} \theta_0 \\ \theta_1 \\ \vdots \\ \theta_{m-1} \end{bmatrix} = \begin{bmatrix} 1 \\ 0 \\ \vdots \\ 0 \end{bmatrix}$$

Verify that (28) is the minimum variance boundary kernel of order 2.

[Gasser and Müller, 1979]

**11.** Consider the estimator $\mu_{\lambda d}(t)$ defined in (45) for the dth derivative of $\mu$ at t. Show that if $\mu$ is in $C^m[0,1]$ $(m > d)$, (12)-(13) hold, $n\to\infty$, $\lambda\to 0$ and $n\lambda \to \infty$, then

$$E(\mu_{\lambda d}(t) - E\mu_{\lambda d}(t))^2 \sim \frac{\sigma^2}{n\lambda^{2d+1}} \int_{-1}^{1} u^m K(u)du$$

and

$$E\mu_{\lambda d}(t) - \mu^{(d)}(t) \sim \lambda^{m-d}\mu^{(m)}(t)\int_{-1}^{1} u^m K(u)du/m! \quad .$$

Use these expressions to obtain an asymptotic form for the risk $E(\mu_{\lambda d}(t) - \mu^{(d)}(t))^2$. Find the asymptotically optimal bandwidth and risk.

[Georgiev, 1984, Müller, 1984b].

**12.** One method for reducing the bias of a kernel estimator is the following:

i) Compute the residuals $e_j(\lambda) = y_j - \mu_\lambda(t_j)$.

ii) Then compute

$$e_\lambda(t) = \frac{1}{n\lambda}\sum_{j=1}^{n} K(\lambda^{-1}(t - t_j))e_j(\lambda).$$

iii) Now correct the original estimator,

$$\mu_\lambda(t) = \frac{1}{n\lambda}\sum_{j=1}^{n} K(\lambda^{-1}(t - t_j))y_j \quad ,$$

using $e_\lambda(t)$ to obtain

$$\tilde{\mu}_\lambda(t) = \mu_\lambda(t) + e_\lambda(t) .$$

This process if called *twicing.*

Show that, apart from boundary effects, twicing amounts to use of the kernel

$$K_2(u) = 2K(u) - \int_0^1 K(u - s)K(s)ds.$$

Verify that if K is a kernel of order m then

$$\int_{-1}^1 u^j K_2(u)du = 0 , \quad j=1,\ldots,2m - 1,$$

$\int_{-1}^1 K_2(u)du = 1$ and $\int_{-1}^1 u^{2m} K_2(u) \neq 0$. In light of this result discuss the efficacy of twicing compared to the use of higher order kernels.

[Stuetzle and Mittal, 1979].

**13.** Another approach to the removal of boundary bias can be obtained through the use of a generalized jackknife estimator for $\mu(t)$. This estimator will have the form

$$\tilde{\mu}_\lambda(t) = \mu_\lambda(t) + \beta[\mu_\lambda(t) - \mu_{\alpha\lambda}(t)]$$

with $\alpha > 0$ and $\beta$ selected appropriately.

Alter estimator (6d) to where it makes sense at the boundary by defining

$$\mu_\lambda(t) = \frac{1}{\lambda} \sum_{j=1}^n y_j \int_{s_{j-1}}^{s_j} K(\lambda^{-1}(t - s))ds / \frac{1}{\lambda} \sum_{j=1}^n \int_{s_{j-1}}^{s_j} K(\lambda^{-1}(t - s))ds.$$

If $t = q\lambda$, $0 < q < 1$, show that as $n\to\infty$, $\lambda\to 0$, $n\lambda\to\infty$,

$$E\mu_\lambda(t) - \mu(t) \sim -\lambda\mu'(t)M_1(q)/M_0(q) + \frac{\lambda^2}{2}\mu''(t)M_2(q)/M_0(q)$$

with $M_j(q) = \int_{-1}^q u^j K(u)du, j=0,1,2.$ Now show that the choice

$$\beta = A(q)/[\alpha A(q/\alpha) - A(q)]$$

with

$$A(q) = M_1(q)/M_0(q)$$

results in the estimator $\tilde{\mu}_\lambda(t)$ having bias whose leading term in $\lambda$ is of the order $\lambda^2$.

[Rice, 1984a; See also Schucany and Sommers, 1977]

**14.** Show that $\tilde{\mu}_\lambda(t)$ in Exercise 13 is also a kernel estimator and derive the form of its kernel. A natural way to choose $\alpha$ is to select it so that the estimator is an average over an interval of width $2\lambda$. This corresponds to taking $\alpha = 2 - q$. Draw graphs of the kernel for $\tilde{\mu}_\lambda(t)$ under this choice of $\alpha$ and for several choices of q.

[Rice, 1984a]

**15.** Prove equation (36). [Hint: Observe that

$$s_j - s_{j-1} = \frac{t_{j+1} - t_{j-1}}{2}$$

$$= \{P^{-1}(\tfrac{j+1}{n}) - P^{-1}(\tfrac{j-1}{n})\}/2$$

$$= \frac{1}{np(\xi_j)} + O(n^{-2})$$

for $P(s) = \int_0^s p(t)dt$ and $j - 1 \le n\xi_j \le j + 1$.]

**16.** Compute the optimal bandwidth and optimal risk (30)-(31) for the quadratic kernel when

$$\mu(x) = \begin{cases} \alpha x^2 + \beta x + \gamma, \\ \frac{\alpha}{6}x^3 + \frac{\beta}{2}x^2 + \gamma x + \delta, \\ \frac{1}{\alpha^2}e^{-\alpha x} + \frac{1}{\beta}e^{\beta x}. \end{cases}$$

**17.** Since kernel estimators are biased the risk provides a better measure of performance than the estimator's variance. Thus, it might be preferable to "standardize" $\mu_\lambda(t)$ using

$$\frac{\mu_\lambda(t) - \mu(t)}{R_n(\lambda)^{1/2}}$$

with $R_n(\lambda) = n^{-1}\Sigma_{j=1}^{n} E(\mu_\lambda(t_j) - \mu(t_j))^2$. Explore the asymptotic distribution of this quantity for estimator (6d). What happens if you assume that $n\to\infty$, $\lambda\to 0$, $n\lambda\to\infty$ and $n\lambda^5\to 0$? What happens if you take $\lambda=\lambda_{opt}$ from (31)? Discuss the problems that occur for the latter choice of $\lambda$ in terms of confidence interval construction and compare them to the problems which arise with intervals (33) for this same optimal $\lambda$. Which would you prefer to use with optimal $\lambda$, intervals (33) or

$$\mu_\lambda(t) \pm Z_{\alpha/2} R_n(\lambda_{opt})^{1/2} ,$$

if such a choice were possible.

**18.** A heuristic approach to interval estimation for estimator (6d) might borrow from results in regression analysis and use

$$\mu_\lambda(t_j) \pm Z_{\alpha/2}\sigma\{\frac{1}{\lambda}\int_{s_{j-1}}^{s_j} K(\lambda^{-1}(t_j - s))ds\}^{1/2}$$

to estimate $\mu(t_j)$. Under the conditions of Section 4.3 the integral can be approximated by $K(0)/n\lambda$. Establish that when $\lambda = \lambda_{opt}$ and K is the quadratic kernel, $\sigma^2 K(0)/n\lambda_{opt} \sim R_n(\lambda_{opt})$. In view of Problem 17, what might this suggest for interval estimates based on optimal bandwidth choices?

**19.** Among estimators (6a) - (6e), which are invariant under a linear transformation of the data? More precisely, if $\tilde{y}_i = a + by_i$ and $\tilde{\mu}_\lambda$ is the estimator for the $\tilde{y}_i$, which of the estimators satisfy $\tilde{\mu}_\lambda = a + b\mu_\lambda$? Why might such a property be desirable?

**20.** Consider Clark's estimator (6e) with the triangular kernel. Show that the resulting estimator is a cubic polynomial over any interval with endpoints $t_j$ and $t_j \pm \lambda$. Establish that the function has two continuous derivatives with discontinuities in its third derivative at points $t_j$, $t_j \pm \lambda$. This shows that the Clark estimator is a cubic spline.

[Clark, 1977].

**21.** Suppose that $\underline{t} = (t_1,t_2)'$, i.e., $\underline{t}$ is bivariate. Let K satisfy (4a) - (4d) and define the estimator

$$\mu_\lambda(\underline{t}) = \frac{1}{n\lambda^2}\sum_{j=1}^{n} y_j K(\lambda^{-1}(t_1 - t_{j1}))K(\lambda^{-1}(t_2 - t_{j2}))$$

for the regression function. Give a formal asymptotic expression for $R_n(\underline{t};\lambda) = E(\mu(\underline{t}) - \mu_\lambda(\underline{t}))^2$ for fixed $\underline{t}$. Find the resulting asymptotically optimal bandwidth and risk.

**22.** Show that under the conditions in Section 4.3 and, assuming appropriate modifications at the boundary,

$$\sigma_\lambda^2 = \sum_{j=1}^{n} (\frac{1}{\lambda}\int_{s_{j-1}}^{s_j} K(\lambda^{-1}(t_j - s)\ )ds)(y_j - \mu_\lambda(t_j))^2$$

is a consistent estimator of $\sigma^2$.

[Härdle and Gasser, 1984]

**23.** Suppose we observe data from the model

$$y_{ij} = \mu(t_i) + \varepsilon_{ij}, \quad i = 1,\ldots,n, \quad j = 1,\ldots,n_i\ .$$

Discuss how one can construct a parallel of the kernel estimator for this case.

[Hint: Use property (1.4).]

**24.** The data sets below represent the growth data for the boy and girl used to obtain Figure 4.7. The reported measurements are standing heights in millimeters.

For these data

a) Fit both data sets using a kernel estimator with

$$K(u) = 15(7u^4 - 10u^2 + 3)/32$$

and bandwidth $\lambda = 3.0$. Renormalize the estimators at the boundary region so that the weights sum to one. Compare the growth curve estimates for the boy and girl.

b) Estimate the acceleration curve $\mu''$ for the boy and girl using results from Section 4.8 with $d = 2$ ,

$$K(u) = 315(77u^6 - 135u^4 + 63u^2 - 5)$$

and $\lambda = 3.8$. Again use these to compare growth patterns for the boy and girl. It may help to refer back to the discussion in Section 4.8 concerning the estimation of the velocity curve for these data and, also, the discussion in Section 4.9 concerning growth curve analysis.

c) Which seems to provide the most information about growth in boys and girls: their growth or acceleration curves? Explain your choice.

| **Age** | **Boy** | **Girl** | **Age** | **Boy** | **Girl** |
|---|---|---|---|---|---|
| 1 month | 534 | 525 | 13.0 years | 1627 | 1552 |
| 3 months | 601 | 584 | 13.5 years | 1669 | 1584 |
| 6 months | 685 | 652 | 14.0 years | 1710 | 1600 |
| 9 months | 729 | 696 | 14.5 years | 1762 | 1607 |
| 1 year | 762 | 736 | 15.0 years | 1809 | 1620 |
| 1.5 years | 822 | 798 | 15.5 years | 1845 | 1626 |
| 2.0 years | 885 | 843 | 16.0 years | 1868 | 1631 |
| 3.0 years | 984 | 930 | 16.5 years | 1883 | 1632 |
| 4.0 years | 1064 | 1014 | 17.0 years | 1894 | 1636 |
| 5.0 years | 1144 | 1085 | 17.5 years | 1900 | 1639 |
| 6.0 years | 1205 | 1135 | 18.0 years | 1909 | 1639 |
| 7.0 years | 1273 | 1200 | 18.5 years | 1913 | —[a] |
| 8.0 years | 1346 | 1266 | 19.0 years | 1913 | 1639 |
| 9.0 years | 1406 | 1316 | 20.0 years | 1915 | 1639 |
| 9.5 years | —[a] | 1344 | | | |
| 10.0 years | 1461 | 1365 | | | |
| 10.5 years | —[a] | 1387 | | | |
| 11.0 years | 1511 | 1423 | | | |
| 11.5 years | 1540 | 1452 | | | |
| 12.0 years | 1569 | 1493 | | | |
| 12.5 years | 1599 | 1524 | | | |

**Source: Data kindly provided by Dr. L. Molinari.**

[a]**Height readings were not available**

**25.** The following data was simulated from the model

$$y_i = \mu(t_i) + \varepsilon_i, \quad i = 1,\ldots,50,$$

with $t_i = (i - 1)/49$ ,

$$\mu(t) = 4.26[\exp(-3.25t) - 4\exp(-6.5t) + 3\exp(-9.75t)]$$

and normal, independent errors with $\sigma^2 = .0025$.

| i | $y_i$ | i | $y_i$ |
|---|---|---|---|
| 1 | -.12485 | 26 | .37580 |
| 2 | -.49933 | 27 | .29548 |
| 3 | -.75620 | 28 | .32839 |
| 4 | -.85224 | 29 | .33951 |
| 5 | -.95771 | 30 | .32649 |
| 6 | -.97225 | 31 | .34362 |
| 7 | -1.03010 | 32 | .25369 |
| 8 | - .89801 | 33 | .30101 |
| 9 | - .82403 | 34 | .26411 |
| 10 | - .71138 | 35 | .22756 |
| 11 | - .53767 | 36 | .25173 |
| 12 | - .48401 | 37 | .20410 |
| 13 | - .42717 | 38 | .19765 |
| 14 | - .23142 | 39 | .20465 |
| 15 | - .17717 | 40 | .17002 |
| 16 | - .03395 | 41 | .15311 |
| 17 | .02761 | 42 | .23946 |
| 18 | - .00263 | 43 | .18240 |
| 19 | .05597 | 44 | .18709 |
| 20 | .11243 | 45 | .21141 |
| 21 | .20459 | 46 | .11498 |
| 22 | .21536 | 47 | .05041 |
| 23 | .25541 | 48 | .16087 |
| 24 | .21113 | 49 | .20821 |
| 25 | .17790 | 50 | .07203 |

a) Fit this data by a kernel estimator using the uniform kernel and GCV to select the bandwidth (be sure to adjust your estimator for boundary effects). Compare the fitted curve to the true curve.

b) Compute confidence intervals for $\mu$ at the $t_i$, possibly using a smaller bandwidth than the one obtained from GCV and the estimator $\hat{\sigma}^2$ of $\sigma^2$ discuss in Section 4.6. Discuss the coverage performance of these estimates and the behavior of $\hat{\sigma}^2$ as an estimator of $\sigma^2$.

c) Another possible estimator of $\sigma^2$ is (Rice, 1986)

$$\hat{\sigma}_R^2 = \sum_{j=1}^{n-1} (y_{j+1} - y_j)^2 / 2(n - 1)$$

Compute this estimator and compare it to $\hat{\sigma}^2$ computed in Part b).

# 5
# Smoothing Splines

## 5.1 Introduction

The topic of the next two chapters is smoothing splines. These are the estimators of $\mu$ discussed in Chapter 1 which derive from the penalized least-squares criterion (1.7). We will find that these estimators are closely related to classical Fourier series, polynomial regression and kernel estimators and, consequently, share many of their properties.

To motivate the smoothing spline concept, suppose that $\mu \in W_2^m[a,b]$. To estimate $\mu$ we want an estimator which fits the data well but, at the same time, has some degree of smoothness. The natural measure of smoothness associated with a function $f \in W_2^m[a,b]$ is $\int_a^b f^{(m)}(t)^2dt$ while a standard measure of goodness-of-fit to the data is $n^{-1}\Sigma_{j=1}^n(y_j - f(t_j))^2$. Thus, one method of combining these two criterion functions to obtain an overall performance measure is to assess the quality of f by its values for

$$(1 - q)n^{-1}\sum_{j=1}^{n}(y_j - f(t_j))^2 + q\int_a^b f^{(m)}(t)^2dt$$

for some $0 < q < 1$. An optimal estimator could then be obtained by minimizing this functional over $f \in W_2^m[a,b]$. However, by setting $\lambda = q/(1 - q)$ we see that this is equivalent to estimating $\mu$ by the function $\mu_\lambda$ which minimizes

$$n^{-1}\sum_{j=1}^{n}(y_j - f(t_j))^2 + \lambda\int_a^b f^{(m)}(t)^2dt\ , \quad \lambda > 0\ , \quad (1)$$

over $f \in W_2^m[a,b]$. The minimizing function, $\mu_\lambda$, is the smoothing spline estimator of the regression function to be studied in this chapter. In view of the above discussion this estimator is seen to derive from a rather natural approach to estimation which incorporates our desires for smoothness and goodness-of-fit directly into the estimation process.

The parameter $\lambda$ $(= q/(1 - q))$ in (1) governs the tradeoff between smoothness and goodness-of-fit. When $\lambda$ is large (q near one) this places a premium on smoothness and heavily penalizes estimators with large mth derivatives. Conversely, a small value of $\lambda$ (q near 0) corresponds to more emphasis on goodness-of-fit. These properties have prompted $\lambda$ to be called the *smoothing parameter*.

Smoothing splines owe their origin to Whittaker (1923) whose work on methods of graduating data stimulated Schoenberg (1964) to derive the smoothing spline estimator. Later, Reinsch (1967) gave an independent derivation for the case of $m = 2$.

It appears that smoothing splines were generally regarded as numerical analysis tools until extensive research, pioneered by Grace Wahba, showed they had useful statistical properties and deserved consideration as a method for performing nonparametric regression analysis. It has now become clear that smoothing splines, and their variants, provide extremely flexible data analysis tools. As a result of this, they are becoming more and more popular in data analysis and have found applications in such diverse areas as analysis of growth data, medicine, remote sensing experiments and economics. For a review of various topics concerning spline smoothing and its applications see Wegman and Wright (1983) and Eubank (1984b).

We begin our study of smoothing splines in the next section by discussing various motivations for these estimators. The topics of existence, computation, etc. for the minimizer of (1) are delayed until Section 5.3. Then, in Section 5.4, we examine the problem of selecting a value for the smoothing parameter. Smoothing splines are also generalized Bayes estimators. This fact is established in Section 5.5 and utilized to motivate one approach to constructing interval estimates and conducting diagnostic analysis. Further study of smoothing splines, including asymptotic analysis, is undertaken in Chapter 6.

### 5.2 Smoothing Splines and Polynomial Regression

The objective of this section is to provide motivation for the use of the smoothing spline estimator of $\mu$ derived from the estimation criterion (1). Thus we will postpone, for the moment, questions concerning existence, uniqueness, etc. of the estimator, and, throughout the section, tacitly assume that $\mu_\lambda$ is well defined.

Consider again the by now familiar regression model

$$y_j = \mu(t_j) + \varepsilon_j , \qquad j = 1,\ldots,n, \tag{2}$$

where $a \le t_1 < \ldots < t_n \le b$ and the $\varepsilon_j$ are zero mean, uncorrelated random errors with common variance $\sigma^2$. Assume that $\mu$ is smooth in the sense that $\mu \in W_2^m[a,b]$.

It will be helpful to rewrite model (2) in an alternative form. Since $\mu \in W_2^m[a,b]$ it follows from Taylor's Theorem (Section 3.5) that there are constants $\theta_0,\ldots,\theta_{m-1}$ such that

$$\mu(t) = \sum_{i=0}^{m-1} \theta_i t^i + \mathrm{Rem}(t)$$

with

$$\mathrm{Rem}(t) = [(m-1)!]^{-1}\int_a^b \mu^{(m)}(x)(t-x)_+^{m-1}dx \quad .$$

Thus (2) is equivalent to

$$y_j = \sum_{i=0}^{m-1} \theta_i t_j^i + \mathrm{Rem}(t_j) + \varepsilon_j , \qquad j = 1,\ldots,n. \tag{3}$$

If the $\mathrm{Rem}(t_j)$ can all be assumed to be small in (3), it would seem reasonable to estimate $\mu$ using an mth order polynomial regression estimator. However, if such an assumption cannot be made, this raises questions about the applicability of the polynomial regression approach.

We will now show that criterion (1) can be viewed as corresponding to an estimation strategy which allows for the possibility that some of the $\mathrm{Rem}(t_j)$ may be appreciably different from zero. To see this, first apply the Cauchy-Schwarz inequality to obtain

$$\mathrm{Rem}(t)^2 \le \frac{(b-a)^{2m-1}}{(2m-1)[(m-1)!]^2} J_m(\mu) \quad ,$$

where

$$J_m(\mu) = \int_a^b \mu^{(m)}(t)^2 dt. \tag{4}$$

Thus, there is a constant C, depending on m but not on $\mu$, such that

$$\sup_{t\varepsilon[a,b]} |\mathrm{Rem}(t)| \le C\, J_m(\mu)^{1/2} \quad .$$

In particular $|\mathrm{Rem}(t_j)| \le C\, J_m(\mu)^{1/2}$, $j=1,\ldots,n$. The size of $J_m(\mu)$ therefore has a direct bearing on the possible magnitudes of the remainder terms $\mathrm{Rem}(t_1),\ldots,\mathrm{Rem}(t_n)$. More precisely, a specification such as $J_m(\mu) \le \rho$, for some constant $\rho > 0$, can be viewed as a statement of one's belief about the extent to which polynomial regression is suitable for estimation of $\mu$.

As a particular example consider the case when m=2. In that instance $J_2(\mu)$ vanishes if and only if $\mu$ is linear. A condition such as $J_2(\mu) \le \rho$ then governs how far $\mu$ is allowed to depart from a linear model. Note that $\mu''$ represents the rate of change of the slope (i.e., derivative) of the regression curve. Thus if $J_2(\mu)$ is known to be small this indicates that $\mu$ is almost linear in the sense that its slope does not change rapidly.

Suppose that, perhaps as a result of experience or prior information, it is possible to place a bound on the size of the remainders by assuming that

$$J_m(\mu) \le \rho \tag{5}$$

for some $\rho \ge 0$. A model of the form (3) along with the specification (5) now becomes an approximate linear model in the sense of Sacks and Ylvisaker (1978). (See also Whitney, 1986, for a somewhat more general development.) Tailoring this terminology to our situation suggests the name "approximate polynomial regression model" for (3) with the side condition (5).

If (5) can be assumed to hold, this condition should be integrated into the estimation procedure. One method of accomplishing this would be to estimate $\mu$ by the minimizer of

$$\mathrm{MSE}(f) = n^{-1} \sum_{j=1}^{n} (y_j - f(t_j))^2 \tag{6}$$

over all function $f \in W_2^m[a,b]$ which satisfy (5). By introducing a Lagrange multiplier to incorporate the constraint, we find this is equivalent to minimizing $\mathrm{MSE}(f) + \lambda(J_m(f) - \rho)$. This is essentially criterion (1). The relationship between the estimators obtained from (5)-(6) and (1) is made precise by the following theorem.

**Theorem 5.1 (Schoenberg, 1964).** Assume that $n \geq m$ and let $\mu(\cdot;\rho)$ be the minimizer of (6) in $W_2^m[a,b]$ subject to the constraint (5). Let $\mu_\lambda$ denote the minimizer of (1) in $W_2^m[a,b]$. Then, there is a computable constant $\rho_0$ such that the two families of functions

$\{\mu(\cdot;\rho)\colon 0 \leq \rho \leq \rho_0\}$ and $\{\mu_\lambda(\cdot)\colon 0 \leq \lambda \leq \infty\}$

are identical in the sense that for any value of $\lambda$ there is a unique $\rho$ such that $\mu_\lambda(\cdot) = \mu(\cdot;\rho)$ and conversely. If $\rho \leq \rho_0$, then $J(\mu(\cdot;\rho)) = \rho$.

Theorem 5.1 has the consequence that the solution to the constrained minimization problem posed by (5)-(6) is a smoothing spline estimator of $\mu$ corresponding to some value of $\lambda$. On the other hand, the choice of a particular value for $\lambda$ corresponds to the assumption that $J_m(\mu) \leq \rho_\lambda$ with $\rho_\lambda = J_m(\mu_\lambda)$ and, hence, reflects our beliefs about the magnitude of the $\mathrm{Rem}(t_j)$.

Notice that if $f \in W_2^m[a,b]$ then $J_m(f) = 0$ if and only if f is a polynomial of order m. This means that by specifying $\rho = 0$ minimization of (6) subject to (5) results in a polynomial regression estimator for $\mu$. Thus, we see that smoothing splines are actually an extension of polynomial regression aimed at guarding against the departure of the regression curve from an idealized polynomial regression model.

The constrained minimization problem which stems from (5)-(6) provides an estimation criterion which incorporates our assumption that $J_m(\mu) \le \rho$ into the construction of an estimator for $\mu$. However, it does not, *a priori*, insure the estimator of any optimal statistical properties. An alternative approach which provides such assurances has been formulated by Speckman (1979) and Li (1982).

Suppose that model (3) holds and it is desired to obtain an estimator of $\mu$ at some fixed point $t \in [a,b]$. Let us restrict attention to linear estimators, i.e., estimators which can be written as $\Sigma_{j=1}^{n} \ell_j y_j$ for constants $\ell_1,\ldots,\ell_n$ which do not involve the $y_j$. Given any such estimator of $\mu(t)$ its performance can be assessed by the risk

$$E(\mu(t) - \sum_{j=1}^{n} \ell_j y_j)^2 \quad .$$

When (5) is known to hold the worst possible case occurs if the true regression function is one that gives the bound

$$\sup_{J_m(\mu)\le\rho} E(\mu(t) - \sum_{j=1}^{n} \ell_j y_j)^2 \quad . \tag{7}$$

To protect against this possibility we could choose the $\ell_j$ to minimize (7). The result is a minimax estimator of $\mu(t)$. Concerning this estimator we state the following theorem.

**Theorem 5.2. (Speckman, 1979, Li, 1982).** The minimax estimator of $\mu(t)$ obtained by minimizing (7) is $\mu_\lambda(t)$, where $\mu_\lambda$ is the smoothing spline estimator obtained by minimizing (1) over functions in $W_2^m[a,b]$ with $\lambda = \sigma^2/n\rho$.

The conclusion of Theorem 5.2 is that, given $\lambda$, the minimizer of criterion (1) provides a minimax estimator of $\mu(t)$ for the particular value $\rho = \sigma^2/n\lambda$. Thus smoothing splines provide a type of worst case protection against departures from a polynomial regression model.

Theorems 5.1 and 5.2 provide motivations for the use of smoothing splines by establishing that they are a natural extension of polynomial regression. One might ask, however, if they provide an improvement over polynomial regression in the sense of reduced risk. The answer to this is in the affirmative. It can be shown (Exercise 1) that, under certain restrictions, there always exists a value of the smoothing parameter such that a smoothing spline will have smaller risk than the corresponding mth order polynomial regression estimator.

## 5.3 Derivation and Computation of the Estimator

The conclusion to be drawn from the previous section is that smoothing splines are a natural extension of polynomial regression estimators which provide superior performance from the viewpoint of their estimation risk. This fact alone suffices to recommend their study. Other optimality properties and motivations will also be discovered in subsequent analysis.

The objective of the present section is to derive a closed form expression for the smoothing spline estimator of $\mu$. We will eventually establish that there is a unique minimizer of (1) over $W_2^m[a,b]$. The minimizing function is a special type of function known as a spline. More precisely, the minimizer is what is called a natural spline. This prompts us to first take a brief detour from our stated goal and introduce the concepts of splines and natural splines.

### 5.3.1 Splines and Natural Splines

Piecewise polynomials play an important role in approximation theory and statistics. This is because their segmented nature gives them more flexibility than polynomials and thereby allows them to adjust more effectively to the local characteristics of a function or data. An important type of piecewise polynomial is what is known as a spline. Splines, as defined here, are piecewise polynomials subject to a maximum number of continuity constraints. However, this terminology is not universally accepted and some consider any piecewise polynomial to be a spline. For an in depth account of splines and their properties, which goes far beyond what is given here, the reader is referred to de Boor (1978).

For our purposes we will define a *spline* of order r with *knots* at $\xi_1,\ldots,\xi_k$ to be any function s of the form (cf. (3.51))

$$s(t) = \sum_{i=0}^{r-1} \theta_i t^i + \sum_{i=1}^{k} \delta_i (t - \xi_i)_+^{r-1} \tag{8}$$

for some set of real coefficients $\theta_0,\ldots,\theta_{r-1}$, $\delta_1,\ldots,\delta_k$. A simple calculation (Exercise 2) reveals this definition is equivalent to the specification that

i) s is a piecewise polynomial of order r on any subinterval $[\xi_i,\xi_{i+1})$,

ii) s has r - 2 continuous derivatives and

iii) s has an (r - 1)st derivative that is a step function with jumps at $\xi_1,\ldots,\xi_k$.

Thus, a spline is a piecewise polynomial whose different polynomial segments have been joined (tied) together at the knots $\xi_1,\ldots,\xi_k$ in a fashion which insures certain continuity properties.

Notice that any piecewise polynomial of order r with r - 1, rather than r - 2, continuous derivatives would actually be a single polynomial of order r. This gives the conclusion that a spline is the smoothest possible piecewise polynomial which still retains a segmented nature.

To appreciate the properties of splines it is helpful to draw a few pictures of their basic component: the plus-function $(t - \xi)_+^{r-1}$ (Exercise 4). Figure 5.1 provides pictures of several examples of splines: a linear spline ($r = 2$) with $k = 1$ knot at $\xi_1 = .5$, a quadratic spline ($r = 3$) with $k = 1$ knot at $\xi_1 = .5$ and a cubic spline ($r = 4$) with two knots at $\xi_1 = .33$ and $\xi_2 = .66$. Observe that for $r \geq 3$ the location of the knots is no longer obvious by visual inspection alone.

Let $S^r(\xi_1,\ldots,\xi_k)$ denote the set of all functions of the form (8). $S^r(\xi_1,\ldots,\xi_k)$ is a vector space in the sense that sums of functions in $S^r(\xi_1,\ldots,\xi_k)$ remain in the set, etc. (Exercise 4). Since the functions $1,t,\ldots,t^{r-1}$, $(t - \xi_1)_+^{r-1},\ldots,(t - \xi_k)_+^{r-1}$ are linearly independent (Exercise 4) it follows that $S^r(\xi_1,\ldots,\xi_k)$ has dimension $k + r$.

Of particular importance for this chapter is the set of natural splines of order $r = 2m$ with $k = n$ knots at the $t_j$. A spline function is a *natural spline* of order $2m$ with knots at $t_1,\ldots,t_n$ if, in addition to the properties implied by i) - iii), it satisfies

iv) s is a polynomial of order m outside of $[t_1,t_n]$.

The name natural spline stems from the fact that, as a result of iv), s satisfies the so called natural boundary conditions

$$s^{(j)}(a) = s^{(j)}(b) = 0 , \quad j = m,\ldots,2m - 1. \tag{9}$$

Let $NS^{2m}(t_1,\ldots,t_n)$ denote the collection of all natural splines of order $2m$ with knots at the $t_j$. $NS^{2m}(t_1,\ldots,t_n)$ is a subspace of the vector space $S^{2m}(t_1,\ldots,t_n)$ obtained by placing the $2m$ linear restricions arising from property iv) on the coefficients in (8). In particular, we see that in order for s to be a natural spline we must have

$$\theta_m = \cdots = \theta_{2m - 1} = 0 \tag{10}$$

in (8) since s must be a polynomial of order m for $t < t_1$. One

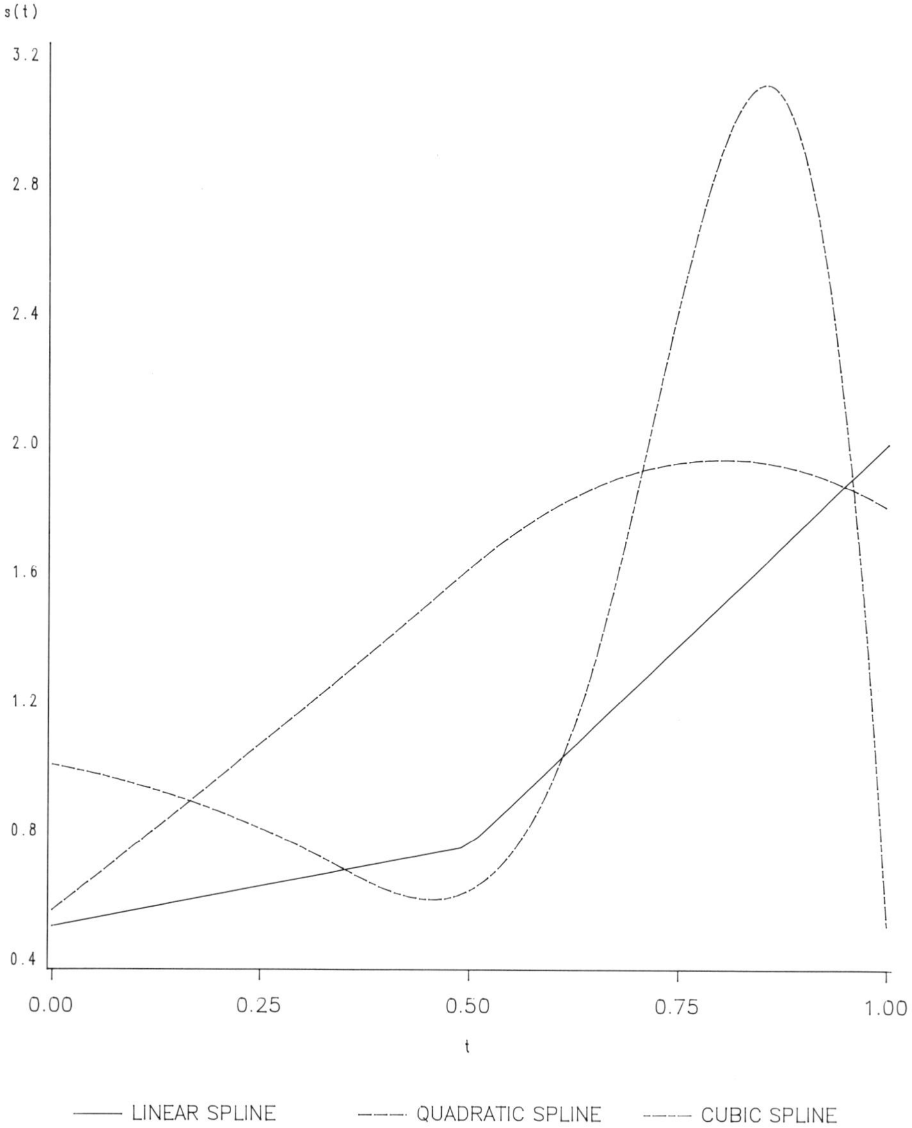


**Figure 5.1** Examples of spline functions.

may verify that $NS^{2m}(t_1,\ldots,t_n)$ has dimension n.

A useful property of natural splines that is required for the next section is provided by the following lemma.

**Lemma 5.1 (Lyche and Schumaker, 1973).** Let $x_1,\ldots,x_n$ be a basis for $NS^{2m}(t_1,\ldots,t_n)$. Then, there are coefficients $\theta_{0j},\ldots,\theta_{m-1,j},\delta_{1j},\ldots,\delta_{nj}$ such that

$$x_j(t) = \sum_{i=0}^{m-1} \theta_{ij} t^i + \sum_{i=1}^{n} \delta_{ij}(t - t_i)_+^{2m-1} \quad . \tag{11}$$

If $s(t) = \Sigma_{j=1}^n \beta_j x_j(t)$ and $f \in W_2^m[a,b]$ then

$$\int_a^b f^{(m)}(t) s^{(m)}(t)dt = (-1)^m (2m - 1)! \sum_{i=1}^{n} f(t_i) \sum_{j=1}^{n} \beta_j \delta_{ij} . \tag{12}$$

*Proof.* The first part of the lemma is merely a restatement of the fact that $NS^{2m}(t_1,\ldots,t_n) \subset S^{2m}(t_1,\ldots,t_n)$ and that (10) holds. Thus we need only verify (12). To accomplish this first observe that $s^{(m+j)}$, $j=0,\ldots,m-1$, vanishes outside of $[t_1,t_n]$ as a result of iv). Then set $t_0 = a$ and $t_{n+1} = b$ and integrate by parts to obtain

$$\int_a^b f^{(m)}(t) s^{(m)}(t)dt = (-1)^{m-1} \int_a^b f'(t) s^{(2m-1)}(t)dt$$

$$= (-1)^{m-1} \sum_{i=0}^{n} \int_{t_i}^{t_{i+1}} f'(t) s^{(2m-1)}(t)dt$$

$$= (-1)^{m-1} \sum_{i=1}^{n-1} \int_{t_i}^{t_{i+1}} f'(t) s^{(2m-1)}(t)dt$$

$$= (-1)^{m-1} \sum_{i=1}^{n-1} \Big\{ [f(t_{i+1}) - f(t_i)]$$

$$\times \Big( \sum_{j=1}^{n} \beta_j (2m - 1)! \sum_{r=1}^{i} \delta_{rj} \Big) \Big\}$$

$$= (-1)^{m-1}(2m-1)! \sum_{j=1}^{n} \left[ \beta_j \{ \sum_{i=2}^{n} f(t_i) \sum_{r=1}^{i-1} \delta_{rj} - \sum_{i=1}^{n-1} f(t_i) \sum_{r=1}^{i} \delta_{rj} \} \right]$$

$$= (-1)^{m}(2m-1)! \sum_{j=1}^{n} \beta_j \sum_{i=1}^{n} f(t_i)\delta_{ij} \quad .$$

In the last step we have used the fact that for $t > t_n$

$$x_j^{(2m-1)}(t) = (2m-1)! \sum_{j=1}^{n} \delta_{ij} = 0$$

as a result of property iv). □

### 5.3.2 Form of the Estimator

We are now prepared to derive a form for the smoothing spline estimator. To avoid clouding the issue it will be useful to first deal with the case where the $t_j$ are distinct, i.e., $a \le t_1 < \dots < t_n \le b$. Also we will assume that a value for $\lambda$ has been selected and therefore treat the smoothing parameter value as fixed. All these restrictions will be relaxed presently.

To motivate the derivation which follows, consider the criterion

$$MSE(f) + \lambda J_m(f)$$

which is to be minimized and let $\mu_\lambda$ be a minimizer in $W_2^m[a,b]$. Then, for any other function $f \in W_2^m[a,b]$ and any constant $\delta$ define

$$\Phi(\mu_\lambda, f; \delta) = (MSE(\mu_\lambda + \delta f) + \lambda J_m(\mu_\lambda + \delta f))/2. \tag{13}$$

Since $\mu_\lambda$ is a minimizer it must be that

$$\Phi(\mu_\lambda, f; 0) \le \Phi(\mu_\lambda, f; \delta)$$

for any $\delta \neq 0$. Clearly (13) is a differentiable function of $\delta$ so it follows that

$$\phi(\mu_\lambda, f) = \left. \frac{d\Phi(\mu_\lambda, f; \delta)}{d\delta} \right|_{\delta=0} = 0 \quad . \tag{14}$$

This is a condition that $\mu_\lambda$ must satisfy if it is a minimizer. To establish Theorem 5.3 below we essentially construct a function which satisfies (14) and then show that it is the unique minimizer of our estimation criterion.

For $f_1, f_2 \in W_2^m[a,b]$ the functional $2\phi(f_1, f_2)$ is called the Gâteaux derivative of $MSE(f) + \lambda J_m(f)$ at $f_1$ in the direction $f_2$. Like an ordinary derivative, the Gâteaux derivative must vanish at a minimum. For discussions of Gâteaux and related functional derivatives see Luenberger (1969) or Tapia and Thompson (1978).

**Theorem 5.3.** Let $x_1, \ldots, x_n$ be a basis for $NS^{2m}(t_1, \ldots, t_n)$ and suppose that $n \geq m$. For fixed $0 < \lambda < \infty$ there is a unique minimizer, $\mu_\lambda$, of (1) in $W_2^m[a,b]$. Moreover, $\mu_\lambda \in NS^{2m}(t_1, \ldots, t_n)$ and therefore has the form $\Sigma_{j=1}^n \beta_{\lambda j} x_j$. The coefficients $\underline{\beta}_\lambda = (\beta_{\lambda 1}, \ldots, \beta_{\lambda n})'$ are the solution to

$$(X + n\lambda G)\underline{\beta}_\lambda = \underline{y} \quad , \tag{15a}$$

where

$$X = \{x_j(t_i)\}_{i,j=1,n} \tag{15b}$$

and

$$G = \{(-1)^m (2m-1)!\,\delta_{ij}\}_{i,j=1,n} \tag{15c}$$

with $\delta_{ij}$ the coefficients defined in (11) for the basis functions $x_1, \ldots, x_n$.

*Proof.* Following the reasoning given at the outset we first note that for any two functions $f_1, f_2 \in W_2^m[a,b]$, $\phi(f_1, f_2)$ in (14) is

$$\phi(f_1, f_2) = -n^{-1} \sum_{i=1}^n f_2(t_i)[y_i - f_1(t_i)] + \lambda \int_a^b f_1^{(m)}(t) f_2^{(m)}(t)dt.$$

Thus, a necessary condition for a function $f_1$ to minimize (1) is that

$$\sum_{i=1}^{n} f_2(t_i)[y_i - f_1(t_i)]/n\lambda = \int_a^b f_1^{(m)}(t) f_2^{(m)}(t)dt$$

for all $f_2 \in W_2^m[a,b]$.

If $f_1$ were a natural spline then we could write $f_1 = \Sigma_{j=1}^n \beta_j x_j$ for some set of coefficients $\beta_1,\ldots,\beta_n$. Using Lemma 5.1 our necessary condition then translates to

$$\sum_{i=1}^{n} f_2(t_i)[y_i - \sum_{j=1}^{n} \beta_j x_j(t_i)]/n\lambda = (-1)^m(2m-1)! \sum_{i=1}^{n} f_2(t_i) \sum_{j=1}^{n} \beta_j \delta_{ij}.$$

Since this must hold for all functions $f_2$ it follows that the condition is equivalent to the $\beta_j$ satisfying

$$\sum_{j=1}^{n} \{x_j(t_i) + n\lambda(-1)^m(2m-1)!\delta_{ij}\}\beta_j = y_i$$

which is recognized as equation (15a). If we can establish that (15) has a unique solution we will then have shown that there is a natural spline which satisfies the necessary condition (14) for minimizing (1).

To verify that the coefficient vector $\underline{\beta}_\lambda$ is uniquely defined by (15) use the result that a system of equations $A\underline{x} = \underline{b}$ has a unique solution if and only if $\underline{x} = \underline{0}$ is the only solution to $A\underline{x} = \underline{0}$. Thus take $\underline{y} = \underline{0}$ in (15) and note that (15) will always have a solution in this instance. Consequently, as $\mu_\lambda \in W_2^m[a,b]$ and also satisfies the necessary condition (14) for minimization of (1) in this case

$$\phi(\mu_\lambda,\mu_\lambda) = n^{-1} \sum_{j=1}^{n} \mu_\lambda(t_j)^2 + \lambda\int_a^b \mu_\lambda^{(m)}(t)^2 dt = 0 .$$

Thus $\mu_\lambda^{(m)}(t) = 0$ almost everywhere and $\mu_\lambda(t_j)^2 = 0$, $j = 1,\ldots,n$. This means that $\mu_\lambda$ is a polynomial of order m which vanishes at $n \geq m$ points and, hence, that $\mu_\lambda(t) \equiv 0$. However, since $x_1,\ldots,x_n$ are a basis for $NS^{2m}(t_1,\ldots,t_n)$, they are linearly independent which has the implication that $\mu_\lambda(t) = \Sigma_{j=1}^n \beta_{\lambda j} x_j(t) \equiv 0$ if and only if $\beta_{\lambda 1} = \ldots = \beta_{\lambda n} = 0$. We conclude from this that (15) has a unique solution.

It now remains to show that $\mu_\lambda = \Sigma_{j=1}^n \beta_{\lambda j} x_j$ actually minimizes (1). Using the fact that $\mu_\lambda$ satisfies $\phi(\mu_\lambda, f) = 0$ for $f \varepsilon W_2^m[a,b]$ and a little algebra we find that

$$n^{-1} \sum_{j=1}^n (y_j - f(t_j))^2 + \lambda \int_a^b f^{(m)}(t)^2 dt$$

$$= n^{-1} \sum_{j=1}^n (y_j - \mu_\lambda(t_j))^2 + \lambda \int_a^b \mu_\lambda^{(m)}(t)^2 dt$$

$$+ 2\phi(\mu_\lambda, \mu_\lambda - f) + n^{-1} \sum_{j=1}^n (\mu_\lambda(t_j) - f(t_j))^2$$

$$+ \lambda \int_a^b (\mu_\lambda^{(m)}(t) - f^{(m)}(t))^2 dt$$

$$\geq n^{-1} \sum_{j=1}^n (y_j - \mu_\lambda(t_j))^2 + \lambda \int_a^b \mu_\lambda^{(m)}(t)^2 dt , \tag{16}$$

since $\phi(\mu_\lambda, \mu_\lambda - f) = 0$ (because $\mu_\lambda - f \varepsilon W_2^m[a,b]$) and all other terms in the sum must be nonnegative. This says that no f in $W_2^m[a,b]$ can make (1) smaller than $\mu_\lambda$. Thus, $\mu_\lambda$ is a minimizer of (1).

To see that $\mu_\lambda$ is the unique minimizer, suppose that $\mu_\lambda$ and f both have the same value for this criterion. Then, from (16), it follows that

$$\int_a^b (\mu_\lambda^{(m)}(t) - f^{(m)}(t))^2 dt = 0$$

and

$$\sum_{j=1}^n (\mu_\lambda(t_j) - f(t_j))^2 = 0 .$$

The first of these relations has the consequence that the mth derivative of $\mu_\lambda$ - f vanishes almost everywhere. This means $\mu_\lambda$ - f must be a polynomial of order m. The second equality entails that $\mu_\lambda(t_j) - f(t_j) = 0$, j=1,...,n. Since $n \geq m$, $\mu_\lambda \equiv f$ and the minimizer must be unique. □

The approach we have taken to proving Theorem 5.3 is due to Lyche and Schumaker (1973). While somewhat tedious, it has the advantage of relying only on integration by parts and Taylor's Theorem rather than more advanced mathematical machinery. There are many other ways to establish this result. Of particular importance are those which employ reproducing kernel Hilbert space (RKHS) theory. A general approach to characterizing the solution to minimization problems, such as the one corresponding to criterion (1), that is based on RKHS methods can be found in Kimeldorf and Wahba (1971).

Theorem 5.3 assures us of the existence and uniqueness of the smoothing spline estimator $\mu_\lambda$ and also gives us a prescription for its computation. Specifically, the smoothing spline estimator of $\mu_\lambda$ is the 2mth order natural spline

$$\mu_\lambda(t) = \sum_{j=1}^{n} \beta_{\lambda j} x_j(t)$$

with the $\beta_{\lambda j}$ computed from (15).

There are two extreme cases for λ which were not included in the theorem: λ = 0 and λ = ∞. In either of these cases there is some question as to what is really meant by criterion (1).

By letting λ = 0 in (15a) we see this case to be equivalent to finding the natural spline which satisfies $\Sigma_{j=1}^{n} \beta_{0j} x_j(t_i) = y_i$, i=1,...,n. The result is what is known as a *natural spline interpolant* to the $y_i$. By following along the lines of the proof of Theorem 5.3 one can establish the following result about its properties.

**Corollary 5.1.** For $n \ge m$, the natural spline interpolant to $y_1,\ldots,y_n$ is the unique minimizer of $J_m(f)$ over $W_2^m[a,b]$ subject to the condition $f(t_j) = y_j$, $j=1,\ldots,n$.

We will have further use for natural spline interpolants when we discuss the selection of basis functions in Section 5.3.3.

If we consider the other extreme, $\lambda = \infty$, one finds that $\mu_\infty = \lim_{\lambda\to\infty}\mu_\lambda$ is just the mth order polynomial regression estimator of $\mu$. A rigorous proof of this fact will be given subsequently. However, the validity of the claim is clear from the fact that as $\lambda$ grows large in (1) this places more and more weight on smoothness, eventually driving $J_m(f)$ to zero. Thus, for $\lambda = \infty$ criterion (1) is equivalent to minimizing MSE(f) subject to $J_m(f) = 0$ which, as was noted in Section 5.2, results in polynomial regression. Thus, for example, the cubic smoothing spline which arises from the choice $m = 2$ is a generalization of the linear regression estimator for $\mu$.

There is an alternative way to write equations (15) which is instructive. Premultiply (15a) by X' to obtain

$$(X'X + n\lambda X'G)\underline{\beta}_\lambda = X'\underline{y} .$$

Now X'G has typical element $(-1)^m(2m-1)!\Sigma_{r=1}^n x_r(t_i)\delta_{ij}$. But, according to Lemma 5.1, this is just $\int_a^b x_i^{(m)}(t)x_j^{(m)}(t)dt$. Thus we have the following corollary.

**Corollary 5.2.** If $n \ge m$ the minimizer of (1) is $\mu_\lambda = \Sigma_{j=1}^n \beta_{\lambda j}x_j$, where $\underline{\beta}_\lambda = (\beta_{\lambda 1},\ldots,\beta_{\lambda n})'$ is the solution to

$$(X'X + n\lambda\Omega)\underline{\beta}_\lambda = X'\underline{y} \tag{17a}$$

with

$$\Omega = \{\int_a^b x_i^{(m)}(t)x_j^{(m)}(t)dt\}_{i,j=1,n}. \tag{17b}$$

Equations (17) resemble the normal equations (A.3) from

linear regression and therefore may look more familiar than (15). In fact, by recalling our work in Chapter 2, we see that the smoothing spline estimator can be viewed as a type of generalized ridge regression estimator. However, unlike most ridge estimators, the matrix which is added to X'X depends on the basis functions in this case.

Using (17) the vector of fitted values is seen to have the form

$$\underline{\mu}_\lambda = (\mu_\lambda(t_1),\ldots,\mu_\lambda(t_n))' = H(\lambda)\underline{y}\ , \tag{18a}$$

where

$$H(\lambda) = X(X'X + n\lambda\Omega)^{-1}X'. \tag{18b}$$

By analogy with linear regression terminology we will call $H(\lambda)$ the *hat matrix*. It is clear that $H(\lambda)$ is symmetric and positive definite and, hence, smoothing splines fall into the class of estimators discussed in Chapter 2. This will have important consequences for the selection of $\lambda$.

The use of criterion (1) for estimation purposes is appropriate if each observation has the same variance, $\sigma^2$. If this is not the case the criterion should be modified by weighting observations inversely proportional to their standard deviations. A general criterion arising from such considerations is

$$n^{-1}\sum_{j=1}^{n} w_j(y_j - f(t_j))^2 + \lambda\int_a^b f^{(m)}(t)^2 dt\ , \tag{19}$$

with weights $w_j > 0$. By taking $w_j = \{\mathrm{Var}\ y_j\}^{-1}$, (19) gives an estimation criterion suitable for regression analysis when the observations have unequal variances. This, of course, assumes that the individual observation variances are known. Silverman (1985) has proposed a method for estimating the variances when this is not the case.

Concerning minimization of criterion (18) we state the following result.

**Theorem 5.4.** Under the notation and conditions of Theorem 5.2, there is a unique minimizer of (19) in $W_2^m[a,b]$. It is the element of $NS^{2m}(t_1,\ldots,t_n)$ whose basis element coefficients are determined by

$$(X + n\lambda G_w)\underline{\beta}_\lambda = \underline{y} \quad ,$$

where

$$G_w = \{(-1)^m(2m - 1)!\delta_{ij}/w_i\}_{i,j=1,n} \quad .$$

*Proof.* Exercise 6.

Theorem 5.4 provides a key to determining how to handle repeated observations. Suppose that there are $n_j \geq 1$ observations at $t_j$ and let $\bar{y}_j$ be their average. Then minimization of criterion (1) is equivalent to minimizing

$$N^{-1}\sum_{j=1}^{n} n_j(\bar{y}_j - f(t_j))^2 + \lambda\int_a^b f^{(m)}(t)^2 dt \quad ,$$

where $N = \Sigma_{j=1}^n n_j$. Consequently, the smoothing spline estimator in this case is a natural spline with coefficients found by solving

$$(X + N\lambda\bar{G})\underline{\beta}_\lambda = \bar{\underline{y}} \quad ,$$

where $\bar{\underline{y}} = (\bar{y}_1,\ldots,\bar{y}_n)'$ and $\bar{G}$ is $G_w$ in Theorem 5.4 with $w_i = n_i$.

### 5.3.3 Selecting a Basis

In Theorem 5.3 it was assumed that a suitable basis for the natural splines was at one's command. There are, of course, many ways of constructing such a basis. In this section we discuss several bases which have appeared in the literature in the specific context of spline smoothing.

We will concentrate on three natural spline bases. The first of these arises from work by Kimeldorf and Wahba (1970a,b) concerning the connection between smoothing splines and prediction from certain stochastic processes. (This aspect of spline smoothing will be explored in Section 5.5). The Kimeldorf-Wahba basis is a useful theoretical tool that leads naturally into the Anselone-Laurent-Reinsch basis which is particularly well suited for computational purposes. The other basis we will consider was developed by Demmler and Reinsch (1975). It also gives a powerful theoretical tool which makes many of the properties of smoothing splines transparent. There are many other useful bases besides those to be considered here; some of these are explored in the exercises. (Exercises 7-8).

**The Kimeldorf-Wahba Basis**

Let

$$Q(s,t) = [(m-1)!]^{-2}\int_a^b (s-u)_+^{m-1}(t-u)_+^{m-1}\,du, \quad a \le s,\ t \le b, \tag{20}$$

and define the associated $n \times n$ matrix

$$Q_n = \{Q(t_i,t_j)\}_{i,j=1,n} \; . \tag{21}$$

The function $Q$ in (20) is actually a covariance kernel for a particular stochastic process related to the Wiener process (see Section 5.5). Among other things this means that the matrix $Q_n$ is positive definite. We will also need the $n \times m$ matrix

$$T = \{t_i^{j-1}\}_{\substack{i=1,n \\ j=1,m}} \tag{22}$$

and $n \times n$ matrix

$$M(\lambda) = Q_n + n\lambda I \; . \tag{23}$$

A natural spline basis, used by Kimeldorf and Wahba (1970a,b) in their work on smoothing splines, is provided by

$$(x_1(t),\ldots,x_n(t)) = (1,t,\ldots,t^{m-1},Q(t,t_1),\ldots,Q(t,t_n))\begin{bmatrix} F \\ \tilde{G} \end{bmatrix} \quad (24a)$$

with

$$F = (T'M(\lambda)^{-1}T)^{-1}T'M(\lambda)^{-1} \quad (24b)$$

and

$$\tilde{G} = M(\lambda)^{-1} - M(\lambda)^{-1}T(T'M(\lambda)^{-1}T)^{-1}T'M(\lambda)^{-1} . \quad (24c)$$

It is not entirely obvious that the $x_j$ defined in this fashion are, in fact, a basis. We must, therefore, first attend to the task of verifying this property.

Observe that, through repeated integration by parts and use of the binomial theorem, it is possible to write

$$(Q(t,t_1),\ldots,Q(t,t_n))$$

$$= \underline{v}(t)'T' + \frac{(-1)^m}{(2m-1)!}((t-t_1)_+^{2m-1},\ldots,(t-t_n)_+^{2m-1}) ,$$

for an $m \times 1$ vector $\underline{v}(t)$ that involves $t$ but not the $t_i$. Since $T'\tilde{G} = 0$, the $x_j$ can then be reexpressed as

$$(x_1(t),\ldots,x_n(t)) = (1,t,\ldots,t^{m-1},(t-t_1)_+^{2m-1},\ldots,(t-t_n)_+^{2m-1})$$

$$\times \begin{bmatrix} F \\ \dfrac{(-1)^m}{(2m-1)!}\tilde{G} \end{bmatrix} .$$

The functions $1,t,\ldots,t^{m-1},(t-t_1)_+^{2m-1},\ldots,(t-t_n)_+^{2m-1}$ are part of a basis for splines of order $2m$ and hence are linearly independent. Thus, $x_1,\ldots,x_n$ will be linearly independent if the matrix $\begin{bmatrix} F \\ \tilde{G} \end{bmatrix}$ has rank $n$. To verify that this is, in fact, the case, merely note that $F$ has rank $m$, $\tilde{G}$ has rank $n - m$ and $\tilde{G}M(\lambda)F' = 0$.

We now know that $x_1,\ldots,x_n$ are linearly independent. Thus if the $x_j$ are in $NS^{2m}(t_1,\ldots,t_n)$ they will necessarily form a basis.

To verify that $x_j$ is a natural spline it is sufficient to show that it satisfies the natural boundary conditions $x_j^{(m+i)}(a) = x_j^{(m+i)}(b) = 0$, $i=0,\ldots,m-1$. Writing

$$(x_1^{(m+i)}(t),\ldots,x_n^{(m+i)}(t)) = (2m-1)\ldots(m-i)((t-t_1)_+^{m-i-1},\ldots,$$
$$(t-t_n)_+^{m-i-1})\tilde{G}$$

we see that the lower boundary conditions are obviously satisfied. For the upper endpoint use the binomial theorem to express the quantities $(b-t_j)_+^{m-i-1}$ as

$$((b-t_1)_+^{m-i-1},\ldots,(b-t_n)_+^{m-i-1}) = \underline{v}_i'T', \quad i = 0,\ldots,m-1,$$

and employ the fact that $T'\tilde{G} = 0$.

We have now shown that $x_1,\ldots,x_n$ defined in (24) are a basis for $NS^{2m}(t_1,\ldots,t_n)$. The coefficients $\underline{\beta}_\lambda$ defined in (15) are found to have a particularly simple form under this basis.

The matrices X and G in (15b)-(15c) can be read off immediately from (25). It is seen that

$$X = TF + Q_n\tilde{G}$$

and

$$G = (-1)^m(2m-1)!\{(-1)^m\tilde{G}/(2m-1)!\} = \tilde{G}.$$

Thus,

$$\begin{aligned} X + n\lambda G &= TF + Q_n\tilde{G} + n\lambda\tilde{G} \\ &= TF + M(\lambda)\tilde{G} = I, \end{aligned} \tag{26}$$

which means $\underline{\beta}_\lambda = \underline{y}$. So the coefficients under this basis are just the response values themselves.

The basis functions defined by (24) are cumbersome and not suited to computation. Wahba (1978) has utilized the fact that $\tilde{G}T = 0$ to obtain alternative basis functions which are more amenable to use in calculations.

First note that since $\tilde{G}T = 0$ this means that the matrix $\tilde{G}$ consists of n vectors belonging to the subspace of $\mathbb{R}^n$ orthogonal to the columns of T. As this subspace has dimension n - m (because T has m linearly independent columns), we can construct a corresponding set of basis vectors $\underline{u}_1,\ldots,\underline{u}_{n-m}$. Let U be the n × (n - m) matrix defined by $U = [\underline{u}_1,\ldots,\underline{u}_{n-m}]$ and note that

$$U'T = 0 . \tag{27}$$

Wahba (1978) observed that, since $\underline{u}_1,\ldots,\underline{u}_{n-m}$ form a basis for the columns of $\tilde{G}$, there is a (n - m) × n matrix A such that

$$\tilde{G} = UA.$$

To solve for A premultiply this equation by $U'M(\lambda)$ and apply (27) to obtain

$$U' = U'M(\lambda)UA.$$

Since $U'M(\lambda)U$ is nonsingular $A = (U'M(\lambda)U)^{-1}U'$ which reveals that

$$\tilde{G} = U(U'M(\lambda)U)^{-1}U'. \tag{28}$$

Now define new basis functions $1,t,\ldots,t^{m-1},g_1(t),\ldots,g_{n-m}(t)$ with

$$(g_1(t),\ldots,g_{n-m}(t)) = (Q(t,t_1),\ldots,Q(t,t_n))U.$$

Since $\underline{\beta}_\lambda = \underline{y}$ we have

$$\begin{aligned}
\mu_\lambda(t) &= \sum_{j=1}^{n} \beta_{\lambda j} x_j(t) = \sum_{j=1}^{n} y_j x_j(t) \\
&= (1,t,\ldots,t^{m-1},Q(t,t_1),\ldots,Q(t,t_n))\begin{bmatrix} F \\ \tilde{G} \end{bmatrix} \underline{y} \\
&= (1,t,\ldots,t^{m-1},g_1(t),\ldots,g_{n-m}(t))\begin{bmatrix} F \\ (U'M(\lambda)U)^{-1}U' \end{bmatrix} \underline{y} \\
&= \sum_{j=0}^{m-1} \theta_{\lambda j} t^j + \sum_{j=1}^{n-m} \gamma_{\lambda j} g_j(t),
\end{aligned} \tag{29a}$$

where

$$\underline{\theta}_\lambda = (\theta_{\lambda 0}, \ldots, \theta_{\lambda, m-1})' = F\underline{y} \tag{29b}$$

and

$$\underline{\gamma}_\lambda = (\gamma_{\lambda 1}, \ldots, \gamma_{\lambda, n-m})' = (U'M(\lambda)U)^{-1}U'\underline{y} \quad . \tag{29c}$$

To obtain an expression for $\mu_\lambda$ it therefore suffices to obtain values for $\underline{\theta}_\lambda$ and $\underline{\gamma}_\lambda$.

To compute $\underline{\gamma}_\lambda$ we can solve the system of equations

$$(U'Q_nU + n\lambda U'U)\underline{\gamma}_\lambda = U'\underline{y}. \tag{29c'}$$

The trick is to choose U in a fashion which makes the system simple to solve.

Wahba (1978) proposed choosing U to satisfy

$$U'U = I \ .$$

Since $T(T'T)^{-1}T'U = 0$, $\underline{u}_1, \ldots, \underline{u}_{n-m}$ are, in this case, a set of eigenvectors corresponding to the n - m zero eigenvalues of the idempotent matrix $T(T'T)^{-1}T'$. Another choice for U will be considered subsequently.

Once $\underline{\gamma}_\lambda$ has been computed the solution can be used to aid in solving for the polynomial coefficients $\underline{\theta}_\lambda$. This follows upon noting that

$$\begin{aligned} T\underline{\theta}_\lambda &= TF\underline{y} \\ &= [I - M(\lambda)\tilde{G}]\underline{y} \\ &= \underline{y} - M(\lambda)U\underline{\gamma}_\lambda \end{aligned}$$

so that

$$(T'T)\underline{\theta}_\lambda = T'(\underline{y} - M(\lambda)U\underline{\gamma}_\lambda) \ . \tag{29b'}$$

These are just the normal equations for regression of $\underline{y} - M(\lambda)U\underline{\gamma}_\lambda$ onto polynomials of order m. Consequently, standard methods can be used to compute $\underline{\theta}_\lambda$.

The vector of fitted values $\underline{\mu}_\lambda = (\mu_\lambda(t_1),\ldots,\mu_\lambda(t_n))'$ has a particularly simple form under this representation for $\mu_\lambda$. It is found that

$$\underline{\mu}_\lambda = \underline{y} - n\lambda U(U'M(\lambda)U)^{-1}U'\underline{y} \quad . \tag{30a}$$

(Exercise 9). Consequently, the hat matrix (18b) can be represented as

$$H(\lambda) = I - n\lambda U(U'M(\lambda)U)^{-1}U'. \tag{30b}$$

**The Anselone - Laurent - Reinsch Basis**

A symmetric $n \times n$ matrix $A = \{a_{ij}\}_{i,j=1,n}$ is said to be $2k + 1$ banded if $a_{ij} = 0$ whenever $|i - j| > k$. The value k is called the half band width. For example, a diagonal matrix corresponds to the case of $k = 0$ and a tridiagonal matrix results when $k = 1$.

The solution of systems of equations involving band matrices can be accomplished much more rapidly than the general case. It can be shown that if A is symmetric, positive definite and $2k + 1$ banded, the equation system $A\underline{x} = \underline{b}$ can be solved in $O(k^2n)$ arithmetic operations (see, e.g., Franklin, 1968).

In view of the computational advantages intrinsic to band matrices it would be helpful if we could choose U in (27) in a manner which results in band structures for some of the matrices in (29). A choice for U with this property has been considered by Anselone and Laurent (1968) and Reinsch (1967, 1971). To discuss their choice it will be helpful to first introduce the concept of divided differences. For a more complete discussion of divided differences than the one given here see de Boor (1978).

Assuming that $t_1,\ldots,t_n$ are distinct, the mth order *divided difference* of a function g at $t_i$ is defined recursively by

$$[t_i,\ldots,t_{i+m}]g = \frac{[t_{i+1},\ldots,t_{i+m}]g - [t_i,\ldots,t_{i+m-1}]g}{t_{i+m} - t_i}$$

with

$$[t_i]g = g(t_i)$$

used to initiate the recursion. For example, the first order divided difference is

$$[t_i,t_{i+1}]g = \frac{g(t_{i+1}) - g(t_i)}{t_{i+1} - t_i}$$

and the second order divided difference is

$$[t_i,t_{i+1},t_{i+2}]g = \frac{[t_{i+1},t_{i+2}]g - [t_i,t_{i+1}]g}{t_{i+2} - t_i}$$

$$= (t_{i+2} - t_i)^{-1}\{(t_{i+2} - t_{i+1})^{-1}g(t_{i+2})$$

$$- [(t_{i+2} - t_{i+1})^{-1} + (t_{i+1} - t_i)^{-1}]g(t_{i+1})$$

$$+ (t_{i+1} - t_i)^{-1}g(t_i)\} .$$

The mth order *normalized divided difference* is defined to be

$$(t_{i+m} - t_i)[t_i,\ldots,t_{i+m}]g = [t_{i+1},\ldots,t_{i+m}]g - [t_i,\ldots,t_{i+m-1}]g. \qquad (31)$$

An important property of mth order divided differences and normalized divided differences is that they vanish if g is a polynomial of order m.

The *mth order normalized divided difference coefficients* are the coefficients of the $g(t_j)$ in (31). For example, in the case of m=1 $g(t_i)$ and $g(t_{i+1})$ both have normalized divided difference coefficients of one with the remaining $g(t_j)$ having zero coefficients. Similarly for m=2, all $g(t_j)$ have zero coefficients except for $g(t_i)$, $g(t_{i+1})$,$g(t_{i+2})$ whose coefficients are

$$\alpha_{0,2}^{[i]} = (t_{i+1} - t_i)^{-1} ,$$

$$\alpha_{1,2}^{[i]} = -[(t_{i+2} - t_{i+1})^{-1} + (t_{i+1} - t_i)^{-1}] ,$$

and

$$\alpha_{2,2}^{[i]} = (t_{i+2} - t_{i+1})^{-1} ,$$

respectively.

Let $\alpha_{0,m}^{[i]},\ldots,\alpha_{m,m}^{[i]}$ denote the nonzero mth order normalized divided difference coefficients at $t_i$. Then, Anselone and Laurent (1968) and Reinsch (1967, 1971) select U' to have ith row

$$\underline{u}_i' = (\underbrace{0,\ldots,0}_{i-1},\alpha_{0,m}^{[i]},\ldots,\alpha_{m,m}^{[i]},\underbrace{0,\ldots,0}_{n-m-i}), \quad i = 1,\ldots,n - m . \tag{32}$$

Since $[t_i,\ldots,t_{i+m}]g = 0$ if g is a polynomial of order m, it follows that U'T = 0 so that U satisfies (27), the condition required for representation (29) to hold.

To see the implications of this choice first note that

$$\underline{u}_i'\underline{u}_j = 0$$

if $|i - j| > m$. Thus, U'U is 2m + 1 banded.

Now, examination of $U'Q_nU$ shows that its typical entry is

$$\begin{aligned}
\underline{u}_i'Q_n\underline{u}_j &= \sum_{r=0}^{m} \sum_{\ell=0}^{m} \alpha_{r,m}^{[i]} \alpha_{\ell,m}^{[j]} Q(t_{i+r},t_{j+\ell}) \\
&= [(m - 1)!]^{-2}\int_a^b \left[ \left( \sum_{r=0}^{m} \alpha_{r,m}^{[i]}(t_{i+r} - u)_+^{m-1}\right) \right. \\
&\qquad \left. \times \left( \sum_{\ell=0}^{m} \alpha_{\ell,m}^{[j]}(t_{j+\ell} - u)_+^{m-1}\right) \right]du \\
&= [(m - 1)!]^{-2}\int_a^b N_{i,m}(u)N_{j,m}(u)du,
\end{aligned}$$

where

$$N_{k,m}(\cdot) = \sum_{r=0}^{m} \alpha_{r,m}^{[k]}(t_{k+r} - \cdot)_+^{m-1}$$

$$= (t_{k+m} - t_k)[t_k, \ldots, t_{k+m}](t - \cdot)_+^{m-1}. \tag{33}$$

The function $N_{k,m}$ is a special kind of spline known as a *B-spline.* These functions have a local support property that is easy to demonstrate. First, it is clear that $N_{k,m}(u)$ vanishes for $u > t_{k+m}$, since all the functions in the sum vanish in this case. Suppose, on the other hand, that $u < t_k$. Fix u and note that in this case $(t - u)_+^{m-1}$ is a polynomial of order m in t for $t > u$. Consequently, its mth order divided difference must vanish and we are led to the conclusion that $N_{k,m}(u) = 0$ if $u \notin [t_k, t_{k+m}]$.

The local support characteristics of the $N_{k,m}$ ensures that

$$\int_a^b N_{i,m}(u) N_{j,m}(u) du = 0$$

if $|i - j| > m - 1$. Thus $\underline{u}_i' Q_n \underline{u}_j = 0$ if $|i - j| > m - 1$ and $U'Q_nU$ is $2m - 1$ banded. This means, for example, that in the case of cubic smoothing splines, $m = 2$, $U'Q_nU$ will be tridiagonal and $U'U$ will be a five diagonal matrix.

Because $U'Q_nU + n\lambda U'U$ is $2m + 1$ banded it follows that $\underline{\gamma}_\lambda$ in (29c') can be computed in $O(m^2n)$ calculations and (29b') will require at most $O(m^3)$ operations for its solution (See, e.g., Franklin, 1968). Thus, by using U as defined in (32) efficient methods for computing the smoothing spline estimator can be developed using algorithms for solving equation systems involving band matrices (see, e.g., Dongarra et al., 1979). For discussion of some specific algorithms for computing $\mu_\lambda$ see Reinsch (1967), Shiau (1985) and Hutchinson and de Hoog (1985, 1987).

**The Demmler - Reinsch Basis**

Another basis for spline smoothing has been studied by Demmler and Reinsch (1975). This basis gives a canonical form for the smoothing spline which makes many of its properties transparent. The basis stems from certain properties of natural spline interpolants.

Let $\underline{\nu} = (\nu_1,\ldots,\nu_n)'$ be a vector of constants. An expression for the natural spline interpolant to $\underline{\nu}$ can be derived from Theorem 5.3, Corollary 5.1, (24), (27) and (32). If we let $s_{\underline{\nu}}$ denote the resulting natural spline, it is found that

$$s_{\underline{\nu}}(t) = (1,\ldots,t^{m-1},Q(t,t_1),\ldots,Q(t,t_n))\begin{bmatrix} F_0 \\ UW^{-1}U' \end{bmatrix}\underline{\nu} \quad ,$$

where

$$F_0 = (T'Q_n^{-1}T)^{-1}T'Q_n^{-1} \quad ,$$

$$W = U'Q_nU$$

and U' is defined in (32).

The smoothing spline estimator is actually a spline of interpolation since

$$\underline{\mu}_\lambda = s_{\underline{\mu}_\lambda}$$

with $\underline{\mu}_\lambda = (\mu_\lambda(t_1),\ldots,\mu_\lambda(t_n))'$ the vector of fitted values. This fact motivated Demmler and Reinsch (1975) to study the properties of the matrix $UW^{-1}U'$.

Let $\underline{1}$, $\underline{t}$, $\underline{t}^2,\ldots,\underline{t}^{m-1}$ denote the columns of T and apply the Gram-Schmidt process (c.f. Example 3.2) successively to obtain m mutually orthogonal vectors of unit length $\underline{x}_1,\ldots,\underline{x}_m$. The first vector will be

$$\underline{x}_1 = n^{-1/2}\underline{1} \quad ,$$

the second

$$\underline{x}_2 = [\underline{t} - (\underline{t}'\underline{x}_1)\underline{x}_1]/|\underline{t} - (\underline{t}'\underline{x}_1)\underline{x}_1| \quad ,$$

etc., where $|\cdot|$ denotes the usual Euclidean norm. The vectors $\underline{x}_1,\ldots,\underline{x}_m$ satisfy

$$UW^{-1}U'\underline{x}_i = \underline{0} \;, \quad i = 1,\ldots,m,$$

and are, therefore, the eigenvectors corresponding to the m zero eigenvalues of $UW^{-1}U'$. Let $\underline{x}_{m+1},\ldots,\underline{x}_n$ denote the remaining eigenvectors, with corresponding positive eigenvalues $\delta_{m+1} \le \ldots \le \delta_n$, and define the matrix

$$X = [\underline{x}_1,\ldots,\underline{x}_n] \; .$$

Demmler and Reinsch then choose their basis for $NS^{2m}(t_1,\ldots,t_n)$ to be

$$(x_1(t),\ldots,x_n(t))$$

$$= (1,t,\ldots,t^{m-1},Q(t,t_1),\ldots,Q(t,t_n))\begin{bmatrix} F_0X \\ UW^{-1}U'X \end{bmatrix} . \qquad (34)$$

In words, the $x_i$ are the natural interpolating splines to the $\underline{x}_i$. Clearly the $x_i$ are all elements of $NS^{2m}(t_1,\ldots,t_n)$. In addition, if there exists coefficients $\underline{c}' = (c_1,\ldots,c_n)$ such that $\Sigma_{j=1}^n c_j x_j(t) \equiv 0$ this implies that $\Sigma_{j=1}^n c_j x_j(t_i) = 0$, $i = 1,\ldots,n$, which, in turn, requires $X\underline{c} = \underline{0}$. Since X is orthogonal this mean $\underline{c} = \underline{0}$ and we must conclude that the $x_j$ are, in fact, a basis for the natural splines with knots as $t_1,\ldots,t_n$.

Under the Demmler-Reinsch basis (34) the form of $\mu_\lambda$ can be deduced from Corollary 5.3. Writing $\mu_\lambda(t) = \Sigma_{j=1}^n \beta_{\lambda j} x_j(t)$ the coefficients $\underline{\beta}_\lambda = (\beta_{\lambda 1},\ldots,\beta_{\lambda n})'$ are determined by

$$[X'X + (n\lambda)(-1)^m(2m-1)!\frac{(-1)^m}{(2m-1)!}X'UW^{-1}U'X]\underline{\beta}_\lambda = X'\underline{y} \qquad .$$

Letting $\delta_1 = \ldots \delta_m = 0$, define

$$\Delta = \mathrm{diag}(\delta_1,\ldots,\delta_m,\delta_{m+1},\ldots,\delta_n)$$

and observe that $X'X = I$ and

$$X'UW^{-1}U'X = \Delta.$$

The equation for $\underline{\beta}_\lambda$ can now be rewritten as

$$\underline{\beta}_\lambda = \Delta(\lambda)X'\underline{y} \ ,$$

where

$$\Delta(\lambda) = \mathrm{diag}((1 + n\lambda\delta_1)^{-1},\ldots,(1 + n\lambda\delta_n)^{-1}) \ .$$

Thus $\mu_\lambda$ takes on the simple form

$$\mu_\lambda(t) = \sum_{j=1}^{n} \frac{\underline{x}_j'\underline{y}}{1+n\lambda\delta_j} x_j(t) = \sum_{j=1}^{m} \underline{x}_j'\underline{y}\ x_j(t) + \sum_{j=m+1}^{n} \frac{\underline{x}_j'\underline{y}}{1+n\lambda\delta_j} x_j(t). \qquad (35)$$

Recalling how $\underline{x}_1,\ldots,\underline{x}_m$ were constructed, it is now clear that the first term in (35), $\Sigma_{j=1}^m \underline{x}_j'\underline{y}\ x_j(t)$, is the mth order polynomial regression estimator of $\mu(t)$. By letting $\lambda \to \infty$ in (35) it then becomes clear that $\lim_{\lambda\to\infty}\mu_\lambda = \mu_\infty$ is the estimator of $\mu$ from polynomial regression. This verifies a claim to that effect made in Section 3.2.2.

Demmler and Reinsch (1975) studied the behavior of the eigenvectors $\underline{x}_1,\ldots,\underline{x}_n$ for $UW^{-1}U'$. Let $\nu(\underline{x})$ denote the number of sign changes in the vector $\underline{x}$ with an arbitrary sign being attributed to a zero component. They then showed that

$$\nu(\underline{x}_i) = i - 1, \qquad i = 1,\ldots,n \ .$$

This says that those $\underline{x}_i$ which correspond to larger eigenvalues tend to have more variable components than those for smaller eigenvalues. The sign changes of the $\underline{x}_i$ also provide information about the oscillation properties of the basis elements $x_1,\ldots,x_n$ since $x_i$ must change signs at least as often as $\underline{x}_i$. Thus, basis elements with higher indices (equivalently those which correspond to larger eigenvalues) would be expected to oscillate more rapidly than the others.

Our original model had $y_j = \mu(t_j) + \varepsilon_j$, $j = 1,\ldots,n$, where $\mu$ was a smooth regression function and the $\varepsilon_j$ were random noise. The fact that $\mu$ is smooth allows us to think of $\underline{y}$ as consisting of a slowly varying component, namely the $\mu(t_j)$, and a high frequency error component. If the eigenvalue-eigenvector decomposition of $UW^{-1}U'$ is viewed as a type of discrete Fourier series, the high frequency portion of $\underline{y}$ should be captured by the basis elements corresponding to those $\underline{x}_i$ having many sign changes. Examination of (35) now indicates that $\mu_\lambda$ is an estimator which tends to damp out high frequency parts of the data since the weight $(1 + n\lambda\delta_i)^{-1}$ will be small for $\underline{x}_i$ which correspond to large eigenvalues.

A plot of several of the Demmler-Reinsch basis functions is shown in Figure 5.2 for the case of cubic smoothing splines ($m = 2$) with $n = 25$ and equally spaced data over $[0,1]$, i.e., $t_j = j/n$, $j=1,\ldots,n$. In this case

$$x_1(t) \equiv n^{-1/2}$$

and

$$x_2(t) = 12^{1/2}(t - (n + 1)/2n)/(n - 1/n)^{1/2}.$$

The remaining $x_i$ are natural interpolating splines to the $\underline{x}_i$ corresponding to certain of the nonzero eigenvalues of $UW^{-1}U'$. Notice the similarity between the behavior of these functions and what one would expect from cosine functions. This reinforces the conclusions drawn from our analogy with Fourier series estimation. It further suggests that smoothing splines are very nearly a variant of classical Fourier series estimators which smooth by damping, rather than deleting, the Fourier coefficients for higher frequency components. Recall that such a property was established for kernel estimators in the case of a periodic model. We will see in the next chapter that the same type of conclusion can be drawn for smoothing splines in that case.

## 5.4 Selecting $\lambda$ and m

To this point we have effectively dodged the issue of how to select the values of $\lambda$ and m in (1). It turns out that the choice of both parameters can be guided by the data. Let us begin by first dealing with the selection of the smoothing parameter.

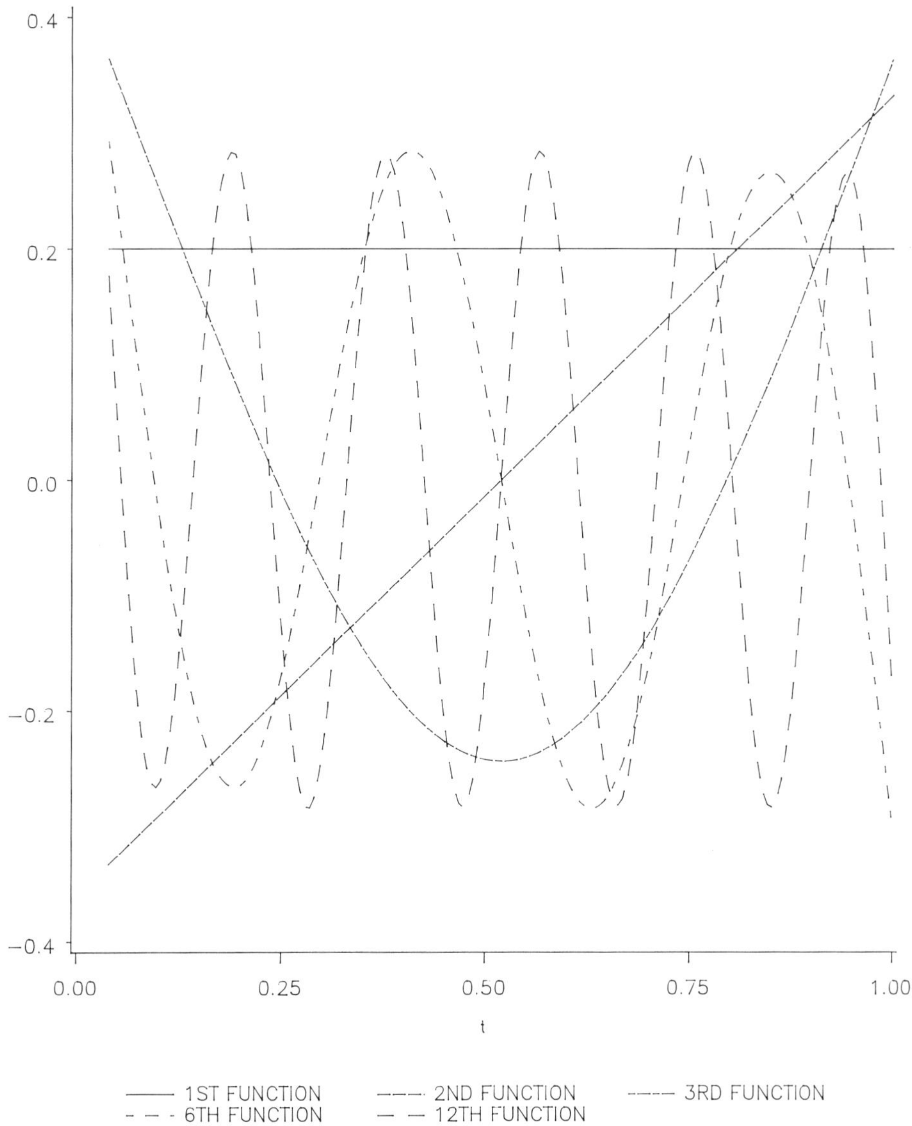


**Figure 5.2** Examples of the Demmler-Reinsch basis functions.

As previously noted the selection of a value for $\lambda$ reflects the premium that is being placed on an estimator's smoothness as measured by $J_m(f)$. This interpretation can be made somewhat more precise by use of Theorem 5.1.

Theorem 5.1 states that given any value of $\lambda$ there is a corresponding value of $\rho$ such that $\mu_\lambda(\cdot) = \mu(\cdot;\rho)$, where $\mu(\cdot;\rho)$ minimizes (6) subject to (5). Now $\mu_\lambda$ is the minimizer of (1) in $W_2^m[a,b]$ and the functions $\mu(\cdot;\rho)$ are all in $W_2^m[a,b]$. Thus, for any fixed $\lambda$, the corresponding value of $\rho$ must be the minimizer with respect to $\rho$ of

$$\mathrm{MSE}(\mu(\cdot;\rho)) + \lambda J_m(\mu(\cdot;\rho)) = \mathrm{MSE}(\mu(\cdot;\rho)) + \lambda\rho.$$

To obtain the last expression we have used the part of Theorem 5.1 which indicates that $J_m(\mu(\cdot;\rho)) = \rho$. $\mathrm{MSE}(\mu(\cdot;\rho))$ is a differentiable function of $\rho$ (Schoenberg, 1964), so we can differentiate our criterion to obtain

$$\lambda = -d\mathrm{MSE}(\mu(\cdot;\rho))/d\rho .$$

Thus, as might be expected from its interpretation as a Lagrange multiplier, $\lambda$ measures the change in the MSE as a function of the smoothness $\rho$. Consequently, $\lambda$ can be viewed as a knob that we twiddle to determine the desired balance between smoothness and mean squared error.

There are a variety of methods which can be used to effect a choice for $\lambda$. One of these is trial and error. Arbitrary values of $\lambda$ are selected until one is found which gives a visually satisfactory fit to the data. This can be time consuming and has the disadvantage that no optimality properties can be attributed to a value selected in this manner. Thus, it may be preferable to use some data driven method for selecting $\lambda$. If nothing else, a data driven method will provide a starting value for $\lambda$ which can then be used for further fine tuning.

If we judge the quality of a choice for $\lambda$ by the corresponding value for the loss

$$L_n(\lambda) = n^{-1} \sum_{j=1}^{n} (\mu(t_j) - \mu_\lambda(t_j))^2 \tag{36}$$

or the risk

$$R_n(\lambda) = EL_n(\lambda) , \tag{37}$$

then our work in Chapter 2 can be used to suggest methods for selecting $\lambda$ from the data. For example, if $\sigma^2$ is known or an estimator of the variance is available, we might choose $\lambda$ as the minimizer of the unbiased risk estimate in (2.14).

Another approach to selecting $\lambda$ which was studied in Chapter 2 is the minimization of the cross-validation criterion

$$CV(\lambda) = n^{-1} \sum_{j=1}^{n} (y_j - \mu_{\lambda(j)}(t_j))^2 ,$$

where $\mu_{\lambda(j)}$ is the estimator arising from (1) when the jth observation has been deleted from the data. Recall from Chapter 2 that $CV(\lambda)$ provides an estimator of the prediction risk which does not require knowledge of $\sigma^2$. Another criterion with this property is the generalized cross-validation (GCV) criterion

$$GCV_n(\lambda) = MSE(\lambda)/[n^{-1}tr(I - H(\lambda))]^2 , \tag{38}$$

where $H(\lambda)$ is the hat matrix defined in (18b).

It should be noted that a data-driven choice for $\lambda$ also provides a value for $\rho$. Specifically, if $\hat{\lambda}$ is a value of $\lambda$ selected by, e.g., one of the above methods then $\hat{\rho} = J_m(\mu_{\hat{\lambda}})$ gives the value for $\rho$ that is suggested by the data.

There is a close relationship between $CV(\lambda)$ and $GCV_n(\lambda)$ for smoothing splines. The connection can be established through use of the important deletion lemma due to Craven and Wahba (1979).

**Lemma 5.2.** Let $\mu_{\lambda(j)}(\cdot;z)$ denote the minimizer of

$$n^{-1}\{\sum_{\substack{i=1 \\ i \neq j}}^{n} (y_i - f(t_i))^2 + (z - f(t_j))^2\} + \lambda J_m(f)$$

over $f \in W_2^m[a,b]$. Then,

$$\mu_{\lambda(j)}(\cdot) = \mu_{\lambda(j)}(\cdot;\mu_{\lambda(j)}(t_j)) \tag{39a}$$

and

$$\mu_{\lambda(j)}(t_j) = (\mu_\lambda(t_j) - h_{jj}(\lambda)y_j)/(1 - h_{jj}(\lambda)) , \tag{39b}$$

where $h_{jj}(\lambda)$ is the jth diagonal element of $H(\lambda)$ in (18b).

*Proof.* To prove this lemma begin by noting that the definition of $\mu_{\lambda(j)}$ entails that

$$n^{-1} \sum_{i \neq j} (y_i - \mu_{\lambda(j)}(t_i))^2 + \lambda\, J_m(\mu_{\lambda(j)})$$

$$= n^{-1}\{ \sum_{i \neq j} (y_i - \mu_{\lambda(j)}(t_i))^2 + (\mu_{\lambda(j)}(t_j) - \mu_{\lambda(j)}(t_j))^2\} + \lambda J_m(\mu_{\lambda(j)})$$

$$\leq n^{-1}\{ \sum_{i \neq j} (y_i - f(t_i))^2 + (\mu_{\lambda(j)}(t_j) - f(t_j))^2\} + \lambda\, J_m(f)$$

for any $f \in W_2^m[a,b]$. Thus (39a) is established.

To verify (39b) note that $\mu_{\lambda(j)}(\cdot;z)$ is the smoothing spline fit to the data where $y_j$ has been replaced by $z$. In particular then, $\mu_{\lambda(j)}(t_j;z) = \Sigma_{i \neq j} h_{ji}(\lambda)y_i + h_{jj}(\lambda)z$ with $h_{j1}(\lambda),\ldots,h_{jn}(\lambda)$ the elements of the jth row of $H(\lambda)$. Since this is linear in $z$ we can expand about any point $z_0$ to obtain

$$\mu_{\lambda(j)}(t_j;z) = \mu_{\lambda(j)}(t_j;z_0) + h_{jj}(\lambda)(z - z_0) .$$

The result now follows by taking $z_0 = y_j$, $z = \mu_{\lambda(j)}(t_j)$ and using (39a) along with the identity $\mu_{\lambda(j)}(t_j;y_j) = \mu_\lambda(t_j)$. □

A particular consequence of Lemma 5.2 is that

$$y_j - \mu_{\lambda(j)}(t_j) = (y_j - \mu_\lambda(t_j))/(1 - h_{jj}(\lambda)).$$

Using this in the CV criterion gives

$$CV(\lambda) = n^{-1} \sum_{j=1}^{n} (y_j - \mu_\lambda(t_j))^2/(1 - h_{jj}(\lambda))^2 .$$

In contrast,

$$GCV_n(\lambda) = n^{-1} \sum_{j=1}^{n} (y_j - \mu_\lambda(t_j))^2 / (1 - n^{-1} \sum_{j=1}^{n} h_{jj}(\lambda))^2 .$$

Thus, for smoothing splines, $GCV_n(\lambda)$ is a variant of $CV(\lambda)$ with the individual leverage values, $h_{11}(\lambda),\ldots,h_{nn}(\lambda)$, replaced by their average $n^{-1}\Sigma_{j=1}^{n} h_{jj}(\lambda)$. We saw results similar to this in Chapter 2 for the case of variable selection problems and ridge regression.

Wahba and Wold (1975a) were the first to propose the use of cross-validation for selection of $\lambda$ in spline smoothing. Then Wahba (1977b, c) and Craven and Wahba (1979) introduced the GCV concept in the context of spline smoothing. GCV now appears to be recognized as the method of choice for smoothing parameter selection. This is due, in part, to various optimality properties which have been established for this criterion. Some of these will now be discussed.

An early landmark result concerning optimality of GCV was established by Craven and Wahba (1979). They showed that if $\mu \in W_2^m[a,b]$ and the $t_j$ are the n-tiles of a positive density on $[a,b]$ then there is a sequence of minimizers $\{\tilde{\lambda}_n\}$ of $EGCV_n(\lambda)$ such that

$$\frac{R_n(\tilde{\lambda}_n)}{\inf\limits_{0\le\lambda\le\infty} R_n(\lambda)} \to 1 \tag{40}$$

as $n \to \infty$. (There was a heuristic aspect to their proof in the case of unequally spaced data that was later made rigorous by Speckman, 1981). This result indicates that the minimizer of $EGCV_n(\lambda)$ and $R_n(\lambda)$ have similar properties in terms of minimizing the risk.

The Craven and Wahba result is not all that one might want since, just like the minimizer of $R_n(\lambda)$, the minimizer of $EGCV_n(\lambda)$ is not observable. It would be more satisfactory if a similar property could be established for $\hat{\lambda}_n$ the minimizer of $GCV_n(\lambda)$.

One of the first results of this type is due to Speckman (1982). He worked with a constrained sequence of minimizers $\tilde{\lambda}_n$ obtained by minimizing $GCV_n(\lambda)$ over $\lambda \geq a_n$, where the $a_n$ are positive constants tending to zero. Speckman was able to show that under certain regularity conditions

$$\frac{L_n(\tilde{\lambda}_n)}{\inf\limits_{0\leq\lambda\leq\infty} L_n(\lambda)} \xrightarrow{P} 1$$

as $n\to\infty$. Cox (1984b) later showed that it was possible to weaken a normality assumption utilized by Speckman in his work. Finally, the restriction to a constrained minimizing sequence was lifted by Li (1986) who showed that

$$\frac{L_n(\hat{\lambda}_n)}{\inf\limits_{0\leq\lambda\leq\infty} L_n(\lambda)} \xrightarrow{P} 1 ,$$

provided $\mu$ is not a polynomial of order m.

The above results suggest that GCV provides a method for selecting a value of $\lambda$ which will produce a near minimal value for the risk in large samples. Li (1985) has also been able to attach some consistency properties to the cross-validated estimator $\mu_{\hat{\lambda}_n}$ of $\mu$ . His work, along with work by Utreras (1979, 1981b, 1983, 1986b) and Speckman (1981) on the properties of the eigenvalues of the matrix $UW^{-1}U'$ utilized to construct the Demmler-Reinsch basis, leads to the conclusion that

$$L_n(\hat{\lambda}_n) \xrightarrow{P} 0$$

as $n \to \infty$, under reasonable restrictions.

It is always necessary to use caution in transferring conclusions drawn from asymptotic theory to finite samples. However, the asymptotic properties of GCV should bolster our faith in its ability to guide the selection of $\lambda$. Simulation experiments reported in Craven and Wahba (1979), Silverman (1984a) and

Wahba (1985) suggest that GCV is a reliable method of selecting $\lambda$ for samples of reasonable size, e.g., $n \geq 100$.

There has been considerable interest in the computational problems associated with the GCV criterion. A useful computational form for $GCV_n(\lambda)$ can be derived from results in Section 5.3.3. Notice from (29c) and (30a) that

$$\begin{aligned} MSE(\lambda) &= n^{-1} \sum_{j=1}^{n} (y_j - \mu_\lambda(t_j))^2 \\ &= n^{-1}(\underline{y} - \underline{\mu}_\lambda)'(\underline{y} - \underline{\mu}_\lambda) \\ &= n\lambda^2 \underline{\gamma}_\lambda' U'U\underline{\gamma}_\lambda \end{aligned}$$

and, from (30b), that

$$tr(I - H(\lambda)) = n\lambda \; trU(U'M(\lambda)U)^{-1}U' \; .$$

Thus

$$GCV_n(\lambda) = n\underline{\gamma}_\lambda' U'U\underline{\gamma}_\lambda / \{trU(U'M(\lambda)U)^{-1}U'\}^2.$$

As previously noted the numerator of this expression can be computed in $O(m^2 n)$ calculations under the Anselone-Laurent-Reinsch basis. The problems which arise stem from the computation of its denominator.

The original algorithm utilized by Craven and Wahba (1979) for computing $GCV_n(\lambda)$ was quite time consuming and required $O(n^3)$ arithmetic operations. This computational burden was reduced by algorithms proposed by Utreras (1979,1981b) which utilized approximations to the eigenvalues of $U(U'M(\lambda)U)^{-1}U'$ that allowed him to compute its trace and, hence, the denominator of $GCV_n(\lambda)$, in $O(n^2)$ calculations in general and only $O(n)$ operations if the $t_j$ are equally spaced. Another approximate $O(n)$ algorithm, stemming from similar considerations, that is applicable to both equally and unequally spaced data, has been proposed by Silverman

(1984a). Recently Shiau (1985) has developed an exact $O(n)$ algorithm for the equally spaced case that does not require eigenvalue approximations.

Hutchinson and de Hoog (1985, 1987) have developed $O(n)$ algorithms that are applicable to both equally and unequally spaced $t_j$. To motivate the Hutchinson and de Hoog (1985) approach note that

$$\operatorname{tr} U(U'M(\lambda)U)^{-1}U' = \operatorname{tr} U'U(U'M(\lambda)U)^{-1}.$$

We have already seen that $U'U$ is $2m + 1$ banded. Thus, computation of the denominator of $GCV_n(\lambda)$ involves computing the diagonal elements of the product of a $2m + 1$ banded matrix $U'U$ with the full matrix $(U'M(\lambda)U)^{-1}$. However, these will involve only the $2m + 1$ central bands of $(U'M(\lambda)U)^{-1}$. Hutchinson and de Hoog (1985) therefore concentrated on computing the $2m + 1$ central bands of $(U'M(\lambda)U)^{-1}$ and were able to devise an algorithm to accomplish this in $O(m^2n)$ operations. Their approach utilizes only the band structure of $U'M(\lambda)U$ and is therefore independent of the spacing of the design points. In latter work, Hutchinson and de Hoog (1987), an alternative $O(n)$ algorithm was developed which appears to solve some of the problems arising from ill-conditioning that occurred with their first algorithm. Another $O(n)$ algorithm for GCV that is applicable to a general class of problems which include spline smoothing is provided by Ansley and Kohn (1987). For a discussion of how to deal with repeated values see Ansley and Kohn (1986). Computational aspects of more general GCV problems are treated in Bates and Wahba (1983). The description of a general set of Fortran-77 code for a variety of GCV problems can be found in Bates, Lindstrom, Wahba and Yandell (1987).

Given an efficient method for computing $GCV_n(\lambda)$, such as those of Hutchinson and de Hoog (1985, 1987) or Ansley and Kohn (1987), it is a fairly simple matter to actually minimize the GCV criterion. There are numerous algorithms for minimizing nonlinear functions which can be used for this purpose. A method which does not require the use of a derivative is to be preferred. One method which seems to have been successful for this purpose is the golden section algorithm (see, e.g., Kennedy and Gentle, 1980). Many times it may be feasible to actually use a global search to find the minimizer of the GCV function. This is usually done by computing $GCV_n(\lambda)$ at small increments of log $\lambda$ and has the advantage of providing more information about the GCV function than other more automatic methods. Since it is quite easy to plot $GCV_n(\lambda)$ along with such a method one can identify local minima and other characteristics of interest.

Criterion (1) also involves the parameter m that must be chosen in some fashion. A particular choice for m may derive from prior beliefs concerning the nature of $\mu$. If, for example, we feel that $\mu$ should be approximately linear, then $m = 2$ is the value which should be used. If $\mu$ is believed to be nearly quadratic then we should take $m = 3$, etc. Thus, if a particular polynomial regression model is being considered this provides a natural choice for m. A corresponding choice of $\lambda = \infty$ by the data can then be viewed intuitively as an indication of the appropriateness of the candidate model.

If certain smoothness properties are known or believed to hold for $\mu$ this can also suggest a choice for m. In this regard, a standard choice is $m = 2$ which corresponds to the assumption that $\mu$ is continuously differentiable and has a square integrable second derivative. The resulting estimator is a cubic smoothing spline.

When one is unwilling to choose a value for m it may be selected from the data. To illustrate how this can be done let us alter our notation slightly and use $GCV_n(\lambda,m)$ instead of $GCV_n(\lambda)$ in order to explicitly indicate the criterion's dependence on m. Then, for any particular value of m we can minimize $GCV_n(\lambda,m)$ with respect to $\lambda$. Let $\hat{\lambda}_m$ be the resulting minimizer. An optimal m is then provided by $\hat{m}$, the minimizer of $GCV_n(\hat{\lambda}_m,m)$. For a nice example of this approach to selecting $\lambda$ and m by GCV see Wahba and

Wendelberger (1980). Of course there is nothing special about the GCV criterion and this plan of attack could be used with other criteria as well.

The use of automated methods for simultaneous selection of m and $\lambda$ has been studied by Gamber (1979a) and Speckman (1982). Gamber (1979a) uses the unbiased risk estimate (2.14), assuming $\sigma^2$ to be known, for this purpose. She is able to establish some weak optimality properties of this approach for certain function types. Speckman (1982) restricts m to lie within a finite set of integers M and then, for each m $\varepsilon$ M, obtains a restricted minimizer $\tilde{\lambda}_m$, as discussed above, for the GCV criterion. If we let $L_n(\lambda,m)$ denote the loss for any value of m and $\lambda$, Speckman is able to show that under certain conditions,

$$\frac{L_n(\tilde{\lambda}_{\hat{m}}, \hat{m})}{\min_{m \varepsilon M} L_n(\tilde{\lambda}_m, m)} \overset{P}{\to} 1 ,$$

where $\hat{m}$ is the minimizing value of $GCV_n(\tilde{\lambda}_m,m)$ over m $\varepsilon$ M.

As an illustration of the types of results that can be expected from the use of GCV to select $\lambda$ consider the data shown in Figure 5.3. The data is given in Exercise 21 and consists of the ratios of weights to heights for boys in a sample of families from the twelve north central states (see, Eppright, et. al. 1972, Gallant and Fuller, 1973 and Gallant, 1977). Three fits to the data using cubic smoothing splines are also exhibited in Figure 5.3 corresponding to $\lambda$ = 1380, $\lambda$ = .00138 and $\lambda \doteq 1.38$. The first two fits represent, essentially, linear regression and interlation and are, respectively, too smooth and too rough. The third fit with $\hat{\lambda} \doteq 1.38$ seems about right. This value of $\lambda$ is the one suggested by GCV as can be seen from a plot of the GCV function shown in Figure 5.4. We have also plotted an estimate of the predictive risk obtained from (2.13) with $\sigma^2$ estimated by the Gasser, Sroka, Steinmetz estimator discussed in Section 4.6. Notice the similarities of the two criterion functions and the fact that both are minimized at essentially the same value of $\lambda$.

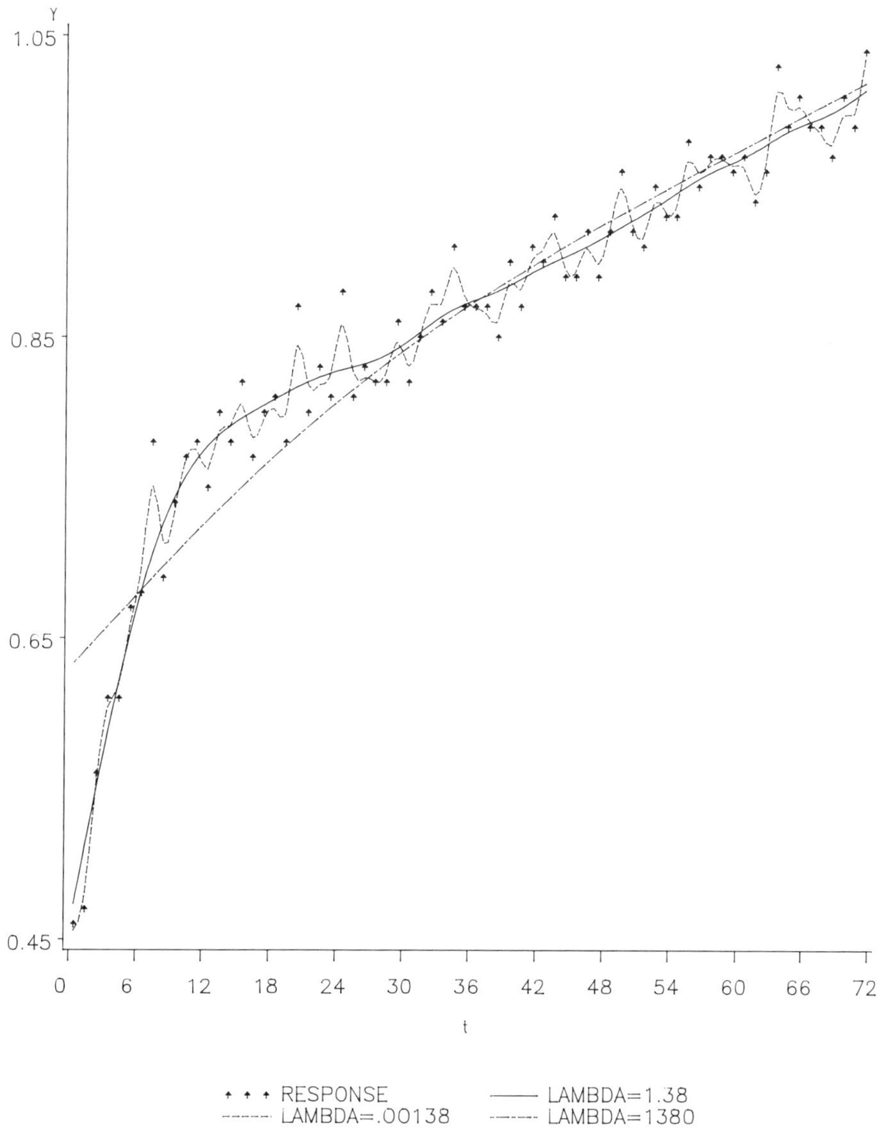


**Figure 5.3** Spline fits to the weight-height ratio.

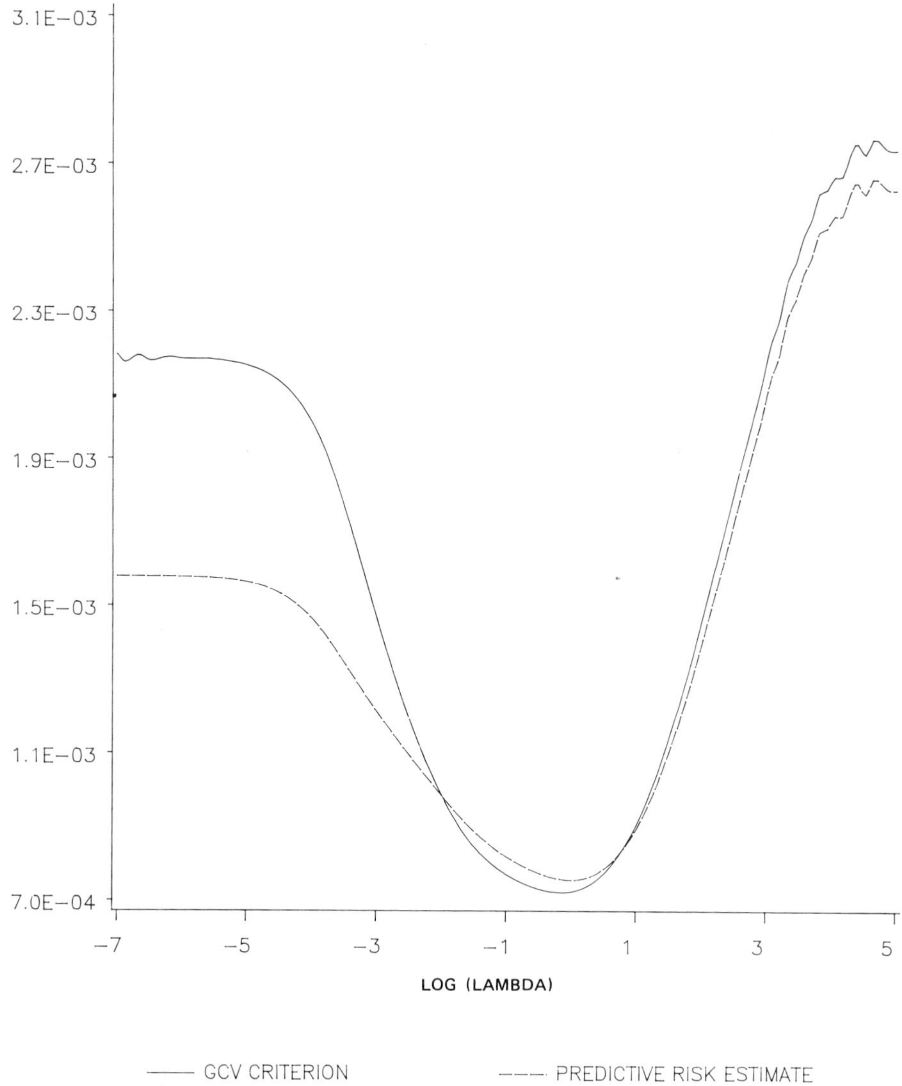


**Figure 5.4** GCV criterion and estimate of the predictive risk for the weight-height ratio data.

## 5.5 Bayesian Interpretations and Inference

In previous discussions we have focused on the properties of $\mu_\lambda$ when $\mu$ is some unknown function in $W_2^m[a,b]$. However, the fact that smoothing splines are a type of generalized ridge regression estimator (see the discussion after Corollary 5.2) means they also have an interpretation as a type of Bayes estimator (see Exercise 2.3). This latter viewpoint provides additional insights into the properties and motivations for the estimators. Therefore, in this section we explore the Bayesian perspective on spline smoothing and show that smoothing splines result from a type of Bayesian polynomial (nonparametric) regression analysis. The Bayesian approach pays some immediate dividends by suggesting various methods for interval estimation and diagnostic analysis.

### 5.5.1 Bayesian Polynomial Regression

A stochastic process is an indexed collection of possibly dependent random variables. There are many familiar examples of stochastic processes. For instance, a sequence of related random variables observed over time such as yearly sales is a stochastic process. In this particular example the random variables are the sales at each year and each random variable is indexed by its corresponding year.

For the developments in this section we require the idea of a normal process on [a,b]. This is a collection of random variables indexed by a continuous variable t taking on values in [a,b]. Let Y(t) be the random variable corresponding to any particular value of t. Then, we denote a stochastic process on [a,b] by $\{Y(t): t \in [a,b]\}$. The process is said to be normal if $\{Y(t_1),\ldots,Y(t_r)\}$ has a multivariate normal distribution for any finite collect of points $t_1,\ldots,t_r$ from [a,b]. To specify a normal process it suffices to specify its *mean value function*

$$\mu(t) = EY(t)$$

and its *covariance function* or *covariance kernel*

$$K(s,v) = \text{Cov}(Y(s), Y(v)).$$

With these preliminaries let us now consider the Bayesian polynomial regression model where the observations $(t_1,y_1),\ldots,(t_n,y_n)$ are obtained by sampling from a stochastic process $\{Y(t)\colon t \in [a,b]\}$ at predetermined points $t_1,\ldots,t_n$. The Y process is assumed to have a special form with

$$Y(t) = \mu(t) + \varepsilon(t), \quad t \in [a,b], \tag{41a}$$

where $\{\varepsilon(t)\colon t \in [a,b]\}$ is a zero mean normal process with

$$\text{Cov}(\varepsilon(s), \varepsilon(v)) = \begin{cases} 0, & v \neq s, \\ \sigma^2, & v = s. \end{cases} \tag{41b}$$

A Bayesian structure is introduced by placing a prior distribution on the mean value function $\mu$ of the Y process. It is assumed that $\{\mu(t)\colon t \in [a,b]\}$ has the same prior distribution as the stochastic process

$$\sum_{j=0}^{m-1} \theta_j t^j + \sigma_s Z(t), \quad \sigma_s > 0, \tag{41c}$$

where $\underline{\theta} = (\theta_0,\ldots,\theta_{m-1})'$ has a zero mean, normal distribution with variance-covariance matrix $\nu I$ and $\{Z(t)\colon t \in [a,b]\}$ is a zero mean normal process with covariance kernel (20). The polynomial coefficients, i.e., the $\theta_j$, and the Z process are also assumed to be uncorrelated. The Z process is what is called an (m - 1) -fold integrated Weiner process (see Parzen, 1962a and Shepp, 1966).

Model (41) seems to be quite a departure from our original model (2) where $\mu$ was some deterministic function in $W_2^m[a,b]$. In essense this model treats $\mu$ as being a polynomial plus a special type of random, rather than deterministic, error function. A line of reasoning which justifies this approach has been advanced by

Blight and Ott (1975). They note that once a deterministic function $\mu$ has been represented as far as seems possible by a polynomial, the experimenter then regards the remainder terms $\text{Rem}(t_1),\ldots,\text{Rem}(t_n)$ as errors with zero predicted values. Although $\text{Rem}(t)$ is deterministic, this belief can be incorporated into the estimation process by assuming that the $\text{Rem}(t_i)$ have a zero mean Gaussian (i.e., normal) prior distribution. The polynomial coefficients can also be assumed to have a normal prior distribution with mean $\bar{\underline{\theta}}$ and variance-covariance matrix $\nu I$. The last assumption can be justified by recalling that the Bayesian estimator for $\underline{\theta}$ in ordinary polynomial regression reduces to the usual least-squares estimator in the limit as $\nu \to \infty$. These considerations lead us to the use of a prior distribution for $\mu$ that coincides with the distribution of

$$\sum_{j=0}^{m-1} \theta_j t^j + \sigma_s Z(t) ,$$

where $\underline{\theta}$ is m-variate normal with $E\underline{\theta} = \bar{\underline{\theta}}$ and $\text{Var } \underline{\theta} = \nu I$. The partiular choices $\bar{\underline{\theta}} = \underline{0}$ and use of covariance kernel (20) for the Z process gives model (41). Other related models can be found in Blight and Ott (1975), Young (1977), O'Hagan (1978) and Steinberg (1983, 1985).

An alternative but related view of model (41) is that we extract a polynomial trend from $\mu$ and then approximate the remainder by a stochastic process. Such practices are not without precedent. Indeed, time series analysts have used this approach with great success to model complicated time series that may include deterministic components. To be more precise consider the discrete time signal-plus-noise model

$$y_t = \mu_t + \varepsilon_t, \qquad t = 0, \pm 1, \ldots,$$

where $\mu_t$ is a signal that may be deterministic, stochastic or both and is uncorrelated with the discrete time white noise process $\{\varepsilon_t: t = 0, \pm 1, \ldots\}$. A white noise process in this setting is a process with a covariance function of the form

$$\mathrm{Cov}(\varepsilon_v,\varepsilon_s) = \begin{cases} 0 , & v \neq s \\ \sigma^2 , & v = s . \end{cases}$$

The problem is to extract the signal from the response and filter out the noise. However, the signal portion of $y_t$ is often so complicated that no parametric model can be suggested for its form. Various flexible approximations are frequently used to handle such cases.

In practice the signal, $\mu_t$, can be adequately represented in many cases by removing a polynomial trend and treating the remainder, $\mathrm{Rem}_t$, as being a multiple of a process such as

$$Z_t = \sum_{j=0}^{p} \phi_j W_{t-j} ,$$

where $\{W_t: t = 0, \pm 1, \ldots\}$ is a discrete time white noise process with unit variance and the $\phi_j$ are real coefficients. The process $\{Z_t: t = 0, \pm 1, \ldots\}$ is called a pth order moving average process.

The estimation strategy which stems from these considerations is that of fitting an approximate model

$$y_t \doteq \sum_{j=0}^{m-1} \theta_j t^j + \sigma_s Z_t + \varepsilon_t, \qquad \sigma_s > 0 .$$

Various methods can then be used to estimate the $\theta_j$ and $\phi_j$.

The previous discussion has an obvious parallel in the case where t is continuous rather than discrete. If the process is (41b), then the approach would be to express $\mu(t)$ as $\sum_{j=0}^{m-1} \theta_j t^j$ + Rem(t) and approximate Rem(t) by a continuous time stochastic process. In the continuous time setting, moving average processes have the form

$$Z(t) = \int_a^b \phi(t - u)\,dW(u),$$

where $\{W(u): a \le u \le b\}$ is the Weiner process and $\phi$ is some specified function. Thus, Rem(t) might be approximated by $\sigma_s Z(t)$.

A particular choice for $\phi$ is $\phi(u) = u_+^{m-1}/(m-1)!$. In this case the Z process has covariance kernel (20) and model (41) is the end result. Thus, model (41) can be viewed as an approximation to model (2) where a stochastic approximation is used for the remainder term. Further discussion of this and related motivations can be found in Ansley and Wecker (1981) and Wecker and Ansley (1983).

We should note that model (41) is nonparametric in the sense that, unlike parametric models, no shapes are ruled out for the function $\mu$. However, it is not equivalent to model (2). The difficulty lies in the fact that if $\mu$ were to have an mth derivative it would be the derivative of a Weiner process. But the Wiener process provides the classic example of a smooth process that is nowhere differentiable with probability one. So, $J_m(\mu)$ fails to exist almost surely. More precisely, the sample paths of process (41c) have zero probability of being in $W_2^m[a,b]$ (for a more rigorous argument see Wahba, 1983). This leads to an interesting estimation paradox to be discussed subsequently.

Let us now determine the estimator of $\mu(t)$ obtained from model (41). The following lemma due to Wahba (1983) will prove helpful for this purpose. Its proof is left for the reader.

**Lemma 5.3.** Suppose that $\underline{y}$, $\underline{\mu}$ and $\underline{\varepsilon}$ are n-variate normal random vectors related by

$$\underline{y} = \underline{\mu} + \underline{\varepsilon} \ ,$$

with $E[\underline{\mu}] = E[\underline{\varepsilon}] = \underline{0}$, $E[\underline{\mu}\ \underline{\mu}'] = \sigma_s^2\Sigma_\mu$, $E[\underline{\varepsilon}\ \underline{\varepsilon}'] = \sigma^2 I$ and $E[\underline{\mu}\ \underline{\varepsilon}'] = 0$. Let $\underline{h}$ be p-variate normal and satisfy $E[\underline{h}] = \underline{0}$, $E[\underline{h}\ \underline{\mu}'] = \sigma_s^2\Sigma_{h\mu}$, $E[\underline{h}\ \underline{\varepsilon}'] = 0$ and $E[\underline{h}\ \underline{h}'] = \sigma_s^2\Sigma_h$. Then,

$$E[\underline{h}|\underline{y}] = \Sigma_{h\mu}(\Sigma_\mu + n\lambda I)^{-1}\underline{y},$$

where $n\lambda = \sigma^2/\sigma_s^2$. Also

$$\mathrm{Var}(\underline{h}|\underline{y}) = \sigma_s^2(\Sigma_h - \Sigma_{h\mu}\Sigma_\mu^{-1}\Sigma_{\mu h})$$

$$+ \sigma^2 \Sigma_{h\mu}\Sigma_\mu^{-1}\tilde{H}(\lambda)\Sigma_\mu^{-1}\Sigma_{\mu h},$$

with $\tilde{H}(\lambda) = \Sigma_\mu(\Sigma_\mu + n\lambda I)^{-1}$.

To apply this lemma we note that the Bayes estimator of $\mu(t)$ under model (41) and quadratic loss will be the posterior mean $E[\mu(t)|\underline{y}]$. Thus, make the identifications $\underline{\mu} = (\mu(t_1),\ldots,\mu(t_n))'$, $\underline{\varepsilon} = (\varepsilon_1,\ldots,\varepsilon_n)'$, $h = \mu(t)$ and define

$$\underline{t} = (1,\ldots,t^{m-1})', \qquad \underline{v} = (1,\ldots,v^{m-1})'$$

$$T = \{t_i^{j-1}\}_{\substack{i=1,n\\ j=1,m}} ,$$

$$\underline{Q}_t = (Q(t,t_1),\ldots,Q(t,t_n))', \quad \underline{Q}_v = (Q(v,t_1),\ldots,Q(v,t_n))'$$

and

$$Q_n = \{Q(t_i,t_j)\}_{i,j=1,n} .$$

Letting $\omega = \nu/\sigma_s^2$, we then find that

$$\Sigma_h = \omega\underline{t}'\underline{t} + Q(t,t) ,$$

$$\Sigma_{h\mu} = \omega\underline{t}'T' + \underline{Q}_t'$$

and

$$\Sigma_\mu = \omega TT' + Q_n .$$

Combining all this information gives the desired result stated formally below.

**Proposition 5.1.** A Bayes estimator of $\mu(t)$ is

$$\mu_{\omega,\lambda}(t) = E[\mu(t)|\underline{y}] = (\omega\underline{t}'T' + \underline{Q}_t')(\omega TT' + M(\lambda))^{-1}\underline{y} \tag{42a}$$

where $\omega = \nu/\sigma_s^2$ ,

$$M(\lambda) = Q_n + n\lambda I ,$$

and $n\lambda = \sigma^2/\sigma_s^2$. The posterior covariance kernel for $\mu(\cdot)$ is

$$\mathrm{Cov}(\mu(t),\mu(v)|\underline{y}) = \sigma_s^2\{Q(t,v) + \omega\underline{t}'\underline{v}$$

$$- (\omega\underline{t}'T' + \underline{Q}_t')(\omega TT' + Q_n)^{-1}(\omega T\underline{v} + \underline{Q}_v)\}$$

$$+ \sigma^2(\omega\underline{t}'T' + \underline{Q}_t')H_\omega(\lambda)(\omega T\underline{v} + \underline{Q}_v) ,$$

where $H_\omega(\lambda) = (\omega TT' + Q_n)(\omega TT' + M(\lambda))^{-1}$.

In order to actually compute this estimator of $\mu$ we must specify the parameters $\nu$ and $\lambda$. In particular, specification of the variance for the polynomial coefficients, $\nu$, would require some knowledge of how close values of $\underline{\theta}$ should be to their mean. In our situation it seems unlikely that any information of this type would be available and, hence, it makes sense to try to dispense with the parameter $\nu$ in some fashion. This can be accomplished by letting $\nu \to \infty$. The result is what is called a diffuse or improper prior for $\underline{\theta}$. Priors of this form can be envisioned as an attempt to place a uniform distribution on $\mathbb{R}^m$ which would entail that all values of $\underline{\theta}$ were equally likely. Of course no distribution can be uniform on $\mathbb{R}^m$ and still integrate to one. Hence, the term improper prior is used. Improper priors are ideally suited to situations where the statistician's knowledge about the parameters is limited relative to that contained in the sample. See DeGroot (1970) for further discussion of improper priors.

Letting $\nu \to \infty$ in Proportion 5.1 gives the following theorem due to Wahba (1978, 1983) (See, also Kimeldorf and Wahba, 1970a,b).

**Theorem 5.5.** If an improper prior distribution ($\nu \to \infty$) is used for the polynomial coefficients in model (41)

$$\mu_\lambda(t) \equiv \lim_{\omega\to\infty} \mu_{\omega,\lambda}(t)$$

$$= \underline{t}'(T'M(\lambda)^{-1}T)^{-1}T'M(\lambda)^{-1}\underline{y}$$

$$+ \underline{Q}_t'M(\lambda)^{-1}[I - T(T'M(\lambda)^{-1}T)^{-1}T'M(\lambda)^{-1}]\underline{y} \ . \tag{43a}$$

Also, the limiting (as $\nu \to \infty$) posterior covariance kernel for $\mu(\cdot)$ is

$$\begin{aligned} \mathrm{Cov}(\mu(v),\mu(t)|\underline{y}) = \sigma_s^2\{Q(v,t) &+ \underline{t}'(T'Q_n^{-1}T)^{-1}\underline{v} \\ &- \underline{Q}_t'F_0'\underline{v} - \underline{t}'F_0\underline{Q}_v - \underline{Q}_t'P\underline{Q}_v\} \\ &+ \sigma^2(\underline{t}'F_0 + \underline{Q}_t'P)H(\lambda)(F_0'\underline{v} + P\underline{Q}_v) \end{aligned} \tag{43b}$$

with

$$F_0 = (T'Q_n^{-1}T)^{-1}T'Q_n^{-1}$$

and

$$P = Q_n^{-1} - Q_n^{-1}TF_0 \ .$$

*Proof.* We will prove only the first part of this theorem. The form for the posterior covariance kernel follows from similar arguments and is left as an exercise.

Recall the well known matrix identity, sometimes called the Sherman-Morrison-Woodbury formula, which states that for matrices A, B, C and D, $(A + BC^{-1}D)^{-1} = A^{-1} - A^{-1}B(C + DA^{-1}B)^{-1}DA^{-1}$, assuming all the necessary inverses and products are well defined. An application of this formula gives

$$\begin{aligned} (\omega TT' + M(\lambda))^{-1} &= M(\lambda)^{-1} \\ &+ M(\lambda)^{-1}T(T'M(\lambda)^{-1}T)^{-1}(I + \omega^{-1}(T'M(\lambda)^{-1}T)^{-1})^{-1}T'M(\lambda)^{-1} \ . \end{aligned}$$

Since for $\nu$ (and, hence, $\omega$) sufficiently large the eigenvalues of $\omega^{-1}(T'M(\lambda)^{-1}T)^{-1}$ will all be less than one, a matrix power series can be used to obtain

$$\begin{aligned} (\omega TT' + M(\lambda))^{-1} &= M(\lambda)^{-1} - M(\lambda)^{-1}T(T'M(\lambda)^{-1}T)^{-1}T'M(\lambda)^{-1} \\ &+ \omega^{-1}M(\lambda)^{-1}T(T'M(\lambda)^{-1}T)^{-2}T'M(\lambda)^{-1} + O(\omega^{-2}) \ . \end{aligned}$$

Thus

$$\lim_{\omega \to \infty} (\omega TT' + M(\lambda))^{-1} = M(\lambda)^{-1}[I - T(T'M(\lambda)^{-1}T)^{-1}T'M(\lambda)^{-1}]$$

and

$$\lim_{\omega \to \infty} \omega T'(\omega TT' + M(\lambda))^{-1} = (T'M(\lambda)^{-1}T)^{-1}T'M(\lambda)^{-1}.$$

Combining these facts with formula (42a) gives (43a). □

By referring back to our discussion of the Kimeldorf-Wahba basis (c.f. (24)) we see that $\mu_\lambda$ defined in (43a) is identical to the smoothing spline estimator discussed throughout this chapter. Consequently, smoothing splines can also be viewed as deriving from the polynomial regression model (41) with a diffuse prior placed on the polynomial coefficients. There are other Bayesian models which lead to smoothing spline estimators: see Shiller (1984) and Silverman (1985).

As previously noted the sample paths (i.e., realized values) of $\mu$ have zero probability of being in $W_2^m[a,b]$ under the Bayesian model. However, $\mu_\lambda$ is most definitely a member of this function space. Consequently, the Bayesian approach gives an estimator of $\mu$ which is not a plausible candidate for the realized regression curve. This is the estimation paradox referred to earlier. It causes some conceptual difficulties and, in particular, means that care must be taken in trying to transfer results from the Bayesian model back to model (2). An alternative model due to Silverman (1985) avoids this estimation paradox by modeling $\mu$ as a natural spline which ensures that the sample paths of $\mu$ are in $W_2^m[a,b]$ (see Exercise 11).

The parameter $\lambda$ has a somewhat different meaning under the Bayesian view of $\mu_\lambda$. However, its basic function as an indicator of how much smoothing is needed remains unchanged. Since $\lambda = \sigma^2/n\sigma_s^2$, it follows that $n\lambda$ is the ratio of the noise and signal variance, provided we consider $\{\sigma_s Z(t): t \in [a,b]\}$ to be a stochastic signal in (41c). If $\sigma_s$ is small relative to $\sigma$ then the signal will tend to dominate the lower frequency portion of the data while the noise will be concentrated at high frequencies. Consequently, smoothing is needed which is reflected by the fact that $\lambda$ will be large in this case. Conversely, if $\sigma$ is small relative to $\sigma_s$ appreciable contributions from the signal will also appear in the higher frequency portion of the data so that a rougher estimate, i.e., smaller value for $\lambda$, is in order.

To conduct inference about $\mu$ from the Bayesian model we must, in general, select or estimate values for the remaining unknown parameters: $\lambda$, $\sigma^2$ and $\sigma_s^2$. If point estimation is all that is of interest then, of course, $\lambda$ is all that must be estimated. However if interval estimates for functionals of $\mu$, other than just the $\mu(t_i)$, are needed we must also estimate $\sigma^2$ and $\sigma_s^2$. Let us first deal with selection of $\lambda$.

### 5.5.2 Selection of λ in the Bayes Model

When no *a priori* value is available for $\lambda$ we could again resort to GCV for its selection. Alternatively, a likelihood based approach might be considered. Procedures of this latter variety have been studied by Anderssen and Bloomfield (1974a,b), Wecker and Ansley (1983) and Wahba (1977b,c, 1985). To derive Wahba's likelihood criterion we must first examine the unconditional or *predictive distribution* of the response vector $\underline{y}$.

If our data is obtained from model (41) then $\underline{y}$ has a normal distribution with mean

$$\underline{\mu} = (\mu(t_1),\ldots,\mu(t_n))'$$

and variance-covariance matrix $\sigma^2 I$, conditional on $\underline{\theta}$ and $\underline{Z} = (Z(t_1),\ldots,Z(t_n))'$. Now

$$\underline{\mu} = T\underline{\theta} + \sigma_s \underline{Z}$$

so the unconditional mean of $\underline{y}$ is $E\underline{\mu} = 0$, because $E\underline{\theta} = \underline{0}$ and $E\underline{Z} = \underline{0}$. To find the unconditional variance of $\underline{y}$ observe that

$$\text{Var } \underline{y} = \text{E Var}(\underline{y}|\underline{\theta}, \underline{Z}) + \text{Var E}(\underline{y}|\underline{\theta},\underline{Z})$$

$$= \sigma^2 I + \text{Var}(T\underline{\theta} + \sigma_s \underline{Z})$$

$$= \sigma^2 I + \nu TT' + \sigma_s^2 Q_n$$

$$= \sigma_s^2 (M(\lambda) + \omega TT') \quad . \tag{44}$$

Thus, $\underline{y}$ is n-variate normal with mean $\underline{0}$ and variance-covariance matrix given by (44).

We now wish to let $\omega \to \infty$ as before and utilize the resulting limiting likelihood to obtain a criterion for selecting $\lambda$. To accomplish this we follow Wahba (1985) and first transform from $\underline{y}$ to

$$\begin{bmatrix} \underline{u} \\ \underline{v} \end{bmatrix} = \begin{bmatrix} U' \\ \omega^{-1/2} T' \end{bmatrix} \underline{y} \tag{45a}$$

with U any matrix that satisfies (27). It is then found that

$$\text{Cov}(\underline{u},\underline{v}) = \sigma_s^2 \omega^{-1/2} U'(M(\lambda) + \omega TT')T = \sigma_s^2 \omega^{-1/2} U'M(\lambda)T, \tag{45b}$$

$$\text{Var}(\underline{v}) = \sigma_s^2 \omega^{-1} T'(M(\lambda) + \omega TT')T \tag{45c}$$

and

$$\text{Var } \underline{u} = \sigma_s^2 U'M(\lambda)U \tag{45d}$$

as $U'T = 0$. Now let $\omega \to \infty$ and observe that only the distribution of $\underline{u}$ involves $\lambda$ since

$$\lim_{\omega \to \infty} \text{Cov}(\underline{u}, \underline{v}) = 0$$

and

$$\lim_{\omega \to \infty} \text{Var } \underline{v} = \sigma_s^2 (T'T)^2 \ .$$

Thus, we can restrict attention to the portion of the likelihood corresponding to $\underline{u}$ for the purpose of selecting $\lambda$.

The log of the $\underline{u}$ likelihood is

$$-\frac{1}{2}\log|\sigma_s^2 U'M(\lambda)U| - \frac{1}{2\sigma_s^2}\underline{u}'(U'M(\lambda)U)^{-1}\underline{u}$$

$$= -\frac{n-m}{2}\log\sigma_s^2 - \frac{1}{2}\log|U'M(\lambda)U| - \frac{1}{2\sigma_s^2}\underline{u}'(U'M(\lambda)U)^{-1}\underline{u}\ .$$

From this we see that a maximum likelihood estimator of $\sigma_s^2$ is

$$\tilde{\sigma}_{s\lambda}^2 = \underline{u}'(U'M(\lambda)U)^{-1}\underline{u}/(n - m). \tag{46}$$

Substituting $\tilde{\sigma}_{s\lambda}^2$ back into the likelihood function gives the concentrated likelihood

$$-\frac{n-m}{2}\log\tilde{\sigma}_{s\lambda}^2 - \frac{1}{2}\log|U'M(\lambda)U| - \frac{n-m}{2}\qquad .$$

Maximizing this is equivalent to minimizing

$$\mathrm{GML}_n(\lambda) = \underline{u}'(U'M(\lambda)U)^{-1}\underline{u}/|(U'M(\lambda)U)^{-1}|^{1/(n-m)}$$

$$= \underline{y}'U(U'M(\lambda)U)^{-1}U'\underline{y}/|(U'M(\lambda)U)^{-1}|^{1/(n-m)},$$

which is the generalized maximum likelihood (GML) criterion studied in Wahba (1985).

Both GCV and GML appear to have similar properties for selecting $\lambda$ under the Bayesian formulation. An indication of this is provided by the following theorem due to Wahba (1977b,c,1985) which relates the minimizers of the risk and GML and GCV criteria under this model. For notational simplicity we will write $R(\lambda)$, $\mathrm{GML}(\lambda)$ and $\mathrm{GCV}(\lambda)$ for $R_n(\lambda)$, $\mathrm{GML}_n(\lambda)$ and $\mathrm{GCV}_n(\lambda)$ throughout the theorem and its proof.

**Theorem 5.6.** Let E and $E_\mu$ denote expectations with respect to the distribution of $\underline{\varepsilon} = (\varepsilon_1,\ldots,\varepsilon_n)'$ and the prior distribution of $\underline{\mu} = (\mu(t_1),\ldots,\mu(t_n))'$, respectively. Then $E_\mu R(\lambda)$, $E_\mu \mathrm{EGCV}(\lambda)$ and $E_\mu \mathrm{EGML}(\lambda)$ are all minimized at $\lambda = \sigma^2/n\sigma_s^2$.

*Proof.* Refer to Proposition 5.1 which states that when $\underline{\theta}$ is normal with mean $\underline{0}$ and Var $\underline{\theta} = \nu I$, the optimal vector of estimates for $\underline{\mu}$ is

$$\underline{\mu}_{\lambda,\omega} = H_\omega(\lambda)\underline{y}$$

with

$$H_\omega(\lambda) = \omega TT'(\omega TT' + M(\lambda))^{-1} + Q_n(\omega TT' + M(\lambda))^{-1} \quad ,$$

where $\omega = \nu/\sigma_s^2$ and $M(\lambda) = Q_n + n\lambda I$. Thus denoting the GCV criterion for this estimator by $GCV_\omega(\lambda)$ we see that

$$[n^{-1}tr(I - H_\omega(\lambda))]^2 EGCV_\omega(\lambda) = n^{-1}\underline{\mu}'(I - H_\omega(\lambda))^2\underline{\mu} + \frac{\sigma^2}{n} tr[(I - H_\omega(\lambda))^2]$$

and, since $E\underline{\mu} = \underline{0}$ and Var $\underline{\mu} = \sigma_s^2 Q_n + \nu TT'$,

$$[n^{-1}tr(I - H_\omega(\lambda))]^2 E_\mu EGCV_\omega(\lambda) = \frac{\sigma_s^2}{n} tr[(I - H_\omega(\lambda))^2(Q_n + \omega TT')] + \frac{\sigma^2}{n} tr[(I - H_\omega(\lambda))^2] \ .$$

Using techniques similar to those used in proving Theorem 5.5 it is found that as $\nu$ (and hence $\omega$) approaches infinity

$$tr[(I - H_\omega(\lambda))^2(Q_n + \omega TT')] \rightarrow tr[(I - H(\lambda))^2 Q_n] \quad ,$$

where $H(\lambda)$ is defined in (30b) and we have used the fact that $T'(I - H(\lambda)) = 0$. Consequently under a diffuse prior ($\nu \rightarrow \infty$)

$$[n^{-1}tr(I - H(\lambda))]^2 E_\mu EGCV(\lambda) = \{\frac{\sigma_s^2}{n} tr[(I - H(\lambda))^2 Q_n] + \frac{\sigma^2}{n} tr[(I - H(\lambda))^2]\} \ .$$

To show that $\lambda = \sigma^2/n\sigma_s^2$ minimizes this last expression note that by choosing U so that U'U = I we obtain, as a result of (30b),

$$I - H(\lambda) = (n\lambda)U\Gamma\ \mathrm{diag}((d_1 + n\lambda)^{-1}, \ldots, (d_{n-m} + n\lambda)^{-1})\Gamma'U',$$

where $d_1, \ldots, d_{n-m}$ are the eigenvalues and $\Gamma$ is the matrix of eigenvectors of $U'Q_nU$. This form for $I - H(\lambda)$ allows us to express $E_\mu EGCV(\lambda)$ as

$$E_\mu EGCV(\lambda) = n^2\sigma_s^2 \sum_{i=1}^{n-m} \frac{d_i/n + \sigma^2/n\sigma_s^2}{(d_i/n + \lambda)^2} \Bigg/ \left[\sum_{i=1}^{n-m} (d_i/n + \lambda)^{-1}\right]^2 .$$

When we take $\lambda = \sigma^2/n\sigma_s^2$ this reduces to

$$E_\mu EGCV_n(\sigma^2/n\sigma_s^2) = n^2\sigma_s^2 \Big/ \sum_{i=1}^{n-m} (d_i/n + \sigma^2/n\sigma_s^2)^{-1} .$$

Now consider the difference between $E_\mu EGCV(\lambda)$ and $E_\mu EGCV(\sigma^2/n\sigma_s^2)$ which is found, after a little algebra, to be proportional to

$$\left\{ \sum_{i=1}^{n-m} \frac{d_i/n + \sigma^2/n\sigma_s^2}{(d_i/n + \lambda)^2} \sum_{j=1}^{n-m} (d_j/n + \sigma^2/n\sigma_s^2)^{-1} - \left[\sum_{i=1}^{n-m} (d_i/n + \lambda)^{-1}\right]^2 \right\} .$$

If this quantity can be shown to be nonnegative $\lambda = \sigma^2/n\sigma_s^2$ will have then been established as a minimizer. To see that this is the case let

$$\alpha_i(\lambda) = (d_i/n + \sigma^2/n\sigma_s^2)^{1/2}/(d_i/n + \lambda) ,$$

$$\xi_i = (d_i/n + \sigma^2/n\sigma_s^2)^{-1/2},$$

$i = 1, \ldots, n - m$, and apply the Cauchy-Schwarz inequality to obtain

$$\left[\sum_{i=1}^{n-m}(d_i/n+\lambda)^{-1}\right]^2=\left[\sum_{i=1}^{n-m}\alpha_i(\lambda)\xi_i\right]^2\le\sum_{i=1}^{n-m}\alpha_i(\lambda)^2\sum_{j=1}^{n-m}\xi_j^2$$

$$=\sum_{i=1}^{n-m}\frac{d_i/n+\sigma_s^2/n\sigma^2}{(d_i/n+\lambda)^2}\sum_{j=1}^{n-m}(d_i/n+\sigma^2/n\sigma_s^2)^{-1}.$$

Equality holds in the last expression if and only if

$$\alpha_i(\lambda)=C(\lambda)\xi_i,\qquad i=1,\ldots,n-m,$$

for some constant $C(\lambda)$. This is equivalent to $(d_i/n + \sigma^2/n\sigma_s^2)/(d_i/n + \lambda) = C(\lambda)$. If the $d_i$ do not all coincide, equality can only occur for $\lambda = \sigma^2/n\sigma_s^2$ and the minimizer is unique. In the unlikely event that the $d_i$ are all equal the statement of the theorem still stands since $E_\mu EGCV(\lambda)$ is constant in that case.

The verification that $\sigma^2/n\sigma_s^2$ is also a minimizer of $E_\mu R(\lambda)$ and $E_\mu EGML(\lambda)$ follows along similar lines and is left as an exercise. □

Theorem 5.6 suggests that GCV and GML should perform similarly if the Bayesian model holds. However, the model which is probably of most interest is model (2) with $\mu \in W_2^m[a,b]$. An important application for the Bayesian approach is, as we shall see, to suggest plans of attack for inference in model (2).

Wahba (1985) has investigated the utility of the GML criterion for selecting $\lambda$ when model (2), rather than (41), is believed to hold. She shows that if $\mu \in W_2^m[a,b]$ and $\tilde{\lambda}_n$ and $\bar{\lambda}_n$ are minimizers of $EGCV_n(\lambda)$and $EGML_n(\lambda)$ then

$$1+o(1)\le R_n(\bar{\lambda}_n)/R_n(\tilde{\lambda}_n)<\infty\qquad,$$

where $o(1)$ is positive and decays to zero as $n \to \infty$. Under additional restrictions, such as $\mu \in W_2^m[a,b]$ with $\mu^{(j)}(a) = \mu^{(j)}(b)$, $j = m,\ldots,2m-1$, she is able to show that $R_n(\bar{\lambda}_n)/R_n(\tilde{\lambda}_n)$

becomes arbitrarily large. Since we already know that

$$R_n(\tilde{\lambda}_n)/\inf_{0\le\lambda\le\infty} R_n(\lambda) \to 1$$

as $n\to\infty$, the conclusion to be drawn from Wahba's results is that GML may not perform well in the case of a nonstochastic regression function. Simulation results in Wahba (1985) tend to confirm this.

There are other likelihood based methods for selecting $\lambda$. One of these is due to Wecker and Ansley (1983). They do not place a diffuse prior on the polynomial coefficients but, instead, utilize maximum likelihood techniques to estimate $\underline{\theta}$, $\sigma_s^2$ and $\lambda$. Their criterion for selecting $\lambda$ is found to be (Exercise 13)

$$WA(\lambda) = \underline{y}'U(U'M(\lambda)U)^{-1}U'\underline{y}/|M(\lambda)^{-1}|^{1/n}.$$

See Wahba (1985) for some discussion of their approach.

### 5.5.3 Estimation of $\sigma^2$ and $\sigma_s^2$

Given a value for $\lambda$ there are many potential estimators of $\sigma^2$ and $\sigma_s^2$ which can be justified through the Bayesian framework. For example, Wahba (1985) has suggested that we estimate $\sigma^2$ and $\sigma_s^2$ by

$$\sigma_\lambda^2 = nMSE(\lambda)/tr(I - H(\lambda)) \tag{48}$$

and

$$\sigma_{s\lambda}^2 = J_m(\mu_\lambda)/(trH(\lambda) - m) \quad ,$$

respectively. The first estimator is motivated by the fact that it generalizes the usual estimate of the error variance from linear models: the sum of squared errors divided by the sample size less the number of estimated parameters which is the trace of the identity minus that of the hat matrix in that case (c.f. (A.10)). The estimator $\sigma_{s\lambda}^2$ is just a rescaling of the estimated smoothness criterion $J_m(\mu_\lambda)$.

For known $\lambda$ one finds that both $\sigma_\lambda^2$ and $\sigma_{s\lambda}^2$ are unbiased estimators of their respective parameters. To see this write

$$nMSE(\lambda) = (n\lambda)^2 \underline{u}'(U'M(\lambda)U)^{-1}U'U(U'M(\lambda)U)^{-1}\underline{u}$$

and

$$J_m(\mu_\lambda) = \underline{u}'(U'M(\lambda)U)^{-1}U'Q_nU(U'M(\lambda)U)^{-1}\underline{u}$$

with $\underline{u} = U'\underline{y}$. Now take expectations using Lemma A.1, the predictive distribution of $\underline{u}$ and the fact that $\lambda = \sigma^2/n\sigma_s^2$ to obtain the desired conclusion. A similar argument shows that estimator (46) is unbiased for $\sigma_s^2$ in this case.

In the case where $\lambda$ is also estimated, a maximum likelihood estimator of $\sigma^2$ is $\tilde{\sigma}^2_{s\tilde{\lambda}}$, where $\tilde{\lambda}$ is the GML estimator of $\lambda$ and $\tilde{\sigma}^2_{s\lambda}$ is defined in (46). A corresponding maximum likelihood estimator of $\sigma^2$ is $\tilde{\sigma}^2_{\tilde{\lambda}} = n\tilde{\lambda}\tilde{\sigma}^2_{s\tilde{\lambda}}$. Note that

$$\begin{aligned}(n - m)\tilde{\sigma}^2_{\tilde{\lambda}} &= (n - m)n\tilde{\lambda}\tilde{\sigma}^2_{s\tilde{\lambda}} \\ &= n\tilde{\lambda}\underline{u}'(U'M(\tilde{\lambda})U)^{-1}\underline{u} \\ &= n\tilde{\lambda}\underline{u}'(U'M(\tilde{\lambda})U)^{-1}(U'M(\tilde{\lambda})U)(U'M(\tilde{\lambda})U)^{-1}\underline{u} \\ &= (n\tilde{\lambda})^2\underline{u}'(U'M(\tilde{\lambda})U)^{-1}U'U(U'M(\tilde{\lambda})U)^{-1}\underline{u} \\ &\quad + n\tilde{\lambda}\underline{u}'(U'M(\tilde{\lambda})U)^{-1}U'Q_nU(U'M(\tilde{\lambda})U)^{-1}\underline{u} \\ &= n(MSE(\tilde{\lambda}) + \tilde{\lambda}J_m(\mu_{\tilde{\lambda}})) .\end{aligned}$$

Thus, $\tilde{\sigma}^2_{\tilde{\lambda}}$ is $n/(n - m)$ times our original criterion evaluated at the estimator $\mu_{\tilde{\lambda}}$ of $\mu$. Consequently, this estimator also has intuitive appeal. Shiller (1984) derived a similar "estimator" for $\sigma^2$ by treating $\lambda$ as known and using a noninformative prior for $\sigma^{-2}$. One finds that $\tilde{\sigma}_\lambda^{-2}$ is the posterior mean of $\sigma^{-2}$ in that case.

Simulation results in Wahba (1983) indicate that $\sigma^2_\lambda$ behaves satisfactorily as an estimator of $\sigma^2$ when $\mu$ is nonrandom. Other estimators related to $\sigma_\lambda^2$, $\sigma_{s\lambda}^2$, $\tilde{\sigma}_\lambda^2$ and $\tilde{\sigma}_{s\lambda}^2$ have been studied by Shiau (1985) and Green, Jennison and Seheult (1985).

### 5.5.4 Interval Estimation

There are several methods available for interval estimation using smoothing splines. One of these is based on the Bayesian interpretation of $\mu_\lambda$ and was first studied by Wahba (1983).

Let us concentrate on constructing interval estimates for $\mu(t_1),\ldots,\mu(t_n)$. As a result of Theorem 5.5 we see that

$$\text{Var}(\underline{\mu}|\underline{y}) = \sigma^2 H(\lambda). \tag{49}$$

In particular then

$$\text{Var}(\mu(t_i)|\underline{y}) = \sigma^2 h_{ii}(\lambda) ,$$

where $h_{ii}(\lambda)$ is the ith diagonal element of the hat matrix $H(\lambda)$. Thus, for any $0 < \alpha < 1$

$$P(|\mu(t_i) - \mu_\lambda(t_i)| \le Z_{\alpha/2}[\sigma^2 h_{ii}(\lambda)]^{1/2}) = 1 - \alpha ,$$

from which we obtain the $100(1 - \alpha)\%$ "confidence interval"

$$\mu_\lambda(t_i) \pm Z_{\alpha/2}[\sigma^2 h_{ii}(\lambda)]^{1/2} \tag{50}$$

for $\mu(t_i)$. Strictly speaking this is not a confidence interval but rather a prediction interval since $\mu(t_i)$ is a random variable under the Bayesian model.

When $\sigma^2$ is unknown in (50) it can be estimated using, for example, estimator (48). The computation of the $h_{ii}(\lambda)$ can be accomplished efficiently by using the techniques outlined in Hutchinson and deHoog (1985, 1987).

The confidence interval (50) has the pleasing intuitive property that it directly parallels the form of the interval estimate for the value of $\mu(t_i)$ in linear regression analysis (c.f. (A17)). While such intervals are perfectly valid under the Bayesian model the question of most import is how do intervals such as (50) perform when $\mu$ is nonrandom and in $W_2^m[a,b]$. The answer to this seems to be that they work remarkably well.

One reason the Bayesian confidence intervals may also work well for model (2) has been advanced by Wahba (1983). She conjectured, and gave an argument to the effect that,

$$R_n(\lambda_{opt}) \doteq \sigma^2 \mathrm{tr}H(\lambda_{opt})/n . \tag{51}$$

More precisely her argument suggested that $\sigma^2 C\mathrm{tr}H(\lambda_{opt})/n = R_n(\lambda_{opt})(1 + o(1))$ for some constant C in the interval $[(4m + 1)(2m - 1)/8m^2, 1]$. This result has since been established for the case of periodic smoothing splines by Hall and Titterington (1985) who found that $C = (4m + 1)(2m - 1)/8m^2$. Nychka (1986) assumes that the $\varepsilon$'s have finite eighth moments, $m \geq 2$ and $\mu \in W_2^{2m}[a,b]$ satisfies the natural boundary conditions $\mu^{(j)}(a) = \mu^{(j)}(b) = 0$, $j = m,\ldots,2m - 1$. He then shows that $\sigma_{\tilde{\lambda}}^2 \mathrm{tr}H(\tilde{\lambda})/R_n(\lambda_{opt}) \xrightarrow{P} (4m + 1)(2m - 1)/8m^2$, where $\tilde{\lambda}$ is a (restricted) minimizer of the GCV criterion. The conclusion that can be drawn from this is that if the leverage values are all near their average and the smoothing parameter has been selected in an optimal fashion then the use of the Bayesian intervals should roughly correspond to using the risk $R_n(\lambda)$, rather than the variance of $\mu_\lambda(t_i)$, to assess the accuracy of $\mu_\lambda(t_i)$ as an estimator of $\mu(t_i)$.

In her discussion of Silverman (1985), Wahba points out the implications of (51) in terms of the frequency interpretations of the intervals (50). She notes that for nonrandom $\mu$ when $\lambda$ has been selected optimally the coverage properties of the Bayesian intervals should be viewed as across a curve. Thus, for example, if a sample of 100 observations is taken following model (2) and the corresponding 100 intervals (50) are computed with $\alpha = .05$ and $\lambda$ selected by GCV, then one should have approximately 95 (e.g., 94, 93 or 96, etc.) of the true function values contained in their respective confidence intervals. This type of pattern should persist over repeated sampling. However, it should be anticipated that the interval estimates will tend to fail to cover the *same* points in *different* samples. The points where such failures will occur will usually lie in regions where the bias of the smoothing spline estimator has its largest magnitude. For some illustrations of this property see Wahba (1983).

Another reason for hoping that the Bayesian intervals might work well for estimation of $\mu$ stems from the fact that for many cases in parametric inference the frequentist coverage probabilities for Bayesian intervals tend to their posterior coverage probabilities as $n \to \infty$. One might hope this would also be true for the intervals (50). However, Cox (1986) has shed doubt on this possibility by showing that simultaneous Bayesian intervals for a class of models closely related to model (41) do not possess this property.

An obvious question to ask is why not use an interval estimate of the form $\mu_\lambda(t_i) \pm Z_{\alpha/2}[\text{Var}\ \mu_\lambda(t_i)]^{1/2}$, where Var $\mu_\lambda(t_i)$ is the true variance of $\mu_\lambda(t_i)$ under model (2). The reason such intervals are not satisfactory when model (2) holds is that $\mu_\lambda(t_i)$ is a biased estimator of $\mu(t_i)$ and, hence, its variance tends to understate the expected (squared) magnitude of its departure from the parameter being estimated. Thus, replacement of Var $\mu_\lambda(t_i)$ by something like $\sigma^2 h_{ii}(\lambda)$ which, as a result of (51), should tend to estimate $R_n(\lambda)$ seems like a sound practice since it tends to give wider interval estimates and takes account of the fact that $\mu_\lambda$ is biased. It can actually be shown that the Bayesian intervals are always wider than intervals based on the true variance (Exercise 15).

To investigate the behavior of the intervals in (50) when $\mu \in W_2^m[a,b]$ a small scale simulation was conducted. Data was generated from model (2) with normal errors, equally spaced $t_j$ and regression functions

$$\mu_{(1)}(t) = 4.26\{\exp(-3.25t) - 4\exp(-6.5t) + 3\exp(-9.75t)\}$$

and

$$\mu_{(2)}(t) = \{B_{10,5}(t) + B_{7,7}(t) + B_{5,10}(t)\}/3,$$

where

$$B_{p,q}(t) = \Gamma(p + q)t^{p-1}(1 - t)^{q-1}/\Gamma(p)\Gamma(q), \ 0 \le t \le 1 ,$$

and $\Gamma(\cdot)$ is the gamma function. The function $\mu_{(1)}$ is a rescaled version of a function considered by Wahba and Wold (1975a) whereas $\mu_{(2)}$ was utilized in the simulation study in Wahba (1983). Two sample size configurations were used: m = 50 replicates of sample size n = 80 and m = 100 replicates of n = 40. Four values were chosen for $\sigma$: .2, .4, .6, and .8.

Smoothing splines were fitted to each data set using GCV to select the smoothing parameters. The intervals (50) were then computed with $\alpha$ = .05 and (48) was used to estimate $\sigma^2$. The results summarized in Table 5.1 are the proportion of times the true function values were contained in the interval estimates. Consequently, the entries in Table 5.1 should be compared to the nominal value of .95. Figures 5.5 and 5.6 provide plots of typical simulated data sets for each of the regression functions along with plots of the true and estimated curves and interval estimates. As might be expected, the intervals seem to perform worse where the true curve changes rapidly and the estimator $\mu_\lambda$ will be more biased.

**Table 5.1 Empirical Coverage Levels**

| $\mu_{(1)}$ | m = 100 n = 40 | m = 50 n = 80 |
|---|---|---|
| $\sigma$ = .2 | .944 | .941 |
| $\sigma$ = .4 | .908 | .914 |
| $\sigma$ = .6 | .854 | .876 |
| $\sigma$ = .8 | .850 | .856 |
| $\mu_{(2)}$ | | |
| $\sigma$ = .2 | .911 | .954 |
| $\sigma$ = .4 | .912 | .931 |
| $\sigma$ = .6 | .923 | .931 |
| $\sigma$ = .8 | .933 | .933 |

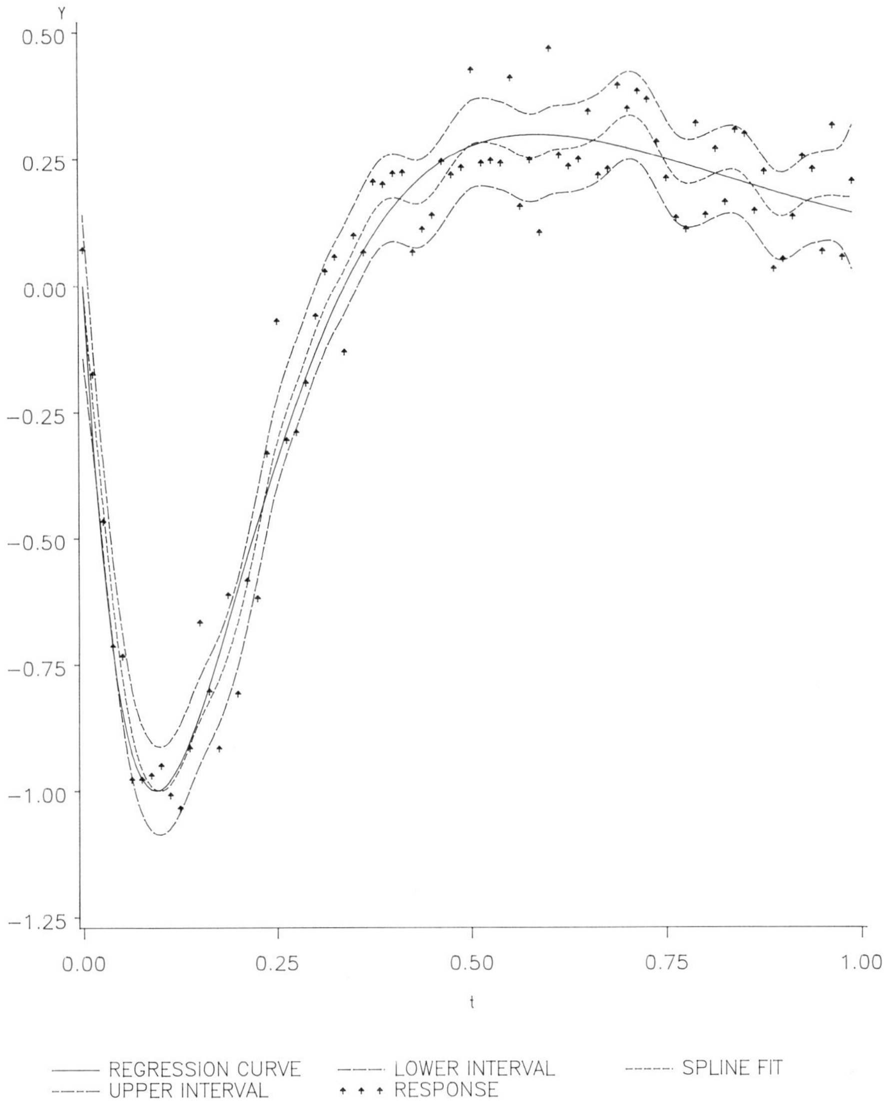


**Figure 5.5** Spline fit and interval estimates for simulated data.

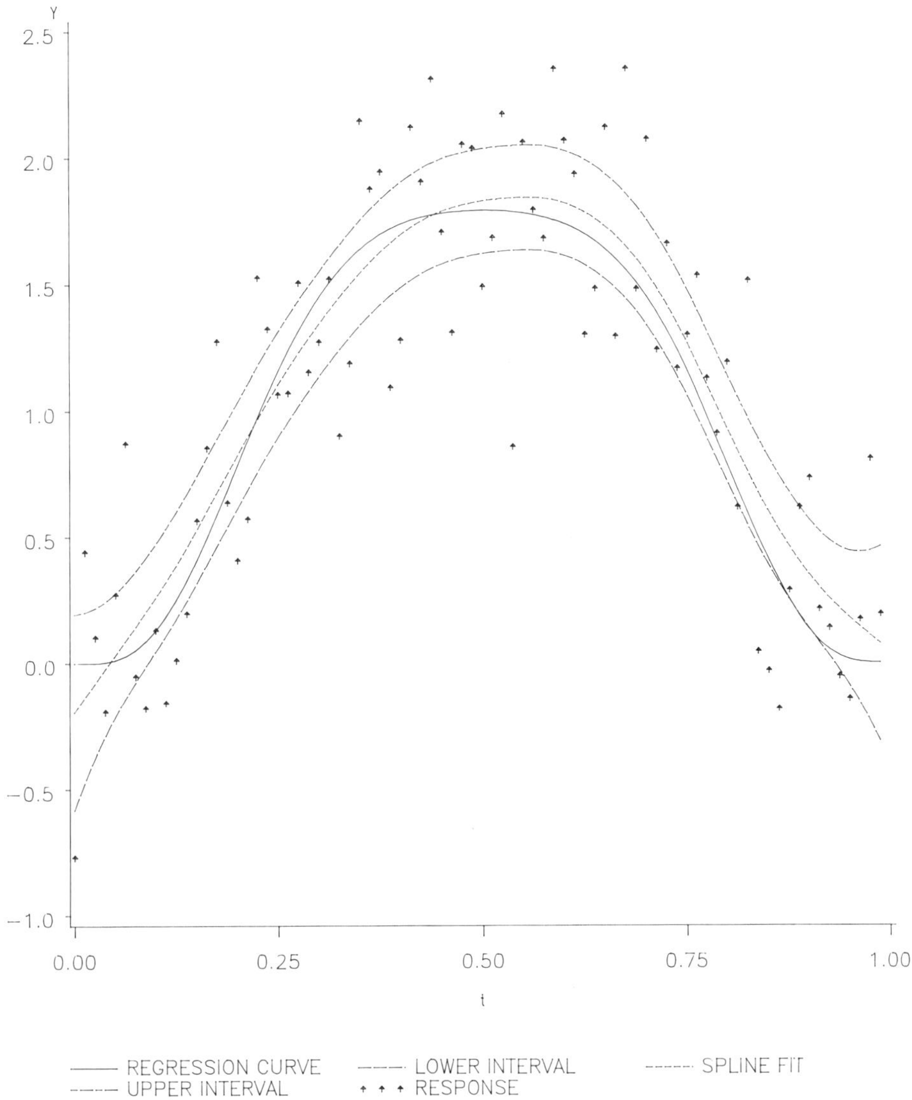


**Figure 5.6** Spline fit and interval estimates for simulated data.

To analyze the data in Table 5.1 one might treat each proportion in the table as stemming from a binomial experiment consisting of 4000 trials and then test the hypothesis that the true underlying coverage probabilities are .95. The assumptions for the binomial are not met here since the coverage probabilities undoubtedly are not the same at all points along the two regression curves. Nonetheless, for comparison purposes we conducted a analysis of this type using the asymptotic normal distribution for a sample proportion to construct a test statistic. It was found that the hypothesis of a 95% coverage level was rejected at the 5% level of significance in all cases except for $\mu_{(1)}$ when m = 100, n = 40, and $\sigma$ = .2 and for $\mu_{(2)}$ when m = 50, n = 80 and $\sigma$ = .2. Similar large sample tests reject the hypotheses that the confidence intervals perform the same for $\mu_{(1)}$ and $\mu_{(2)}$ at each of the different configurations for m,n and $\sigma$ except m = 100, n = 40 and $\sigma$ = .4.

A simulation of this calibre is insufficient to either confirm or reject the utility of an estimation procedure. However, it can indicate certain things which may happen. For example, we see that the intervals (50) may work reasonably well with some functions (e.g., $\mu_{(2)}$) and not so well with others (e.g., $\mu_{(1)}$). It also suggests that the use of such intervals may be best advised in large samples. The latter recommendation is hardly surprising. Other simulations reported by Wahba (1983) give similar, although somewhat better, qualitative results for the method.

There are other approaches to interval estimation besides the Bayesian procedure. Some have proposed the use of jackknife (see Wold, 1974) and bootstrap techniques for construction of interval estimates. Knafl, Sacks and Ylvisaker (1982) develop an approach to interval estimation which explicitly incorporates the bias into the estimation process. They are able to obtain conservative interval estimates using their procedure.

Notice that the intervals in (50) are not confidence bands but are instead aimed at estimating each of the $\mu(t_i)$ individually with confidence level $1 - \alpha$. To obtain intervals which might be expected to hold simultaneously for all the $\mu(t_i)$, Gamber (1979b) has suggested using an approach based on the distribution of the maximum absolute deviation $\max_{1 \leq j \leq n} |\mu(t_j) - \mu_\lambda(t_j)|$. For discussion of a general approach to the construction of confidence bands see Knafl, Sacks and Ylvisaker (1985).

Another approach to simultaneous estimation of $\mu(t_1), \ldots, \mu(t_n)$ has been proposed by Li (1983, 1985). Assuming $\sigma^2$ to be known he employs a Stein estimate for $\underline{\mu}$ of the form

$$\tilde{\underline{\mu}}_\lambda = (\tilde{\mu}_\lambda(t_1), \ldots, \tilde{\mu}_\lambda(t_n))' = \underline{y} - \frac{\sigma^2}{\sigma^2_\lambda}(I - H(\lambda))\underline{y}$$

with $\sigma^2_\lambda$ defined in (48). The risk is estimated by Stein's unbiased risk estimate

$$\begin{aligned} \mathrm{SURE}_n(\lambda) &= \sigma^2 - \sigma^4 \operatorname{tr}(I - H(\lambda))/n\sigma^2_\lambda \\ &= \sigma^2 - \sigma^4/\mathrm{GCV}_n(\lambda) \end{aligned}$$

discussed in Section 2.3. Li is able to show that if the $\varepsilon_i$ are normally distributed then for any $\delta > 0$

$$p_n(\delta) = \sup_{\underline{\mu} \in \mathbb{R}^n} P(|\mathrm{SURE}_n(\hat{\lambda}) - n^{-1} \sum_{j=1}^{n} (\mu(t_j) - \tilde{\mu}_{\hat{\lambda}}(t_j))^2| > \delta) \to 0$$

as $n \to \infty$ when $\hat{\lambda}$ is the GCV estimator of $\lambda$. If $\sigma^2$ is unknown it can be replaced by a consistent estimator in the definition of $\tilde{\underline{\mu}}_\lambda$ and $\mathrm{SURE}_n(\lambda)$ with similar results.

If we choose $\delta = \delta(\alpha)$ so that $p_n(\delta(\alpha)) \leq \alpha$ this gives a conservative $100(1 - \alpha)\%$ confidence region for $\underline{\mu}$: namely

$$\{\underline{\mu}: n^{-1}(\underline{\mu} - \tilde{\underline{\mu}}_{\hat{\lambda}})'(\underline{\mu} - \tilde{\underline{\mu}}_{\hat{\lambda}}) \leq \mathrm{SURE}_n(\lambda) + \delta(\alpha)\} \quad .$$

Li (1983) discusses one approach to choosing $\delta(\alpha)$. However, the development of methods for selecting $\delta(\alpha)$ in practice remains an open problem.

It is interesting to note that Li's method for interval estimation is related to Wahba's Bayesian method. To see this, estimate $\sigma^2$ by $\sigma_\lambda^2$ in SURE ($\lambda$). One finds that the resulting estimator of the risk is $\sigma_\lambda^2 \mathrm{tr}H(\lambda)/n$ which is suggested by (51). A closely related estimator of the risk has also been studied by Nychka (1986).

Throughout our discussion of interval estimates we have focused only on the estimation of $\mu$ at the $t_i$. It may well be that there are other functionals of $\mu$ for which interval estimates may be desired. For example, one may wish to estimate $\mu(t)$ for some $t \neq t_i$ or maybe $\mu'(t)$ at some point $t \in [a,b]$. Bayesian intervals can also be constructed for such cases by using the posterior variance of the functional obtained from the posterior covariance kernel of $\mu$ in Theorem 5.5 (see Wahba, 1983, for details). Unlike the case of estimation of the $\mu(t_i)$ it also becomes necessary to estimate $\sigma_s^2$ in this case. Another Bayesian formulation due to Silverman (1985) avoids the need for estimating $\sigma_s^2$ (Exercise 11) as do jackknife and bootstrap methods. The Knafl, Sacks and Ylvisaker (1982) technique is also applicable to estimation of functionals.

### 5.5.5 Diagnostic Analysis

To assess the quality of a fit to a set of data it is always useful to conduct *a posteriori* analysis involving the examination of residuals, detection of influential data values, etc. In the context of linear models this type of diagnostic analysis has been extensively studied and diagnostic methodology is at an advanced level of development. (See Section A.1.3 of the Appendix). The same cannot be said for other areas in regression. This is true, in particular, of nonparametric regression.

Smoothing splines are a type of nonparametric regression estimator for which the diagnostic problem has received a good bit of attention. Some early comments concerning diagnostic methods for these estimators can be found in Wold (1974). Wendelberger

(1981) also discusses diagnostic analysis for smoothing splines and proposes the use of normal probability plots for the residuals as a possible tool.

It turns out that in the case of smoothing splines it is possible to derive diagnostic techniques by analogy with the linear regression case through use of the Bayesian model (41). The key lies in noting that the vector of residuals

$$\underline{e}_\lambda = (e_\lambda(t_1),\ldots,e_\lambda(t_n))' = \underline{y} - \underline{\mu}_\lambda$$

has variance-covariance matrix

$$\text{Var } \underline{e}_\lambda = \sigma^2(I - H(\lambda)) \tag{52}$$

when the Bayesian model holds. To see this use (30b) and (45a) to write

$$\begin{aligned}\underline{e}_\lambda &= (I - H(\lambda))\underline{y} \\ &= n\lambda U(U'M(\lambda)U)^{-1}\underline{u}\end{aligned}$$

and then apply (45d) noting that $\sigma_s^2 n\lambda = \sigma^2$.

Equation (52) is a direct parallel of the form of the variance-covariance matrix for residuals from linear regression analysis (c.f.(A19)). Thus, one might consider using standardized residuals

$$r_{\lambda i}(\sigma) = e_i(\lambda)/\sigma(1 - h_{ii}(\lambda))^{1/2} \tag{53a}$$

to assess the quality of the fit to $y_i$. When $\sigma^2$ is unknown it may be estimated by (48) to obtain a *studentized residual*

$$r_{\lambda i} = r_{\lambda i}(\sigma_\lambda) \ . \tag{53b}$$

While the exact distribution of $r_{\lambda i}$ is unknown its magnitude might be compared to values from a Student's t-distribution with approximate degrees of freedom given by tr(I - H(λ)). This choice for degrees of freedom is motivated by the fact that in a linear model with p parameters the studentized residuals are compared to values from a t - distribution with n - p degrees of freedom which is the trace of the identity minus that of the hat matrix for that case.

Another choice for an estimator of $\sigma^2$ in (53a) is discussed in Eubank (1984a,1985) and Eubank and Gunst (1986) (see Exercise 16). An alternative type of studentized residual has been proposed by Silverman (1985) where, by analogy with the practice of GCV, the leverage values, $h_{ii}(\lambda)$, are replaced by their average $n^{-1}\mathrm{tr}H(\lambda)$ in (53).

An indication of the validity of studentized residuals in the case where $\mu$ is nonrandom and in $W_2^m[a,b]$ is provided by Table 5.2. Using the data employed to obtain Table 5.1 in Section 5.5.4 studentized residuals (53b) were computed for the corresponding smoothing spline fits with the smoothing parameter selected by GCV. These residuals were then compared to the .05 level two sided critical value for a Student's t distribution with $\mathrm{tr}(I - H(\hat{\lambda}))$ approximate degrees of freedom. What is recorded in the table is the proportion of times the studentized residuals exceeded the critical value for each case. An analysis similar to the one conducted for Table 5.1 was performed on the values in Table 5.2 with the table entries being treated as sample proportions for random samples of size 4000. The hypothesis of a true 5% rejection level was not rejected at the 5% level of significance in any of the m = 50, n = 80 cases. However, the hypothesis was rejected in all of the m = 100 and n = 40 cases except for when $\sigma = .2$. If anything the method seems conservative in the latter instances so we would not expect this approach to lead to the detection of spurious outliers.

As in interval estimation one can argue that the use of $\sigma(1 - h_{ii}(\lambda))^{1/2}$ to rescale $e_i(\lambda)$ is more desirable than the use of its actual standard deviation under model (2) for this purpose. The reasoning is that, unlike the residuals from linear models, $\underline{e}_\lambda$ does not have zero expectation. Thus, a more accurate assessment of the accuracy of $\underline{e}_\lambda$ as an estimator of the zero vector is provided by

**Table 5.2 Empirical Significance Levels for Studentized Residuals**

| $\mu_{(1)}$ | m = 100, n = 40 | m = 50, n = 80 |
|---|---|---|
| $\sigma = .2$ | .0433 | .0473 |
| $\sigma = .4$ | .0390 | .0460 |
| $\sigma = .6$ | .0398 | .0455 |
| $\sigma = .8$ | .0410 | .0455 |
| $\mu_{(2)}$ | | |
| $\sigma = .2$ | .0383 | .0458 |
| $\sigma = .4$ | .0408 | .0458 |
| $\sigma = .6$ | .0418 | .0448 |
| $\sigma = .8$ | .0418 | .0448 |

$$n^{-1}E\underline{e}_\lambda'\underline{e}_\lambda = n^{-1}\sigma^2 tr(I - H(\lambda))^2 + n^{-1}(\underline{\mu} - E\underline{\mu}_\lambda)'(\underline{\mu} - E\underline{\mu}_\lambda)$$

$$= \sigma^2 - 2\sigma^2 trH(\lambda)/n + R(\lambda).$$

However, in view of (51) this can be approximated by $\sigma^2(1 - trH(\lambda)/n)$ when $\lambda$ has been chosen optimally. This gives further credence to Silverman's proposal for studentized residuals and provides a motivation for the use of measures such as (53).

Since

$$\underline{\mu}_\lambda = H(\lambda)\underline{y}$$

this means that

$$\mu_\lambda(t_j) = \sum_{i=1}^{n} h_{ij}(\lambda)y_i$$

and, hence, the jth leverage value $h_{jj}(\lambda)$ tells us how much influence $y_j$ has on its own fit. Consequently, the leverage values also provide diagnostic information about the smoothing spline estimator.

Under the Demmler-Reinsch basis we find that the ijth element of $H(\lambda)$ can be expressed as

$$h_{ij}(\lambda) = \sum_{r=1}^{n} x_{ir}x_{jr}/(1 + n\lambda\delta_r)$$

with $0 = \delta_1 = \ldots = \delta_m < \delta_{m+1} \leq \ldots \leq \delta_n$. Thus by letting $\lambda$ tend to zero and infinity we find that for any $0 < \lambda < \infty$

$$h_{ii}(\infty) = \lim_{\lambda\to\infty} h_{ii}(\lambda) \leq h_{ii}(\lambda) \leq 1.$$

However, $h_{ii}(\infty)$ is the ith leverage value for regression on polynomials of order m. Consequently, observations tend to have higher leverage with smoothing spline fits to data than they do for polynomial regression. This might be anticipated since smoothing splines are more flexible and will therefore conform more to the data than will a polynomial estimator.

There is also a distance interpretation for the $h_{ii}(\lambda)$. Note that

$$\bar{\underline{x}}' = n^{-1}\underline{1}'X = n^{-1/2}(1,0,\ldots,0)$$

under the Demmler-Reinsch basis. Thus,

$$h_{ii}(\lambda) = (\underline{x}_i - \bar{\underline{x}})'\Delta(\lambda)(\underline{x}_i - \bar{\underline{x}}) + n^{-1} \quad ,$$

where $\underline{x}_i$ is the ith row of X and $\Delta(\lambda)$ is defined prior to (35). Consequently, leverage values for smoothing splines measure the distance between the ith observation on the basis functions $x_1, \ldots, x_n$ and their average $\bar{\underline{x}}$ just as they do for linear regression (c.f.(A21)). By analogy with the linear regression case we might say that an observation has high leverage if its leverage value exceeds $3\mathrm{tr}H(\lambda)/n$.

It is also possible to develop parallels of measures which are used to assess the influence of observations in linear regression. For example, a smoothing spline analog of Cook's distance measure is (c.f. (A23)).

$$D_{\lambda i} = r_{\lambda i}(\sigma_\lambda)^2 h_{ii}(\lambda)/(1 - h_{ii}(\lambda))\mathrm{tr}H(\lambda)$$

Thus, one might deem an observation influential if its corresponding Cook's distance exceeds the lower 10% point of an F-distribution with approximate numerator and denominator degrees of freedom provided by $trH(\lambda)$ and $tr(I - H(\lambda))$, respectively (c.f. Section A.1.3 ).

The motivation for using $D_{\lambda i}$ is essentially the same as in the linear regression case. Let $\underline{\mu}_\lambda$ and $\underline{\mu}_{\lambda(i)}$ denote the vectors of fitted values at point $t_1,\ldots,t_n$ obtained from the entire data and from when the ith point has been deleted. As a result of Lemma 5.2

$$\underline{\mu}_\lambda = H(\lambda)\{\underline{y} - (e_i(\lambda)/(1 - h_{ii}(\lambda)))\underline{1}_i\} \quad ,$$

where $\underline{1}_i$ is a vector of all zeros except for a one as its ith component. Identity (49) suggests that we might rescale the difference $\underline{\mu}_\lambda - \underline{\mu}_{\lambda(i)}$ using $\sigma^2 H(\lambda)$ to obtain

$$(\underline{\mu}_\lambda - \underline{\mu}_{\lambda(i)})'(\sigma^2 H(\lambda))^{-1}(\underline{\mu}_\lambda - \underline{\mu}_{\lambda(i)})$$

$$= e_i(\lambda)^2 h_{ii}(\lambda)/(1 - h_{ii}(\lambda))^2\sigma^2 = \sigma_\lambda^2 trH(\lambda)D_{\lambda i}/\sigma^2 \quad .$$

Thus, $D_{\lambda i}$ is an appropriately scaled measure of the influence of $(t_i, y_i)$ on the overall fit.

As an illustration of the use of the diagnostics discussed above for analysis of data by smoothing splines consider the German hyperinflation data shown in Figure 5.7. This data consists of values for the logarithm of the money supply as a function of the logarithm for the premium, or discount, on a forward contract for foreign exchange during the German hyperinflation (see Frenkel, 1977, and Ansley and Wecker, 1981). A cubic smoothing spline was fit to this data with the smoothing parameter value $\hat{\lambda}$ selected by GCV. The resulting curve estimate is also depicted in Figure 5.7. One objective in analyzing this data is to see the additional insight that can be gained by the use of measures such as studentized residuals, leverage values and Cook's distance measures over the use of residuals alone.

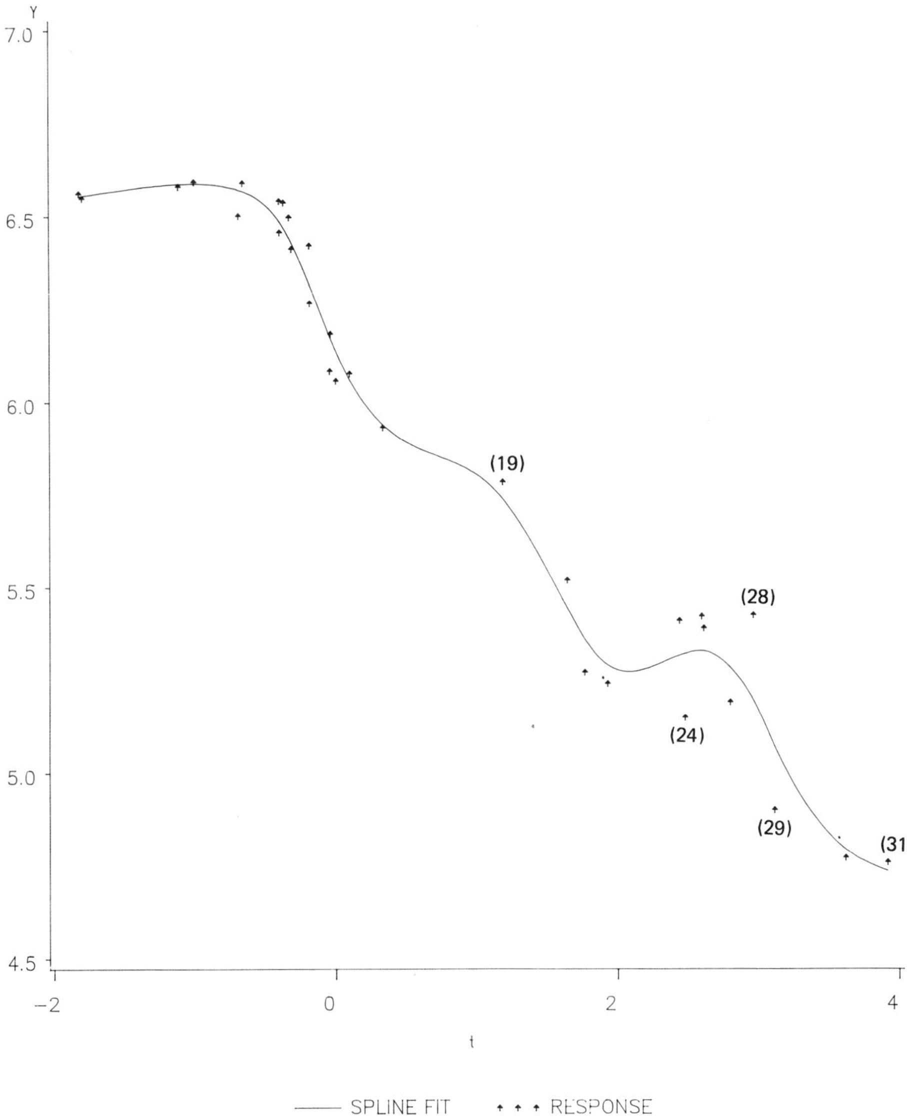


**Figure 5.7** Spline fit to the German hyperinflation data.

Table 5.3 gives a summary of the diagnostics obtained for our fit to the German hyperinflation data. What has been presented is all cases whose studentized residuals exceed two and/or whose leverage values exceed $2\mathrm{tr}H(\hat{\lambda})/n = .67$. These are also all the observations whose Cook's distance measures are of appreciable magnitude. The next largest Cook's distance measure after those in the table is .064, about half the size of that of observation 24. To aid in the analysis the case numbers of the observations in Table 5.3 have also been indicated on Figure 5.7.

Observations 19, 24, 28, 29 and 31 represent the most influential observations for our fit to the German hyperinflation data. However, their influence on the fit stems from different reasons. These reasons can be deduced by examination of their corresponding diagnostics in Table 5.3.

**Table 5.3. Diagnostics for the German Hyperinflation Data**

| Observation Number | $h_{ii}(\hat{\lambda})$ | $r_{\hat{\lambda}i}$ | $D_{\hat{\lambda}i}$ |
|---|---|---|---|
| 19 | .760 | .935 | .268 |
| 24 | .257 | -2.018 | .136 |
| 28 | .335 | 2.802 | .384 |
| 29 | .460 | -2.344 | .455 |
| 31 | .860 | .634 | .239 |

The influence of observations 19 and 31 is primarily due to high leverage. From examination of the data in Figure 5.7 it is clear that case 19 has high leverage resulting from inadequate information about values of the log exchange premium over a portion of its range. This indicates that conclusions drawn about predictions made in this interval will be predicated largely on the information in the 19th observation. The high leverage for observation 31 is reflective of the position of its value for the log exchange premium at the extreme of the data set.

The influence of observations 28 and 29 derives from the fact that they lie significantly far from the fitted curve as measured by their values for $|r_{\hat{\lambda}i}|$. Observation 24 also has a significantly large studentized deleted residual. However, it is not a high leverage point and, as indicated by its lower value for $D_{\hat{\lambda}i}$, is less influential on the fit. This latter observation enforces the point that not all outliers need influence the fit.

To demonstrate the additional utility of the diagnostics in Section 3 over examination of residuals alone, a plot of the residuals has been presented in Figure 5.8. Cases 24, 28 and 29 stand out as outliers. However, examination of their studentized deleted residual values, all of which exceed 2, provides a quantitative confirmation of this intuitive designation for these points. In addition, by consideration of their values for $D_{\hat{\lambda}i}$, we recognize that, of these three points, observations 28 and 29 have the most influence on the fit. Also note that from examination of the residual plot no special importance would be attributed to observations 19 and 31. This is not surprising since high leverage points will always have small residuals. Their influence was clearly revealed, however, by examination of the $D_{\hat{\lambda}i}$ values.

In summary, our analysis of the German hyperinflation data has pointed out several sources of difficulty for the corresponding cross-validated smoothing spline fit that would not have been recognized through examination of residuals alone. Specific remedies would depend on the subject matter area and might include consideration of alternative transformations for the data, downweighting certain points in the smoothing criterion (1) or the use of robust smoothing methodology (see Section 6.2.5) which takes leverage into account. These possibilities will not be pursued here.

More detailed treatments of diagnostic methods for smoothing splines with examples can be found in Eubank (1984a, 1985), Silverman (1985) and Eubank and Gunst (1986). Another example of the use of these techniques is given in Section 6.3.

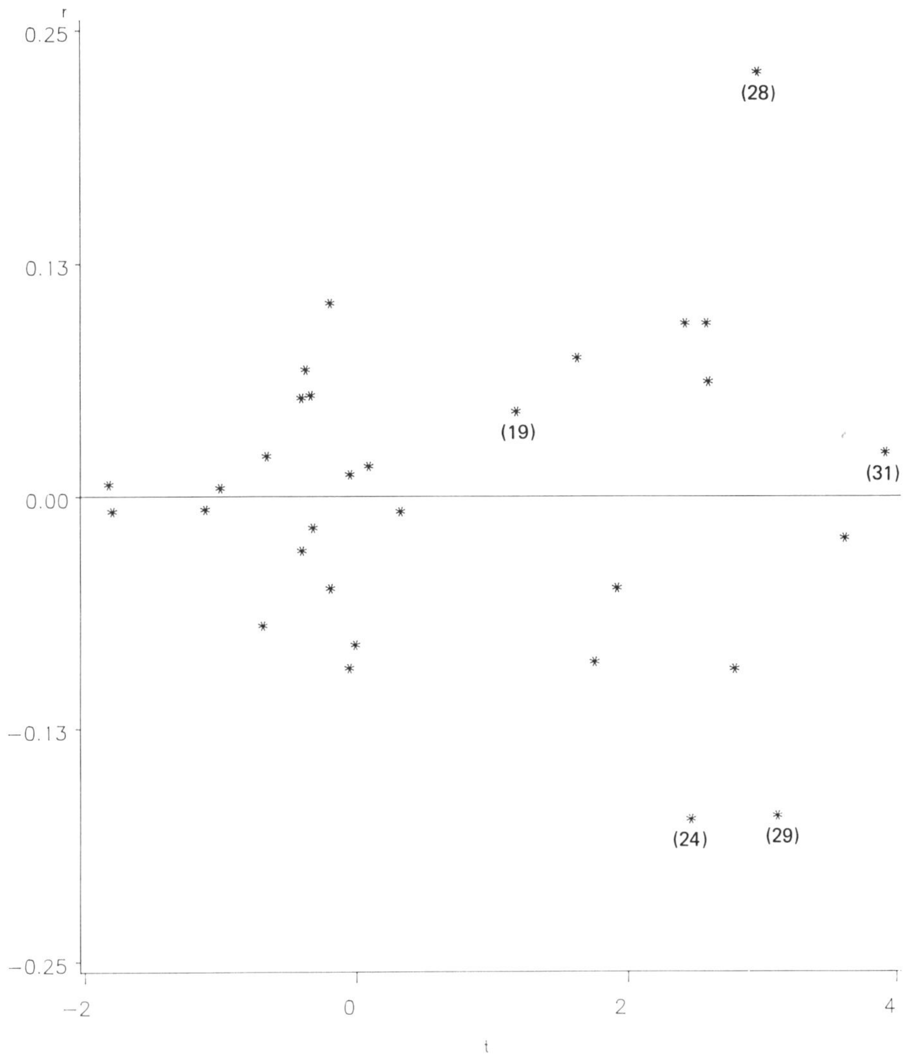


**Figure 5.8** Residuals for the German hyperinflation data.

## EXERCISES

**1.** Let $R_n(\lambda) = n^{-1}\Sigma_{j=1}^{n} E(\mu(t_j) - \mu_\lambda(t_j))^2$. Show that, under appropriate conditions, there exists a value $\lambda_0$ in $(0,\infty)$ such that $R_n(\lambda_0) < R_n(0)$ and $R_n(\lambda_0) < R_n(\infty) = \lim_{\lambda\to\infty} R_n(\lambda)$. [Hint: Use the Demmler-Reinsch basis.]

**2.** Verify that a spline as defined in (8):

a) is a piecewise polynomial of order r on any subinterval between knots,

b) has r - 2 continuous derivatives and

c) has a step function as its (r - 1)st derivative with jumps at the knots.

**3.** Draw pictures of $(t - \xi)_+^j$ for j = 0,1,2 and 3.

**4.** Show that $1, t, \ldots, t^{r-1}, (t - \xi_1)_+^{r-1}, \ldots, (t - \xi_k)_+^{r-1}$ are r + k linearly independent functions. Then show that $S^r(\xi_1, \ldots, \xi_k)$ is a vector space. What is its dimension?

**5.** Prove Corollary 5.1.

**6.** Prove Theorem 5.4.

**7.** Let

$$(x_1(t), \ldots, x_n(t)) = (1, t, \ldots, t^{m-1}, Q(t,t_1), \ldots, Q(t,t_n))$$

$$\times \begin{bmatrix} (T'Q_n^{-1}T)^{-1}T'Q_n^{-1} \\ UW^{-1}U' \end{bmatrix},$$

where the notation which is utilized is that of Section 5.3.3. Show that $x_j$ is the unique element of $NS^{2m}(t_1, \ldots, t_n)$ which satisfies

$$x_j(t_r) = \begin{cases} 0, & r \neq j, \\ 1, & r = j, \end{cases}$$

and conclude from this that the $x_j$ form a basis for $NS^{2m}(t_1, \ldots, t_n)$. What form does $\underline{\beta}_\lambda$ in (15) take under this basis? What form does the hat matrix have under this basis?
[Utreras, 1981]

**8.** A local support basis for natural splines was proposed by Greville (1969). The basis functions are defined by

$$x_j(t) = \begin{cases} [t_1,\ldots,t_{m+j-1}](\cdot - t)_+^{2m-1}, & j = 1,\ldots,m, \\ (t_{j+m} - t_{j-m})[t_{j-m},\ldots,t_{j+m}](\cdot - t)_+^{2m-1}, & j=m+1,\ldots,n-m, \\ (-1)^{n+m-j}[t_{j-m},\ldots,t_n](t - \cdot)_+^{2m-1}, & j = n - m + 1,\ldots,n. \end{cases}$$

Provided $n \geq 2m$ they form a basis for $NS^{2m}(t_1,\ldots,t_n)$. Show that $X$ and $G$ in (15) are $2m - 1$ and $2m + 1$ banded, respectively.
[Hint: Use the local support properties of the $x_j$ to establish the banded nature of $X$. For $G$ show that

(a) the $j$th column of $G$ is $(-1)^{m+1}(2m - 1)!$ times the vector of $(m + j - 1)$st divided difference coefficients at $t_1$ for $j=1,\ldots,m$,

(b) for $j = m + 1,\ldots,n - m$, the $j$th column of $G$ is $(-1)^{m+1}(2m - 1)!$ times the 2$m$th normalized divided difference coefficients at $t_{j-m},\ldots,t_{j+m}$, and

(c) for $j = n - m + 1,\ldots,n$, the $j$th column of $G$ consists of $(-1)^{n-j}(2m - 1)!$ times the $(n + m - j)$th divided difference coefficients over $t_{j-m},\ldots,t_n$.]

[Lyche and Schumaker, 1973, Schumaker, 1980.]

**9.** Verify identity (30).

**10.** Complete the proof of Theorem 5.5.

**11.** Silverman (1985) has proposed the following alternative to model (41). He assumes that

$$y_i = \sum_{j=1}^{n} \beta_j x_j(t_i) + \varepsilon_i ,$$

where the $\varepsilon_j$ are zero mean uncorrelated normal random errors with common variance $\sigma^2$ and $x_1,\ldots,x_n$ are a basis for $NS^{2m}(t_1,\ldots,t_n)$. The vector $\underline{\beta}$ has a prior likelihood proportional to $\exp\{-n\lambda\underline{\beta}'\Omega\underline{\beta}/2\sigma^2\}$, where $\Omega$ is defined in (17b). Show that $\mu_\lambda$ is the posterior mean of $\mu = \Sigma_{j=1}^{n}\beta_j x_j$ under this formulation and derive the posterior variance of the vector $\underline{\mu} = (\mu(t_1),\ldots,\mu(t_n))'$ and

$\mu(t)$. Compare the variance of $\mu(t)$ under this model with the variance under model (41). How do they differ? More generally, compare the variances of any linear functional of $\mu$ under model (41) and Silverman's model. Discuss the implications of this for interval estimation. Find the variance-covariance matrix for the residual vector under Silverman's model.

**12.** Complete the proof of Theorem 5.6.

**13.** Consider the case of model (41) where, instead of our original assumption that $\underline{\theta}$ is stochastic, the vector of polynomial coefficients is assumed to be some fixed but unknown vector in $\mathbb{R}^m$. Find the $\underline{y}$ likelihood for this case and then show that a maximum likelihood estimator of $\lambda$ under this model can be obtained by minimizing

$$WA(\lambda) = \underline{y}'U(U'M(\lambda)U)^{-1}U'\underline{y}/|M(\lambda)^{-1}|^{1/n} .$$

[Wecker and Ansley, 1983]

**14.** Verify identity (49).

**15.** Let $h_{ii}^{(2)}(\lambda)$ be the ith diagonal element of $H(\lambda)^2$. Show that $h_{ii}(\lambda) \geq h_{ii}^{(2)}(\lambda)$ and thereby conclude that the Bayesian confidence intervals (50) for the $\mu(t_i)$ will tend to be wider than intervals which use the true variance of the $\mu_\lambda(t_i)$ under model (2).

**16.** In some cases it may be deemed more appropriate to use an estimator of $\sigma^2$ which does not involve the ith observation in (53). Establish that, when the ith observation is deleted from the data, estimator (48) becomes

$$\sigma^2_{\lambda(i)} = \sum_{j\neq i} (e_j(\lambda) + h_{ji}(\lambda)e_i(\lambda)/(1 - h_{ii}(\lambda)))^2/n\tau_{(i)} ,$$

where

$$n\tau_{(i)} = n - 1 - \sum_{j\neq i} [h_{jj}(\lambda) + h_{ij}(\lambda)^2/(1 - h_{jj}(\lambda))] .$$

[Hint: Use Lemma 5.2.]

[Eubank, 1984a, 1985, Eubank and Gunst, 1986].

**17.** Let $\underline{\mu}_\lambda = (\mu_\lambda(t_1),\ldots,\mu_\lambda(t_n))'$ and let $\underline{\mu}_{\lambda(i)}$ denote the corresponding vector of fitted values for when the ith observation has been deleted from the data.

a) Show that

$$\underline{\mu}_\lambda - \underline{\mu}_{\lambda(i)} = -H(\lambda)[\underline{1}_i e_\lambda(t_i)/(1 - h_{ii}(\lambda))],$$

where $\underline{1}_i$ is the vector of all zeros except for a one as its ith entry.

b) Let $V_c = \{\underline{\ell}: \underline{\ell}'H(\lambda)^{-1}\underline{\ell} \le c\}$ for $c > 0$. Prove that

$$\sup_{\underline{\ell}\in V_c} [\underline{\ell}'(\underline{\mu}_\lambda - \underline{\mu}_{\lambda(i)})]^2 = ch_{ii}(\lambda)[e_i(\lambda)/(1 - h_{ii}(\lambda))]^2 .$$

From a Bayesian viewpoint, what is $V_c$ and what does this identity suggest concerning diagnostic analysis.

c) By choosing c appropriately in b) derive the Cook's distance measure $D_{\lambda i}$ in Section 5.5.5.

[Eubank and Gunst, 1986]

**18.** Suppose that there are $n_i \ge 1$ observations taken at a particular design point $t_i$. Show that the corresponding leverage value for a smoothing spline estimator satisfies $h_{ii}(\lambda) \le n_i^{-1}$.

**19.** Refer to the growth data for the boy from the Zürich growth study in Exercise 4.24. Fit this data using a cubic smoothing spline with GCV used to select λ. Construct Bayesian confidence intervals for the true growth curve and conduct diagnostic analysis of the fit.

**20.** Refer to the simulated data in Exercise 4.25. Fit this data using a cubic smoothing spline with λ selected by GCV. Assess the performance of the Bayesian confidence intervals for the $\mu(t_i)$. Discuss and explain areas where any difficulties arise and compare your findings to those from Exercise 4.25. Also conduct diagnostic analysis and comment on your findings.

**21.** The data given below represent the ratios of weight to height for boys from a study by Eppright, et al. (1972). A fit to this data using a cubic smoothing spline with λ chosen by GCV is shown in Figure 5.3. Discuss the properties of this fit.

For the GCV fit to this data it is found that the GCV estimate of $\lambda$ is $\hat{\lambda} = 1.38$, $\mathrm{tr}H(\hat{\lambda}) \doteq 10$ and $\sigma^2_{\hat{\lambda}} = .00062$. Given along with the data are the corresponding fitted and leverage values. Use this information to construct Bayesian confidence intervals and conduct diagnostic analysis for this data.

| Age (in months) | $y_i$ | $\mu_{\hat{\lambda}}(t_i)$ | $h_{ii}(\hat{\lambda})$ | Age (in months) | $y_i$ | $\mu_{\hat{\lambda}}(t_i)$ | $h_{ii}(\hat{\lambda})$ |
|---|---|---|---|---|---|---|---|
| .5 | .46 | .4736 | .413 | 32.5 | .88 | .8591 | .133 |
| 1.5 | .47 | .5123 | .253 | 33.5 | .86 | .8642 | .133 |
| 2.5 | .56 | .5505 | .178 | 34.5 | .91 | .8686 | .133 |
| 3.5 | .61 | .5873 | .148 | 35.5 | .87 | .8720 | .133 |
| 4.5 | .61 | .6219 | .140 | 36.5 | .87 | .8749 | .133 |
| 5.5 | .67 | .6537 | .139 | 37.5 | .87 | .8778 | .133 |
| 6.5 | .68 | .6820 | .139 | 38.5 | .85 | .8810 | .133 |
| 7.5 | .78 | .7064 | .137 | 39.5 | .90 | .8848 | .133 |
| 8.5 | .69 | .7268 | .137 | 40.5 | .87 | .8888 | .133 |
| 9.5 | .74 | .7437 | .136 | 41.5 | .91 | .8930 | .133 |
| 10.5 | .77 | .7576 | .135 | 42.5 | .90 | .8969 | .133 |
| 11.5 | .78 | .7689 | .134 | 43.5 | .93 | .9005 | .133 |
| 12.5 | .75 | .7779 | .133 | 44.5 | .89 | .9037 | .133 |
| 13.5 | .80 | .7853 | .133 | 45.5 | .89 | .9070 | .133 |
| 14.5 | .78 | .7914 | .133 | 46.5 | .92 | .9106 | .133 |
| 15.5 | .82 | .7964 | .133 | 47.5 | .89 | .9146 | .133 |
| 16.5 | .77 | .8008 | .133 | 48.5 | .92 | .9190 | .133 |
| 17.5 | .80 | .8050 | .133 | 49.5 | .96 | .9234 | .133 |
| 18.5 | .81 | .8091 | .133 | 50.5 | .92 | .9276 | .133 |
| 19.5 | .78 | .8133 | .133 | 51.5 | .91 | .9319 | .133 |
| 20.5 | .87 | .8173 | .133 | 52.5 | .95 | .9365 | .133 |
| 21.5 | .80 | .8207 | .133 | 53.5 | .93 | .9412 | .133 |
| 22.5 | .83 | .8237 | .133 | 54.5 | .93 | .9461 | .133 |
| 23.5 | .81 | .8263 | .133 | 55.5 | .98 | .9509 | .133 |
| 24.5 | .88 | .8286 | .133 | 56.5 | .95 | .9553 | .133 |
| 25.5 | .81 | .8304 | .133 | 57.5 | .97 | .9593 | .133 |
| 26.5 | .83 | .8324 | .133 | 58.5 | .97 | .9630 | .133 |
| 27.5 | .82 | .8350 | .133 | 59.5 | .96 | .9665 | .133 |
| 28.5 | .82 | .8385 | .133 | 60.5 | .97 | .9701 | .134 |
| 29.5 | .86 | .8428 | .133 | 61.5 | .94 | .9741 | .135 |
| 30.5 | .82 | .8479 | .133 | 62.5 | .96 | .9786 | .136 |
| 31.5 | .85 | .8535 | .133 | 63.5 | 1.03 | .9832 | .137 |
| | | | | 64.5 | .99 | .9871 | .138 |
| | | | | 65.5 | 1.01 | .9905 | .139 |
| | | | | 66.6 | .99 | .9934 | .139 |
| | | | | 67.5 | .99 | .9963 | .140 |
| | | | | 68.5 | .99 | .9998 | .148 |
| | | | | 69.5 | 1.01 | 1.0041 | .178 |
| | | | | 70.5 | .99 | 1.0091 | .253 |
| | | | | 71.5 | 1.04 | 1.0145 | .413 |

**22.** Derive explicit expressions for the elements of the matrices U and $U'Q_nU$ in the Anselone-Laurent-Reinsch basis when m=2. In particular give an expression for U'U and $U'Q_nU$ when $t_i = i/n$.

**23.** Derive an expression for the hat matrix corresponding to the estimator derived from (19).

**24.** Consider the case where observations are obtained by sampling at time points $t_1,\ldots,t_n$ from the stochastic process

$$Y(t) = \sum_{j=0}^{m-1} \theta_j t^j + \sigma_s Z(t) + \varepsilon(t), \qquad t \in [a,b], \quad \sigma_s > 0 ,$$

where $\varepsilon(\cdot)$ has zero mean and covariance given by (41b), $Z(\cdot)$ has zero mean and covariance kernel (20) and the $\varepsilon$ and Z processes are uncorrelated. A best linear unbiased predictor (BLUP) of Y(t) based on observations $(t_1,y_1),\ldots,(t_n,y_n)$ is an estimator of the form $\Sigma_{j=1}^n \ell_j y_j$ with coefficients chosen to minimize $\mathrm{Var}(\Sigma_{j=1}^n \ell_j y_j)$ subject to the unbiasedness condition

$$\sum_{j=1}^{n} \ell_j Ey_j = EY(t).$$

Show that the smoothing spline estimator $\mu_\lambda(t)$ with $\lambda = \sigma^2/n\sigma_s^2$ is the BLUP of Y(t).

[Goldberger, 1962, Kimeldorf and Wahba, 1970b]

**25.** To demonstrate the difference between models (1) and (41) consider the sum

$$n^{-1} \sum_{i=1}^{n} \{n(\mu((i+1)/n) - \mu(i/n))\}^2 .$$

When $\mu \in W_2^1[0,1]$ it is easy to argue that this must converge to $\int_0^1 \mu'(t)^2 dt$. Suppose, instead, that $\mu$ is given by (41c) with m=1. Show that the expectation of this sum will diverge in that case and, hence, $E\int_0^1 \mu'(t)^2 dt$ is not defined.

[Wahba, 1983]

# 6

# Smoothing Splines: Extensions and Asymptotic Theory

## 6.1 Introduction

In the previous chapter we studied various properties of smoothing spline estimators. These were estimators suitable for smoothing data from a model of the form

$$y_i = \mu(t_i) + \varepsilon_i , \qquad i = 1,\ldots,n , \tag{1}$$

where the $\varepsilon_i$ are zero mean, uncorrelated random errors and $\mu \in W_2^m[a,b]$. The resulting estimator $\mu_\lambda$ of $\mu$ was obtained by minimizing the penalized least-squares criterion

$$n^{-1} \sum_{j=1}^{n} (y_j - f(t_j))^2 + \lambda J_m(f), \qquad \lambda > 0, \tag{2a}$$

with

$$J_m(f) = \int_a^b f^{(m)}(t)^2 dt \quad , \tag{2b}$$

over $f \in W_2^m[a,b]$. This estimator was found to have various optimality properties which recommend it for use in nonparametric regression analysis.

In the present chapter we continue our study of smoothing spline estimators. In the next section we will see that the basic smoothing spline concept can be extended to affect the solution of a number of other smoothing problems. Then, in Section 6.3, we give a non-rigorous treatment of the asymptotic theory associated with spline smoothing and its variants. Finally, in Section 6.4, some applications of the smoothing techniques presented in Section 6.2 are described.

## 6.2 Extensions

One of the advantages of the penalized least-squares approach which led to spline smoothing is its flexibility for handling other problems. In this section we will demonstrate that criterion (2) can be easily modified to handle a variety of other problems. This results in several interesting variants of the smoothing spline concept which have proved useful in practice. We begin our study of smoothing spline extensions by considering method of regularization estimators.

### 6.2.1 The Method of Regularization

In our previous discussions we have concentrated on a model of the form

$$y_i = \mu_i + \varepsilon_i \ , \qquad i = 1,\ldots,n,$$

where the unknown mean values $\mu_1,\ldots,\mu_n$ were the objects of interest. For example, in model (1) we had $\mu_i = \mu(t_i)$ for some regression function $\mu$ and our emphasis is on the recovery of the function $\mu$ from the data.

A more general formulation than model (1) can be obtained by assuming that the $\mu_i$ are obtained by transformations applied to a function $\eta \in W_2^m[a,b]$. For simplicity these transformations will be assumed to be bounded linear functionals.

A transformation, L, defined on $W_2^m[a,b]$ is said to be a functional if it maps $W_2^m[a,b]$ into some subset of the real line, i.e., it is real valued. A functional L is linear if, for $\alpha_1, \alpha_2 \in \mathbb{R}$ and $f_1, f_2 \in W_2^m[a,b]$, $L(\alpha_1 f_1 + \alpha_2 f_2) = \alpha_1 Lf_1 + \alpha_2 Lf_2$. A linear functional is bounded if there is a positive constant C such that

$$|Lf| \le C||f||_m$$

for all $f \in W_2^m[a,b]$, where $||\cdot||_m$ is any of a variety of possible norms for $W_2^m[a,b]$ (e.g., $||f||_m^2 = \Sigma_{j=1}^m f(s_j)^2 + J_m(f)$ with $a \leq s_1 < \ldots < s_m \leq b$).

Let $L_1,\ldots,L_n$ be bounded linear functionals. Then, a generalization of model (1) arises by assuming that

$$y_i = L_i\eta + \varepsilon_i \quad , \quad i = 1,\ldots,n \; , \tag{3}$$

for some $\eta \in W_2^m[a,b]$. The objective is to estimate $\eta$ rather than the $\mu_i = L_i\eta$. When the $L_i$ are the point evaluation functionals $L_i\eta = \eta(t_i)$ this reverts back to the original problem of estimating an unknown regression function in $W_2^m[a,b]$.

Point evaluation functionals are but one of many possibilities which may be of interest. A fairly general type of linear functional arises from integration against a kernel function $K(s,t)$ assumed, e.g., to belong to $L_2([a,b] \times [a,b])$. The $L_i$ are then defined by

$$L_i\eta = \mu(t_i) = \int_a^b K(t_i,s)\eta(s)ds \; . \tag{4}$$

Kernels such as

$$K(t,s) = I_{[a,t]}(s) \; ,$$

with $I_{[a,t]}(\cdot)$ the indicator function for the interval $[a,t]$, and

$$K(t,s) = [(m-1)!]^{-1}(t-s)_+^{m-1}$$

correspond to the case where one observes averages of a function and derivative estimation, respectively. When $K(t,s)$ depends on s and t only through their difference it is possible to write

$$\mu(t) = \int_a^b K(t-s)\eta(s)ds$$

for some function $K \in L_2[a,b]$. In this case the regression function is obtained by convolving $\eta$ with K and the problem of estimating $\eta$ is called a *deconvolution* problem.

Problems where the data can be assumed to follow models similar to (3) with the $L_i$ defined by kernel functions as in (4) arise in weather forecasting and stereological microscopy. Situations where averages of a function are observed can be found in computerized tomography and any situation where data have been aggregated, e.g., by geographic region. Deconvolution problems occur in geology and physics. We will discuss some of these application areas in more detail in Section 6.4.

Assume that data have been obtained following model (3). An obvious way to extend the concept of smoothing splines for estimation in this case is to estimate $\eta$ by the minimizer of

$$n^{-1} \sum_{j=1}^{n} (y_j - L_j f)^2 + \lambda \int_a^b f^{(m)}(t)^2 dt \tag{5}$$

over all $f \in W_2^m[a,b]$. The resulting estimator will be denoted by $\eta_\lambda$ and termed a *method of regularization* (MOR) estimator of $\eta$. The idea behind MOR estimators stems from work by Tikhonov (1963a,b) who developed a related method for solving ill-posed integral equations where the solution must be regulated in some fashion to reduce its sensitivity to perturbations in the observed data. For a general reference see Lukas (1980).

To derive the minimizer of (5) we require a parallel of the argument used for Theorem 5.3. The first step is to assume that we have a function $\eta_\lambda$ which minimizes the criterion (5). Let f be an arbitrary function in $W_2^m[a,b]$ and $\delta$ a nonzero constant. Then

$$n^{-1} \sum_{j=1}^{n} (y_j - L_j \eta_\lambda)^2 + \lambda \int_a^b \eta_\lambda^{(m)}(t)^2 dt$$

$$\le n^{-1} \sum_{j=1}^{n} (y_j - L_j \eta_\lambda - \delta L_j f)^2 + \lambda \int_a^b (\eta_\lambda^{(m)}(t) + \delta f^{(m)}(t))^2 dt.$$

Thus, by differentiating the right hand side of this expression with respect to $\delta$ and evaluating the result at $\delta = 0$, a necessary condition for $\eta_\lambda$ to be a minimizer is that

$$(n\lambda)^{-1}\sum_{j=1}^{n}(L_jf)(y_j - L_j\eta_\lambda) = \int_a^b f^{(m)}(t)\eta_\lambda^{(m)}(t)dt \qquad (6)$$

for all $f \in W_2^m[a,b]$.

In the case treated by Theorem 5.3 of point evaluation functionals we were able to find a natural spline which satisfied this necessary condition. The key step was the property of a natural spline, s, detailed in Lemma 5.1, which allowed us to express $\int_a^b f^{(m)}(t)s^{(m)}(t)dt$ as a sum involving values of f and coefficients of basis elements. A type of function with similar properties is needed here.

Assume that $L_i(t - \cdot)_+^{2m-1}$, $i = 1,\ldots,n$, and $t^i$, $i = 0,\ldots,m-1$, are m + n linearly independent functions and consider the collection of functions of the form

$$s(t) = \sum_{i=0}^{m-1}\theta_i t^i + \sum_{i=1}^{n}\delta_i L_i(t - \cdot)_+^{2m-1} \qquad (7a)$$

with the $\delta_i$ restricted to satisfy

$$\sum_{i=1}^{n}\delta_i L_i(1 - \cdot)^j = 0\ , \quad j = 0,\ldots,m-1\ . \qquad (7b)$$

This set of functions is a vector space of dimension n which generalizes the concept of natural splines. To verify this latter claim take $L_i$ to be the point evaluation functions in which case $L_i(t - \cdot)_+^{2m-1} = (t - t_i)_+^{2m-1}$ so that (7a) is clearly a spline of order 2m whose mth through (2m - 1)st derivatives vanish at zero. Condition (7b) insures that these same derivatives will also vanish at one and, hence, that s is a natural spline of order 2m with knots at the $t_j$.

Let $x_1,\ldots,x_n$ be a basis for functions of the form (7) so that any function s of this type can be written as $\Sigma_{j=1}^n \beta_j x_j$ for some set of coefficients $\beta_1,\ldots,\beta_n$. Then, a calculation reveals that (Exercise 1)

$$\int_a^b f^{(m)}(t)s^{(m)}(t)dt = (-1)^m(2m-1)!\sum_{i=1}^{n}L_if\sum_{j=1}^{n}\beta_j\delta_{ij}\ , \qquad (8)$$

where $\delta_{ij}$ is the coefficient of $L_i(t - \cdot)_+^{2m-1}$ in $x_j$. Equation (8) is the analog of (5.12) for the case of general linear functionals.

By combining (6) - (8) with arguments similar to those used for Theorem 5.3 it is now relatively straightforward to obtain a closed form expression for the minimizer, $\eta_\lambda$, of (5). Specifically, we have the following result.

**Theorem 6.1.** Let $x_1,\ldots,x_n$ be a basis for functions of the form (7) and assume that for $f \in W_2^m[a,b]$ the conditions $\Sigma_{i=1}^n (L_i f)^2 = 0$ and $\int_a^b f^{(m)}(t)^2 dt = 0$ together imply that $f = 0$. Then the unique minimizer of (5) is $\eta_\lambda = \Sigma_{j=1}^n \beta_{\lambda j} x_j$, with $\underline{\beta}_\lambda = (\beta_{\lambda 1},\ldots,\beta_{\lambda n})'$ obtained as the unique solution of

$$[L + n\lambda G]\underline{\beta}_\lambda = \underline{y} , \tag{9}$$

where L is the n × n matrix whose typical element is $L_i x_j$ and G is an n × n matrix with typical entry $(-1)^m(2m - 1)!$ times the coefficient of $L_i(t - \cdot)_+^{2m-1}$ in $x_j$. Equation system (9) is equivalent to

$$[L'L + n\lambda\Omega]\underline{\beta}_\lambda = \underline{y} \tag{9'}$$

with $\Omega = \{\int_a^b x_i^{(m)}(t) x_j^{(m)}(t)dt\}_{i,j=1,n}$ .

*Proof.* Exercise 2.

The theorem does not specifically address what happens if we let $\lambda$ tend to zero or infinity. The case of $\lambda = \infty$ is found to correspond to estimation of $\eta$ by a polynomial of order m as in the case of spline smoothing. The other extreme, $\lambda = 0$, results in an estimator of $\eta$ which solves the generalized interpolation problem of minimizing

$$\int_a^b f^{(m)}(t)^2 dt$$

subject to

$$L_i f = y_i, \qquad i = 1,\ldots,n$$

(Exercise 3).

The computation of MOR estimators can be tedious in general. The evaluation of basis functions will usually require numerical methods, such as numerical quadrature, for evaluation of the $L_i(t - \cdot)_+^{2m-1}$. Also, the equation system (9) (or 9') will, depending on the chosen basis, generally not have a band structure that can be exploited to ease the computational burden. Several strategies for computing MOR estimators are discussed by Wahba (1980a, 1982).

To select $\lambda$ for MOR estimators one might again use CV or GCV. The idea behind CV and GCV in this setting is that $\eta_\lambda$ provides estimates of the $\mu_i$: namely,

$$\tilde{\mu}_{\lambda i} = L_i \eta_\lambda \quad .$$

Thus, let $\eta_{\lambda(i)}$ denote the minimizer of

$$n^{-1} \sum_{j \neq i} (y_j - L_j f)^2 + \lambda \int_a^b f^{(m)}(t)^2 dt \ .$$

Then, for CV we would choose $\lambda$ to minimize

$$CV(\lambda) = n^{-1} \sum_{i=1}^{n} (y_i - L_i \eta_{\lambda(i)})^2 \quad .$$

If we let $h_{ii}(\lambda)$, $i=1,\ldots,n$, denote the diagonal elements of the matrix which transforms $\underline{y}$ to $(L_1\eta_\lambda,\ldots,L_n\eta_\lambda)'$ arguments similar to those used for Lemma 5.2 reveal that

$$CV(\lambda) = n^{-1} \sum_{i=1}^{n} (y_i - L_i \eta_\lambda)^2/(1 - h_{ii}(\lambda))^2 \ .$$

The GCV criterion for this setting is

$$GCV(\lambda) = n^{-1} \sum_{i=1}^{n} (y_i - L_i \eta_\lambda)^2/(1 - n^{-1} \sum_{i=1}^{n} h_{ii}(\lambda))^2 \ .$$

As is clear from their definitions both CV and GCV provide estimates of the minimizer of the *range* (of the $L_i$) *risk*

$$RR(\lambda) = n^{-1} \sum_{i=1}^{n} E(\mu_i - L_i\eta_\lambda)^2 .$$

This measure assesses the performance of $\lambda$ by how well the $L_i\eta_\lambda$ conform to the $\mu_i$. However, if $\eta$ is the object of interest it may be preferable to try to estimate the value of $\lambda$ which minimizes, e.g., the integrated *domain* (of the $L_i$) *risk*

$$IDR(\lambda) = \int_a^b E(\eta(t) - \eta_\lambda(t))^2 dt .$$

This is complicated by the fact that there is no direct data on $\eta$ and, hence, there is no parallel of CV or GCV for this criterion. Wahba (1977a, 1980a) has given theoretical and numerical evidence that minimizing $RR(\lambda)$ (or $GCV(\lambda)$) can provide an effective estimate of $\eta$ in some cases. However, Rice (1986c) notes this will not always be true and that in some cases it will be preferable to estimate $IDR(\lambda)$ directly using some variant of the unbiased risk estimate discussed in Chapter 2. See Rice (1986c) and O'Sullivan (1986a) for discussions of how this might be done. The question of whether or not GCV will work for estimation of $\eta$ hinges, to a large extent, on whether the optimal rates of decay for the $\lambda$'s minimizing RR and IDR are the same. There are instances such as derivative estimation where the two will coincide (see, e.g., Rice and Rosenblatt, 1983). There are also cases where they will not.

Equation(9') has the implication that $\underline{\beta}_\lambda$ and, hence, $\eta_\lambda$ are Bayes estimators. The connection with Bayesian estimation goes much deeper than this, however. There is a parallel to model (5.41) which leads to $\eta_\lambda$ as an estimator of $\eta$ when a diffuse prior is placed on polynomials of order m. One can then utilize this framework for the development of inferential and diagnostic techniques and maximum likelihood methods for selecting $\lambda$. A thorough treatment of these topics can be found in Wahba (1983, 1985). The MOR estimator of $\eta$ also has certain minimax properties discussed in Speckman (1979) and Li (1982).

There are various ways to generalize the concept of MOR estimators. One useful formulation allows for the presence of nonlinear as well as linear functionals in model (3). This problem has been explored by O'Sullivan (1983) and O'Sullivan and Wahba (1985). Other modifications include the use of alternative measures of fidelity to the data and smoothness in criterion (5). These and other possibilities are explored in Kimeldorf and Wahba (1970 a,b,1971), O'Sullivan (1983, 1986b), O'Sullivan, Yandell and Raynor (1986), Raynor and Bates (1983), Wahba (1980a, 1982), Weinert (1978), Weinert and Kailath (1974), Weinert and Sidhu (1978), Weinert, Desai and Sidhu (1979) and Weinert, Byrd and Sidhu (1980). An excellent overview of a number of statistical issues related to MOR estimation and the solution of ill-posed problems is given in O'Sullivan (1986a).

### 6.2.2 Constraints

There are many cases where additional information about the properties of a function being estimated is available which can be introduced into the estimation process to obtain improved estimators. Such information can often be appended to criterion (2) or (5) through the introduction of constraints. These constraints may take on a variety of forms, some of which will be discussed here. Attention will be restricted to constrained smoothing splines. The more general situation of constrained MOR estimation follows a similar pattern and is discussed in detail in Wahba (1982).

In some instances, theory or prior experience may dictate the values of the regression function at certain points or, more generally, the values of certain linear functionals of $\mu$. Let $L_j$, $j=1,\ldots,r$, be r linear functionals and $\ell_j$, $j=1,\ldots,r$, a set of specified values. Then, if it is known that $L_j\mu = \ell_j$ estimation should be conducted by minimizing (2) subject to

$$L_j f = \ell_j, \quad j = 1,\ldots,r. \tag{10}$$

The solution to this problem is a function of the form

$$s(t) = \sum_{j=0}^{m-1} \theta_j t^j + \sum_{j=1}^{n} \delta_j (t - t_j)_+^{2m-1}$$

$$+ \sum_{j=1}^{r} \gamma_j L_j (t - \cdot)_+^{2m-1} \quad , \tag{11}$$

where the $\delta_j$ and $\gamma_j$ must satisfy $\Sigma_{j=1}^{n} \delta_j (1 - t_j)^k = \Sigma_{j=1}^{r} \gamma_j L_j (1 - \cdot)^k = 0$, $k = 0,\ldots,m - 1$. An argument which leads to this can be developed by first supposing that $\tilde{f}$ is a minimizer of (2) subject to (10). Then, as a consequence of the discussion following Theorem 6.1, we can find a function $\tilde{s}$ of the form (11) which minimizes $\int_a^b f^{(m)}(t)^2 dt$ over all functions in $W_2^m[a,b]$ which satisfy

$$f(t_j) = \tilde{f}(t_j) \quad , \quad j=1,\ldots,n,$$

and

$$L_j f = \ell_j \quad , \quad j=1,\ldots,r.$$

Thus, $\tilde{s}$ would satisfy the constraints and give a smaller value for criterion (2). Consequently, the minimizing function must have the stated form. An explicit expression for the minimizer can now be obtained by minimizing (2) subject to (10) with respect to the coefficients in (11). (Exercise 4).

For some problems of interest $\mu$ may be known to take on values only within a certain interval or exhibit some type of monotonicity or concavity behavior. For example, if $\mu$ represents a percentage measure, then we have $0 \le \mu(t) \le 1$ for all t. If $\mu$ is a quantile or distribution function of a continuous random variable then $\mu$ must be monotone nondecreasing which is reflected by the fact that $\mu' \ge 0$. A general type of minimization problem stemming from considerations such as the above is the minimization of (2) over some closed convex subset C of $W_2^m[a,b]$. Examples of such subsets include

$$C = \{f \in W_2^m[a,b]: \; p \le f(t) \le q\} \; ,$$

$$C = \{f \in W_2^m[a,b]: \quad f'(t) \geq 0\}$$

and

$$C = \{f \in W_2^m[a,b]: \quad f''(t) \leq 0\}, \quad m > 2,$$

which correspond to functions bounded between constants p and q, functions which are monotone nondecreasing and functions which are concave down, respectively.

Minimization of (2) subject to $f \in C$ is a difficult problem in general. It is possible to show that there is a unique solution. However, characterization of the solution and its computation pose challenging problems. Some results along these lines can be found in Laurent (1969), Wright and Wegman (1980), Wong (1984), Utreras (1984, 1985a) and Irvine, Marin and Smith (1985). As an illustration of the difficulties which arise we mention a result due to Utreras (1985a). He shows that, for m = 2 and n even, the minimizer of (2) subject to $f'(t) \geq 0$ is a spline of order 4 with knots at $t_1,\ldots,t_n$ and at most n + 2 additional knots $\xi_1,\ldots,\xi_{n+2}$. The problem is that the number and location of the $\xi_j$ is unknown. Utreras suggests that an exchange algorithm might be used to locate the additional knots. However, the development of methods for computing splines under convexity constraints remains an open problem in most cases.

It may be possible in some instances to approximate C in a fashion which lends itself to simplified estimator computation. Specifically, let $L_j$, j=1,...,r, be r linear functionals and ($p_j$, $q_j$), j=1,...,r, a collection of constants with $p_j \leq q_j$. Then, suppose that C can be approximated by

$$C_a = \{f \in W_2^m[a,b]: \quad p_j \leq L_j f \leq q_j\} \ . \tag{12}$$

As an example of where such an assumption is reasonable, consider the case where $\mu$ represents a proportion so that

$$C = \{f \in W_2^m[a,b] \ : \ 0 \leq f(t) \leq 1\} \ .$$

One could then define $C_a$ by taking $L_j f = f(s_j)$ and

$$C_a = \{f \in W_2^m[a,b]: \ 0 \le f(s_j) \le 1, \ \ j=1,\ldots,r\}$$

for some sufficiently fine grid of points $0 \le s_1 < \ldots < s_r \le 1$.

Minimization of (2) over a set such as (12) is somewhat more amenable to analysis. One can show that (Exercise 5) the minimizer must be of the form (11). Thus, we need only be concerned with functions of this type and the minimization problem reduces to finding the $\theta$, $\delta$ and $\gamma$ coefficients which minimize (2) subject to the constraints. This is a quadratic programming problem and there are a variety of algorithms which can be used to compute a solution. For some specific proposals see Wahba (1980a, 1982), Villalobos (1983) and Villalobos and Wahba (1983,1985,1987). Wahba (1973) shows that the solutions obtained using $C_a$ will approximate the minimizer over C in many cases of interest.

The selection of $\lambda$ in the case of equality constraints such as (10) falls within the framework of Chapter 2 since the estimator for that case is linear and has a closed form. However, for the other types of constraints we have considered the estimator will generally not have a closed form and iterative techniques will be required. Wahba (1980a, 1982) has proposed an analog of the GCV criterion for this setting (Exercise 6) which seems to work quite well in many cases. (See, e.g., Wahba, 1982, and Villalobos, 1983).

### 6.2.3 Multivariate Smoothing Splines

One of the key assumptions in model (1) is that t is scalar valued. In general, this will not be the case and instead there will be $p > 1$ variables $t_1,\ldots,t_p$ whose relationships with Y are to be studied. Let

$$\underline{t} = (t_1,\ldots,t_p)'$$

and assume that responses $y_1,\ldots,y_n$ have been obtained corresponding to values $\underline{t}_1,\ldots,\underline{t}_n$ of $\underline{t}$ that are related through the model

$$y_j = \mu(\underline{t}_j) + \varepsilon_j \quad , \quad j = 1,\ldots,n \ . \tag{13}$$

The $\varepsilon_j$ are assumed, as before, to be zero mean, uncorrelated errors. The difference between model (1) and (13) is that now the regression function, $\mu$, is allowed to be a function of more than one variable. The question to be addressed is how to extend the smoothing spline concept to incorporate this assumption.

It turns out that there are actually a number of ways to develop multivariate smoothing splines. Various possibilities are suggested by the work of Duchon (1977), Meinguet (1979), Dyn, Wahba and Wong (1979), Dyn and Wahba (1982), Cox (1984a) and Barry (1986). We will concentrate on one of these which is known as the thin plate or Laplacian smoothing spline.

The choice of a generalization of spline smoothing for vector valued $\underline{t}$ depends upon how one decides to extend the smoothness criterion (2b) to the multivariate case. There are a number of reasonable ways this can be done. One possibility is

$$J_{m,p}(f) = \sum_{\alpha_1+\ldots+\alpha_p=m} \{(m!/\alpha_1!\ldots.\alpha_p!)$$

$$\times \int_{-\infty}^{\infty}\ldots\int_{-\infty}^{\infty}(\partial^m f(\underline{t})/\partial t_1^{\alpha_1}\ldots\partial t_p^{\alpha_p})^2 d\underline{t}\} \ .$$

This measure of smoothness is defined, loosely speaking, on the set $W_2^m(\mathbb{R}^p)$ of all functions on $\mathbb{R}^p$ for which all the partial derivatives of total order m are square integrable over $\mathbb{R}^p$. For a more mathematically precise description of $W_2^m(\mathbb{R}^p)$ see Duchon (1977) or Meinguet (1979).

If one can assume that $\mu \in W_2^m(\mathbb{R}^p)$ the obvious generalization of criterion (2) is

$$n^{-1}\sum_{j=1}^{n}(y_j - f(\underline{t}_j))^2 + \lambda J_{m,p}(f) \ . \tag{15}$$

The minimizer of (15) over all $f \in W_2^m(\mathbb{R}^p)$ will be denoted by $\mu_\lambda$ and termed a *Laplacian smoothing spline*.

Wendelberger (1981) has pointed out that, when m=p=2, (15) can be interpreted as a measure of the amount of energy required to bend a thin plate of infinite extent that is connected to the $y_i$ by springs. In that case, $\lambda$ represents a constant related to the flexibility of the plate. In general we may view (15) much like (2) in the sense of providing a combination of measures of fidelity to the data and smoothness.

The smoothing parameter $\lambda$ in (15) regulates the weight given to each of the two components of the criterion. The extreme cases of $\lambda = 0$ and $\lambda = \infty$ are found to correspond to interpolation of the data and regression on polynomials of order m, just as in the univariate case. However, by polynomial regression we now mean regression on multivariate polynomials in $t_1, \ldots, t_p$. Such polynomials will be linear combinations of functions of the form

$$t_1^{\alpha_1} \cdots t_p^{\alpha_p}$$

and will therefore have total order m provided they involve only those functions with $\alpha_1 + \cdots + \alpha_p \le m - 1$.

In general the set of polynomials of total order m will have dimension

$$M = \binom{p + m - 1}{p} .$$

As a specific example take $m = 2$. Then $M = p + 1$ and the corresponding set of polynomials will consist of linear combinations of $1, t_1, t_2, \ldots, t_p$. Consequently, Laplacian smoothing splines provide a generalization of linear regression on the variables $t_1, \ldots, t_p$ in that case.

It is perhaps not immediately clear from (15) that Laplacian smoothing splines are a generalization of our original smoothing spline estimator. This is because when $p = 1$ the smoothness measure (14) still involves integration over the entire line while

(2b) is concerned only with the interval [a,b]. However, the fact is that when p = 1 only functions whose mth derivatives are zero outside of $[t_1, t_n]$ need be considered as minimizers of either (2) or (15), since the corresponding smoothness measures can only be increased if that is not the case. This has the implication that minimization of both (2) and (15) is equivalent to finding the minimizer of

$$n^{-1} \sum_{j=1}^{n} (y_j - f(t_j))^2 + \lambda \int_{t_1}^{t_n} f^{(m)}(t)^2 \, dt.$$

The smoothing spline estimator clearly solves this problem and must therefore coincide with the Laplacian smoothing spline with p = 1.

A closed form expression for the Laplacian smoothing spline estimator has been derived by Wahba and Wendelberger (1980). Let $p_j$, j=1,...,M, be a basis for the polynomials of total order m and assume that 2m - p > 0 and that the n × M matrix

$$T = \{p_j(\underline{t}_i)\}_{\substack{i=1,n \\ j=1,M}}$$

has rank M. Then, (15) has a unique minimizer over $W_2^m(\mathbb{R}^p)$ of the form

$$s(\underline{t}) = \sum_{j=1}^{M} \theta_j p_j(\underline{t}) + \sum_{j=1}^{n} \gamma_j Q_p(\underline{t}, \underline{t}_j) \quad , \tag{16}$$

where

$$Q_p(\underline{t}, \underline{s}) = \begin{cases} |\underline{s} - \underline{t}|^{2m - p} \ln(|\underline{s} - \underline{t}|) \ , & p \text{ even} \ , \\ |\underline{s} - \underline{t}|^{2m - p} \ , & p \text{ odd} \ , \end{cases}$$

and the coefficient vector $\underline{\gamma} = (\gamma_1, \ldots, \gamma_n)'$ must satisfy

$$T'\underline{\gamma} = \underline{0}.$$

This last condition is a multivariate analog of the restriction on natural splines that their mth through (2m - 1)st derivatives must vanish at b.

Since $T'\underline{\gamma} = \underline{0}$ this means that $\underline{\gamma}$ lies in the $n - M$ dimensional subspace of $\mathbb{R}^n$ that is perpendicular to the columns of T. Let $\underline{Z}_{M+1},\ldots,\underline{Z}_n$ be a basis for this subspace and set $Z = [\underline{Z}_{M+1},\ldots,\underline{Z}_n]$. Then

$$\underline{\gamma} = Z\,\underline{\zeta}$$

for some vector $\underline{\zeta}$. Now make the identifications

$$x_j(\underline{t}) = p_j(\underline{t})\ , \quad j = 1,\ldots,M \quad ,$$

$$x_j(\underline{t}) = (Q_p(\underline{t},\underline{t}_1),\ldots,Q_p(\underline{t},\underline{t}_n))\underline{Z}_j, \quad j = M+1,\ldots,n,$$

$$X = \{x_j(\underline{t}_i)\}_{i,j=1,n}$$

and

$$\underline{\beta}' = (\underline{\theta}', \underline{\zeta}')\ .$$

Then any function of the form (16) can be expressed as

$$s(\underline{t}) = \sum_{j=1}^{n} \beta_j x_j(\underline{t})\ .$$

By minimizing (15) over functions of this type we find the Laplacian smoothing spline estimator of $\mu$ to be $\mu_\lambda = \Sigma_{j=1}^n \beta_{\lambda j} x_j$, where $\underline{\beta}_\lambda = (\beta_{\lambda 1},\ldots,\beta_{\lambda n})'$ is the solution of

$$(X'X + n\lambda\Omega)\underline{\beta}_\lambda = X'\underline{y}, \tag{17a}$$

and $\Omega$ denotes the matrix with typical element

$$\sum_{\alpha_1+\cdots+\alpha_p=m} (m!/\alpha_1!\ldots\alpha_p!)\int_{-\infty}^{\infty}\cdots\int_{-\infty}^{\infty} \left\{ \frac{\partial^m x_i(\underline{t})}{\partial t_1^{\alpha_1}\ldots\partial t_p^{\alpha_p}} \times \frac{\partial^m x_j(\underline{t})}{\partial t_1^{\alpha_1}\ldots\partial t_p^{\alpha_p}} \right\} dt\ . \tag{17b}$$

It is clear from (17) that $\mu_\lambda$ is a linear estimator that fits into the framework of Chapter 2. Thus, the techniques described there can be used for selecting $\lambda$ and even m. In particular, GCV appears to work well in this setting (c.f., Wahba and Wendelberger, 1980, Wendelberger, 1981, and Hutchinson and Bischof, 1983). Utreras (1985b) has even shown that the Craven and Wahba result (5.40) extends to the multivariate case.

Estimator (16) can be extended to include the case where

$$y_i = L_i\eta + \varepsilon_i \ , \qquad i=1,\dots,n,$$

for some collection of bounded linear functionals $L_1,\dots,L_n$ and $\eta$ an element of $W_2^m(\mathbb{R}^p)$. This more general development is pursued in Wahba and Wendelberger (1980) and Wahba (1981a). Also, constraints can be appended to the minimization problems. See Villalobos (1983) and Villalobos and Wahba (1983, 1985,1987) for a treatment of this topic. In the case of p = 2 Wahba (1981b) has studied a parallel of Laplacian smoothing splines which are suitable for smoothing data on the sphere.

The Laplacian smoothing spline is also a Bayes estimator. There are several models which lead to this conclusion. A simple approach is to extend the model of Silverman (1985) for the univariate case and assume that $\mu = \Sigma_{j=1}^n \beta_j x_j$ with $\underline{\beta} = (\beta_1,\dots,\beta_n)'$ having a prior log likelihood equal , apart from an additive constant,to $-n\lambda\underline{\beta}'\Omega\underline{\beta}/2$. If $\Omega$ were nonsingular this would correspond to a zero mean normal prior density for $\underline{\beta}$. However, since $\Omega$ has rank n - M, the prior is partially improper giving infinite variance to the polynomial coefficients. If we let $\underline{\mu}_\lambda = (\mu_\lambda(\underline{t}_1),\dots,\mu_\lambda(\underline{t}_n))'$ and $\underline{e}_\lambda = \underline{y} - \underline{\mu}_\lambda$ one finds that identities (5.49) and (5.52) still hold; that is,

$$\mathrm{Var}(\underline{\mu}_\lambda | \underline{y}) = \sigma^2 H(\lambda)$$

and

$$\mathrm{Var}\ \underline{e}_\lambda = \sigma^2(I - H(\lambda)) \quad ,$$

where

$$H(\lambda) = X(X'X + n\lambda\Omega)^{-1}X' \quad . \tag{18}$$

This provides a motivation for using inferential and diagnostic techniques, exactly like those in the univariate case, for multivariate smoothing splines. Discussions and illustrations of this are given in Wendelberger (1981), Wahba (1983), Carmody (1985) and Eubank and Gunst (1986).

### 6.2.4 Partial Splines

A useful generalization of model (1) allows for both a parametric and nonparametric part for the mean values. More specifically, let $z_1,\ldots,z_q$ and $t_1,\ldots,t_p$ be p + q variables which are believed to be related to Y. Define the $y_j$ and $\underline{t}_j$ as before and denote the n observations on $\underline{z} = (z_1,\ldots,z_q)'$ by $\underline{z}_1,\ldots,\underline{z}_n$. Then assume that

$$y_j = \underline{z}_j'\underline{\zeta} + \mu(\underline{t}_j) + \varepsilon_j, \qquad j = 1,\ldots,n , \tag{19}$$

for zero mean uncorrelated random errors $\varepsilon_1,\ldots,\varepsilon_n$ and $\mu \in W_2^m(\mathbb{R}^p)$.

Model (19) treats Y and $\underline{z}$ as if they were linearly related but allows for nonlinearity in $\underline{t}$. There are a number of situations where this type of model might be deemed appropriate. Particularly noteworthy is the problem of analysis of covariance, where $\underline{\zeta}$ represents a vector of treatment effects and $\underline{t}$ a vector of covariates. The classical method of analysis would assume that the covariates are also linearly related to Y. However, as demonstrated by Green, Jennison and Seheult (1985), there are cases where such an assumption may not be suitable and the possibility of nonlinearity in $\underline{t}$ must be entertained.

To estimate $\underline{\zeta}$ and $\mu$ one might use the minimizer of

$$n^{-1}\sum_{j=1}^{n}(y_j - \underline{z}_j'\underline{\zeta} - f(\underline{t}_j))^2 + \lambda J_{m,p}(f) \tag{20}$$

over all $\underline{\zeta} \in \mathbb{R}^q$ and $f \in W_2^m(\mathbb{R}^p)$. Setting $Z' = [\underline{z}_1,\ldots,\underline{z}_n]$ and using the notation of Section 6.2.3, it follows easily from our work there (Exercise 7) that under the assumption that $2m - p > 0$ and that $[T,Z]$ has rank $M + q$ the minimizing function is

$$\underline{z}'\underline{\zeta}_\lambda + \sum_{j=1}^{n} \beta_{\lambda j} x_j(\underline{t}) \tag{21a}$$

with the coefficient vectors $\underline{\zeta}_\lambda$ and $\underline{\beta}_\lambda = (\beta_{\lambda 1}, \ldots, \beta_{\lambda n})'$ given by

$$\underline{\beta}_\lambda = (X'X + n\lambda\Omega)^{-1}X'(\underline{y} - Z\underline{\zeta}_\lambda) \tag{21b}$$

and

$$\underline{\zeta}_\lambda = [Z'(I - H(\lambda))Z]^{-1}Z'(I - H(\lambda))\underline{y} , \tag{21c}$$

where $H(\lambda)$ is defined in (18).

The estimator (21) is known as a *partial spline*. It originated from work by Rice (1982), Engle et al. (1986) and Wahba (1984a,b). See also Wahba's discussion to Huber (1985).

The parameter $\lambda$ in (20) can, of course, be chosen via the methods of Chapter 2. The extreme case of $\lambda = \infty$ is seen to correspond to linear regression of Y on $\underline{z}$ and polynomials of order m in $\underline{t}$. Thus, in particular, when $m = 2$ partial splines provide a generalization of the classical approach to analysis of covariance.

There are several Bayesian models which lead to partial spline estimators. Some of these are explored in Wahba (1984a,b), Green, Jennison and Seheult (1985), Shiller (1985), Shiau (1985), Heckman (1986) and Eubank (1986). Some of the implications of such models for diagnostic and inferential analysis are studied by Shiau (1985), Green, Jennison and Seheult (1985) and Eubank and Gunst (1986).

Partial splines are but one of many possible methods for conducting analysis of covariance when the covariates enter into the model in a nonlinear fashion. Green (1985) gives an overview of several related methods, including (discrete) partial splines, in the context of field trial analysis. Other possible estimators are explored in Wahba (1986), Heckman (1986) and Speckman (1986).

### 6.2.5 Robust Smoothing Splines

In Section 4.11 a robust version of kernel estimation was discussed whose purpose was to guard against deleterious effects of discrepant responses. It is possible to formulate a similar type of estimator from a smoothing spline perspective. If $\rho$ is some nonnegative function that is symmetric about zero, one might choose to estimate $\mu$ by the minimizer of

$$n^{-1} \sum_{j=1}^{n} \rho(y_j - f(t_j)) + \lambda \int_a^b f^{(m)}(t)^2 dt \tag{22}$$

over $f \in W_2^m[a,b]$. We will call this estimator a *robust* or *M-type smoothing spline*. When $\rho(u) = u^2$ the robust smoothing spline and ordinary smoothing spline coincide. Thus, the idea behind (22) is to replace the traditional choice $\rho(u) = u^2$ with another choice that is less sensitive to extreme values and, therefore, provides an estimator which is more resistant to the influence of outliers.

Estimators of $\mu$ derived from criteria such as (22) have been studied by Anderssen, Bloomfield and McNeil (1974), Huber (1979), Utreras (1981a) and Cox (1981, 1983a). Cox (1981, 1983a) shows that if $\rho$ is strictly convex and $n \geq m$, (22) will have a unique minimizer in $W_2^m[a,b]$. It follows from this and Corollary 5.1 that the minimizing function must be a natural spline. Consequently, we need only minimize (22) over functions of the form $\sum_{j=1}^{n} \beta_j x_j$, where the $x_j$ are any basis for $NS^{2m}(t_1, \ldots, t_n)$. If $\rho$ admits a derivative, $\psi = \rho'$, the problem is equivalent to finding coefficients which satisfy

$$-\sum_{i=1}^{n} x_j(t_i)\psi(y_i - \sum_{r=1}^{n} \beta_r x_r(t_i)) + 2\lambda \sum_{r=1}^{n} \beta_r \omega_{jr} = 0, \quad j = 1, \ldots, n,$$

where $\omega_{rj}$ is the rjth element of $\Omega$ in (5.17b). These are a system of nonlinear equations which are the analog of the normal equations (5.17) used to find the natural spline coefficients for ordinary smoothing splines. They will, in general, require iterative methods to compute a solution. Some possible algorithms for this purpose are discussed in Anderssen, Bloomfield and McNeil (1974), Huber (1979) and Utreras (1981a). Choices for $\psi$ (or $\rho$) are discussed in Section 4.11.

The problem of data-driven selection of $\lambda$ for robust smoothing splines has not, as yet, been explored to any great extent. One possible approach has been suggested by Wahba (1980a) who defines a generalization of GCV for nonlinear problems which can be applied to this case. Robust smoothing splines are also Bayes estimators (Cox, 1981) and hence the potential exists for development of diagnostic and inferential techniques which parallel those in Section 5.5 for these estimators.

### 6.2.6 The Penalty Function Method

In Sections 6.2.1 - 6.2.5 we have examined some of the possible generalizations of the basic smoothing spline estimator. These serve to illustrate the flexibility of the penalized least-squares approach to nonparametric estimation. It is therefore no surprise that our discussions have really done little more than scratch the surface of the subject.

A general approach to nonparametric estimation can be based on what might be termed the penalty function method. Under this method an estimate of an unknown function is obtained by optimizing

$$n^{-1} \sum_{i=1}^{n} \rho_i(\underline{y}, f) + \lambda J(f) \tag{23}$$

over $f \in C \subset W$, where $C$ is some subset of a general function space $W$, $J$ is a smoothness measure associated with $W$ and the $\rho_i$ are functionals on $\mathbb{R}^n \times W$. For example, if $\rho_i(\underline{y}, f) = (y_i - f(t_i))^2$, $C = W = W_2^m[a,b]$ and $J = J_m$, the optimal estimator is a smoothing spline. In a similar manner one can obtain all the estimators discussed in this chapter. There are also many other possibilities some of which are discussed in Wegman (1982). We will mention only two.

If the $y_i$ represent a random sample of observations with common density $p \varepsilon W$, a maximum penalized likelihood (MPL) estimator of p can be obtained by minimizing

$$-n^{-1} \sum_{i=1}^{n} \log f(y_i) + \lambda J(f)$$

over

$$C = \{f \varepsilon W: f \geq 0, \ \int f(t)dt = 1\} .$$

This corresponds to the choice of $\rho_i(\underline{y},f) = -\log f(y_i)$ in (23). MPL density estimators represent an important method of density estimation which stems from the work of Good and Gaskins (1971). Tapia and Thompson (1978) give a detailed treatment of MPL density estimation along with conditions for existence and uniqueness of the estimator in certain cases. Examples of cases where a unique MPL density estimator exists include

$$W = W_2^m[a,b] , \qquad J = J_m$$

and

$$W = W_2^2(\mathbb{R}), \quad J(f) = \int_{-\infty}^{\infty} f'(t)^2 dt + \int_{-\infty}^{\infty} f''(t)^2 dt.$$

See Tapia and Thompson (1978) and Silverman (1982) for other examples.

Another application of the penalty function method is to the analysis of generalized linear models (Nelder and Wedderburn, 1972, McCullagh and Nelder, 1983, Dobson, 1983). For this problem $y_1,\ldots,y_n$ represent independent observations with $y_i$ typically having an exponential density of the form

$$p_i(y) = \exp\{yb(c_i) - d(c_i) + e(y)\}$$

for some choice of constants $c_1,\ldots,c_n$ and functions b, d and e. One finds that the mean values satisfy

$$\mu_i = Ey_i = d'(c_i)/b'(c_i) \ .$$

A form of linear model results if one assumes that there are variables $t_1,\ldots,t_p$ and a function g such that

$$g(\mu_i) = \sum_{j=1}^{p} \beta_j t_{ij} \qquad ,$$

where $\beta_1,\ldots,\beta_p$ are unknown constants and $t_{ij}$ is the value of $t_j$ corresponding to $y_i$. To obtain maximum likelihood estimators of the $\beta_j$ one minimizes

$$\sum_{i=1}^{n} (d(c_i) - y_i b(c_i))$$

with respect to the unknown parameters. A disadvantage of this approach is the restriction to a parametric model for the $g(\mu_i)$. A more general formulation would assume that

$$g(\mu_i) = L_i \eta$$

for $\eta \in W$, some collection of functions on $\mathbb{R}^p$, and the $L_i$ linear functionals defined on W. Then, an MPL estimator of $\eta$ could be derived by minimizing

$$n^{-1} \sum_{j=1}^{n} (d(c_i) - y_i b(c_i)) + \lambda J(f)$$

over $f \in W$.

As a specific example of the generalized linear model framework we mention the case of Poisson distributed $y_i$ in which case

$$p_i(y) = c_i^y \exp\{-c_i\}/y!$$

$$= \exp\{y\log c_i - c_i - \log(y!)\} \ .$$

This corresponds to the choice of $b(c_i) = \log c_i$, $d(c_i) = c_i$ and $e(y) = -\log(y!)$. The means are $\mu_i = c_i$. If we assume a "log-linear" model then

$$\mu_i = \exp\{L_i\eta\}$$

which means that $g(u) = \log u$. An MPL estimator for this case would minimize

$$n^{-1}\sum_{i=1}^{n}[\exp\{L_i f\} - y_i L_i f] + \lambda J(f)$$

over $f \in W$.

Conditions for existence and uniqueness of MPL estimators for the generalized linear model and other related frameworks are given in O'Sullivan (1983, 1986)and Cox and O'Sullivan (1985) along with some discussion of the associated computational problems. Other proposals and examples can be found in Raynor and Bates (1983) and O'Sullivan, Yandell and Raynor (1986).

## 6.3 Asymptotic Theory

In this section we study the large sample behavior of the smoothing spline estimator of the regression function. As with our previous studies of Fourier series and kernel estimators, the problems to be addressed include characterizing the asymptotic properties of the risk and asymptotic distribution of the estimator. Along the way we will also point out certain connections between spline smoothing and other nonparametric regression estimators that have been considered.

The asymptotic theory for smoothing splines is quite complicated. However,it has received extensive study and, as a consequence, is fairly well understood. Our objective will be to present the essential elements of this theory rather than to give an exhaustive treatment. A detailed analysis will therefore be undertaken only for the simplest case of periodic smoothing splines. This allows for a simplified treatment which embodies many of the key features of asymptotic analysis for smoothing splines with a minimum of technical details. The periodic case also provides a springboard whose understanding makes what transpires in the general case seem intuitively plausible. Discussion of the general case will be limited to the description of known results and citation of key references for those who desire further reading.

### 6.3.1 Periodic Smoothing Splines

In this section we study the large sample behavior of a variant of the basic smoothing spline estimator that is known as a *periodic smoothing spline*. This estimator arises when $\mu$ can be assumed to lie in $W_{2,per}^m[a,b]$ which, recall, is the subset of $W_2^m[a,b]$ whose elements satisfy the periodic boundary conditions

$$\mu^{(j)}(a) = \mu^{(j)}(b) \ , \quad j = 0,\ldots\ldots,m - 1 \ .$$

Periodic smoothing splines were introduced by Cogburn and Davis (1974) and, while seldom used in practice, provide an important theoretical tool. The analysis to be presented here for these estimators derives from the work of Wahba (1975) and Rice and Rosenblatt (1981). Other references concerning periodic smoothing splines and their applications are provided by Wahba and Wold (1975b) and Wahba (1980b).

To begin let us simplify matters by assuming that $[a,b] = [0,1]$,

$$t_j = (j - 1)/n, \quad j = 1,\ldots\ldots,n \ ,$$

and that $m = 2$. The *periodic smoothing spline* for this case is then found as the minimizer of

$$n^{-1} \sum_{j=1}^{n} (y_j - f(t_j))^2 + \frac{\lambda}{(2\pi)^4} \int_0^1 f''(t)^2 dt \tag{24}$$

over all $f \in W_{2,per}^2[0,1]$. The factor of $(2\pi)^4$ has been introduced into the criterion simply for subsequent notational convenience.

To derive the minimizer of (24) we will use a different tactic from what has been used previously and employ Fourier analysis. The idea is that any function $f \in W_{2,per}^{2}[0,1]$ admits the representation

$$f(t) = \sum_{j=-\infty}^{\infty} b_j e^{2\pi ijt}$$

with $b_j$ the jth Fourier coefficient

$$b_j = \int_0^1 f(t)e^{-2\pi ijt}dt .$$

Thus, to find the minimizer of (24) it suffices to find its corresponding Fourier coefficients. Our first task is therefore to re-express criterion (24) in terms of the Fourier coefficients of f. The minimizing function can then be obtained by minimization with respect to these Fourier coefficients. We give the details for only the case where n is odd and leave the case where n is even as an exercise.

When n is odd the vectors

$$\underline{x}_j = (1, e^{2\pi ij/n},\ldots,e^{2\pi ij(n-1)/n}),\ j=0,\pm1,\ldots,\pm(n-1)/2 ,$$

form an orthogonal basis for $\mathbb{R}^n$. Thus,

$$\underline{y} = \sum_{|j|\le(n-1)/2} \tilde{\beta}_j \underline{x}_j$$

with

$$\tilde{\beta}_j = n^{-1} \sum_{r=1}^{n} y_r e^{-2\pi ij(r - 1)/n} .$$

In particular, this means that

$$y_k = \sum_{|j| \le (n-1)/2} \tilde{\beta}_j e^{2\pi ij(k-1)/n} , \quad k=1,\ldots,n.$$

This same argument along with the Fourier series representation for f, shows that

$$f(t_k) = \sum_{|j|\le(n-1)/2} c_j e^{2\pi ij(k-1)/n} ,$$

where

$$c_j = n^{-1} \sum_{r=0}^{n-1} f(r/n)e^{-2\pi ijr/n}$$

$$= n^{-1} \sum_{r=0}^{n-1} \left( \sum_{j=-\infty}^{\infty} b_j e^{2\pi ijr/n}\right) e^{-2\pi ijr/n}$$

$$= \sum_{s=-\infty}^{\infty} b_{j+sn} .$$

It can now be easily verified that by substituting the above expressions for $y_k$ and $f(t_k)$ into the MSE term of (24) one obtains

$$n^{-1} \sum_{k=1}^{n} (y_k - f(t_k))^2 = \sum_{|j|\le(n-1)/2} |\tilde{\beta}_j - \sum_{s=-\infty}^{\infty} b_{j+sn}|^2 .$$

By using the fact that f(0) = f(1) and f'(0) = f'(1) it can be seen that f" has Fourier coefficients

$$b_j'' = \int_0^1 f''(t)e^{-2\pi ijt}dt$$

$$= -(2\pi j)^2 b_j .$$

Therefore, by Parseval's relation,

$$\int_0^1 f''(t)^2 dt = \sum_{j=-\infty}^{\infty} (b_j'')^2$$

$$= (2\pi)^4 \sum_{j=-\infty}^{\infty} j^4 |b_j|^2$$

$$= (2\pi)^4 \sum_{|j|\le(n-1)/2} \sum_{s=-\infty}^{\infty} (j + sn)^4 |b_{j+sn}|^2 .$$

Combining this with our expression for the MSE reveals that (24) can be written as

$$\sum_{|j|\le(n-1)/2} \{|\tilde{\beta}_j - \sum_{s=-\infty}^{\infty} b_{j+sn}|^2 + \lambda \sum_{s=-\infty}^{\infty} (j + sn)^4 |b_{j+sn}|^2\} . \qquad (24')$$

This form of the criterion has the desired property of depending only on the Fourier coefficients of f.

Observe that (24') is a sum of n nonnegative functions of the $b_k$ and that there is no overlap in the Fourier coefficients which appear in the different terms. Thus, we may minimize each term in the sum separately. Beginning with the term for j = 0 it is seen that

$$|\tilde{\beta}_0 - \sum_{s=-\infty}^{\infty} b_{sn}|^2 + \lambda \sum_{s=-\infty}^{\infty} (sn)^4 |b_{sn}|^2$$

is minimized by taking

$$b_0 = \tilde{\beta}_0 \tag{25a}$$

and

$$b_{sn} = 0 \tag{25b}$$

for all s ≠ 0.

For general j we require the minimizer of

$$|\tilde{\beta}_j - \sum_{s=-\infty}^{\infty} b_{j+sn}|^2 + \lambda \sum_{s=-\infty}^{\infty} (j + sn)^4 |b_{j+sn}|^2 \quad .$$

Differentiation of this expression with respect to $b_{j+sn}$ shows that the optimal coefficients satisfy

$$b_{j+sn} = (\tilde{\beta}_j - \sum_{\ell=-\infty}^{\infty} b_{j+\ell n})/\lambda(j + sn)^4 \ .$$

Now sum this expression on s to obtain

$$\sum_{s=-\infty}^{\infty} b_{j+sn} = \tilde{\beta}_j r_j/(\lambda + r_j) \quad ,$$

where

$$r_j = \sum_{s=-\infty}^{\infty} (j + sn)^{-4} \ .$$

Thus, by substituting back into our original formula for $b_{j+sn}$ we find the minimizing coefficients to be

$$b_{j+sn} = \tilde{\beta}_j/(\lambda + r_j)(j + sn)^4. \tag{25c}$$

To check that the $b_k$ determined in this fashion actually minimize (24') one need only verify that any pertubation of the coefficients (25) results in a function with a larger value for the criterion.

The periodic smoothing spline estimator is now given explicitly, for n odd, by

$$\begin{aligned}\mu_{\lambda p}(t) &= \sum_{|j|\le(n-1)/2} \sum_{s=-\infty}^{\infty} b_{j+sn} e^{2\pi i(j+sn)t} \\ &= \sum_{|j|\le(n-1)/2} (\tilde{\beta}_j/(\lambda + r_j)j^4)e^{2\pi ijt} \\ &+ \sum_{|j|\le(n-1)/2} (\tilde{\beta}_j/(\lambda + r_j)) \sum_{s\ne0}(j + sn)^{-4}e^{2\pi i(j+sn)} \quad .\end{aligned}$$

This is somewhat complicated and it would simplify matters if we could replace this estimator by an asymptotically equivalent estimator which is somewhat more amenable to analysis. The key to seeing how to do this rests on noting that, as a result of a calculation similar to (3.38)

$$\left|\sum_{s\ne0} (j + sn)^{-4}e^{2\pi i(j+sn)}\right| \le \sum_{s\ne0} (j + sn)^4 = 0(n^{-4})$$

and that

$$\begin{aligned}(\lambda + r_j)j^4 &= 1 + \lambda j^4 + j^4 \sum_{s\ne0} (j + sn)^{-4} \\ &= 1 + \lambda j^4 + j^4\, 0(n^{-4}) \quad .\end{aligned}$$

These two results coupled with the fact that the Fourier coefficients for $\mu$ must decay rapidly to zero suggest that

$$\mu_{\lambda p}(t) \doteq \sum_{|j|\le(n-1)/2} (\tilde{\beta}_j/(1 + \lambda j^4))e^{2\pi ijt} \tag{26}$$

for sufficiently large n. This is, in fact, the case and it can be shown that, under restrictions on n and $\lambda$, $\mu_{\lambda p}$ and the approximation defined by (26) have the same limiting distribution and the risks for the original and approximate estimator behave the same asymptotically. The details involved in verifying this are tedious but straightforward, given the discussion leading to (26) and results in Rice and Rosenblatt (1981). We will not present them here.

In all subsequent discussions we will work with the approximation (26) instead of the original estimator. Rather than introduce new terminology and notation the approximation will also be referred to as a periodic smoothing spline and denoted by $\mu_{\lambda p}$, i.e., we now define

$$\mu_{\lambda p}(t) = \sum_{|j|\leq(n-1)/2} (\tilde{\beta}_j/(1 + \lambda j^4))e^{2\pi ijt} \quad . \tag{27}$$

The estimator (27) is clearly closely related to Fourier series estimators. However, it uses all n of the sample Fourier coefficients. This is in contrast to classical Fourier series estimators which have the form

$$\sum_{|j|\leq r} \tilde{\beta}_j e^{2\pi ijt}$$

for some $r \leq (n - 1)/2$. Usually r is much smaller than $(n - 1)/2$ since Fourier series estimators smooth by ignoring those $\tilde{\beta}_j$ with indices of large magnitude that can be viewed as corresponding to high frequency aspects of the data. Rather than deleting or ignoring higher order Fourier coefficients, estimator (27) uses all n coefficients but weights them by a damping factor $(1 + \lambda j^4)^{-1}$. This is similar to what transpires for kernel estimators. Apart from boundary effects, one might anticipate that smoothing splines for the general case should behave similarly (c.f., our discussion of the Demmler-Reinsch basis in Section 5.3.3).

The parameter $\lambda$ in (27) governs the severity of down weighting for higher order Fourier coefficients. The extremes, $\lambda$

= 0 and $\lambda = \infty$ correspond to a Fourier series estimator based on all n terms (which, recall, interpolates the data) and estimation of $\mu$ by the constant function $\tilde{\beta}_0 = n^{-1}\Sigma_{j=1}^{n} y_j$. Thus, smaller values of $\lambda$ represent the inclusion of more information from higher frequencies into the estimator while larger values result in an estimator that is based on the lower frequency content of the data.

Since $\mu_{\lambda p}$ is a type of damped Fourier series estimator one might wonder if it is not also closely related to kernel estimators which have this property in the periodic case. To see that this is so, first write

$$\mu_{\lambda p}(t) = \sum_{|j| \le (n-1)/2} (\tilde{\beta}_j/(1 + \lambda j^4))e^{2\pi i j t}$$

$$= n^{-1}\sum_{r=1}^{n} y_r \sum_{|j| \le (n-1)/2} (1 + \lambda j^4)^{-1} e^{2\pi i j(t - (r-1)/n)} .$$

Then, for n large, we may argue that

$$\sum_{|j| \le (n-1)/2} (1 + \lambda j^4)^{-1} e^{2\pi i j u}$$

$$\doteq \sum_{-\infty < j < \infty} (1 + \lambda j^4)^{-1} e^{2\pi i j u}$$

$$\doteq \int_{-\infty}^{\infty} (1 + \lambda x^4)^{-1} e^{2\pi i x u} dx$$

$$= \lambda^{-1/4} K(u/\lambda^{1/4}) ,$$

where $K(\cdot)$ is the Fourier transform of $(1 + y^4)^{-1}$, given explicitly by

$$K(u) = \frac{\sin(|u|/\sqrt{2} + \pi/4)}{2} e^{-|u|/\sqrt{2}} . \tag{28}$$

A plot of this function is shown in Figure 6.1. While this kernel does not have finite support, it does satisfy the basic properties

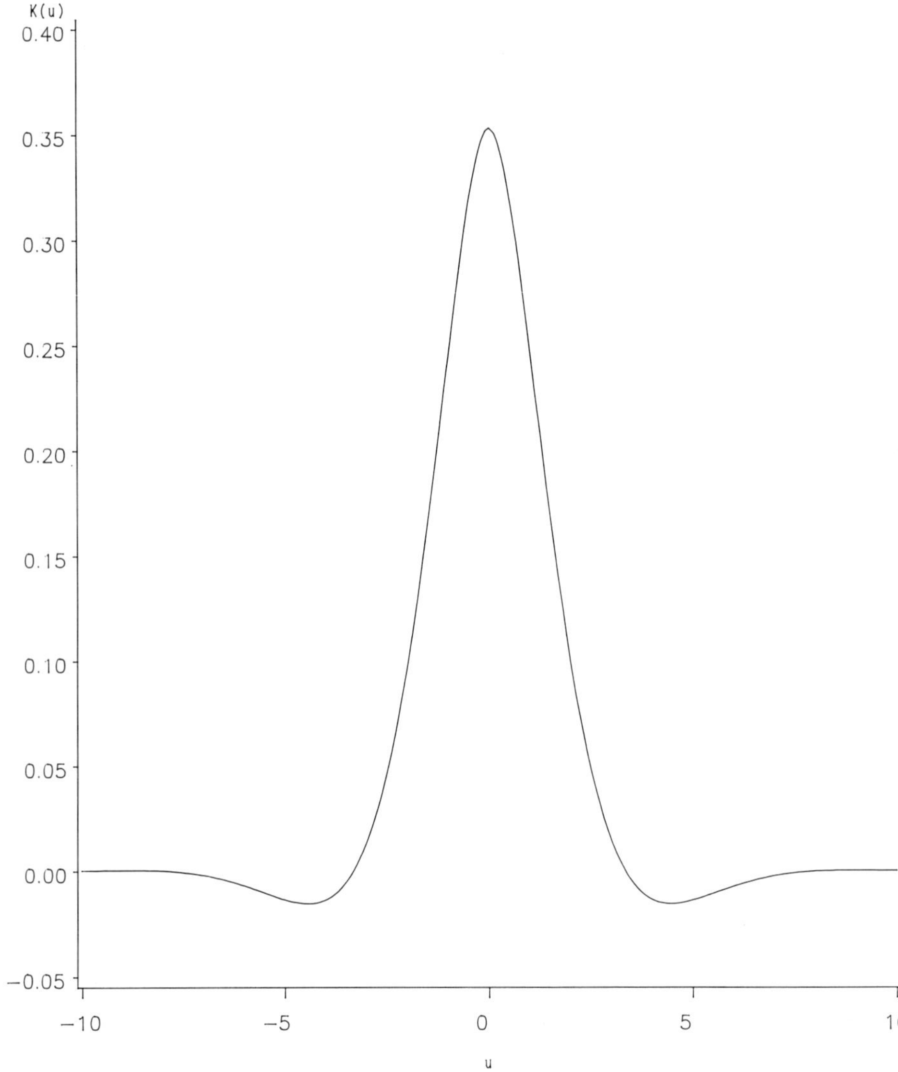


**Figure 6.1** Approximate kernel for spline smoothing.

that are desired for a kernel, such as $\int K(u)du = 1$, $\int uK(u)du = 0$ and $\int K(u)^2 du < \infty$. In addition, it satisfies $\int u^2 K(u)du = \int u^3 K(u)du = 0$ and $\int u^4 K(u)du = -1$. Thus one can think of estimator (27) as corresponding to kernel estimation with a kernel of order four with support on the entire line. The role of the bandwidth is played by $\lambda^{1/4}$. Again, apart from boundary or edge effects, this result might be anticipated to carry over to the general smoothing spline estimator obtained from criterion (2). We will see shortly that this is indeed the case.

Let us now study the (integrated) risk associated with $\mu_{\lambda p}$. Writing $IR_n(\lambda)$ in terms of its variance and squared bias components we have

$$IR_n(\lambda) = \int_0^1 (\mu(t) - E\mu_{\lambda p}(t))^2 dt + \int_0^1 \mathrm{Var}\ \mu_{\lambda p}(t)dt.$$

The asymptotic behavior of each of the two integrals will be examined in turn.

Dealing first with the squared bias component observe that

$$B_n^2(\lambda) = \int_0^1 (\mu(t) - E\mu_{\lambda p}(t))^2 dt$$

$$= \sum_{|r|>(n-1)/2} |\beta_r|^2 + \sum_{|j|\le(n-1)/2} |\beta_j - E\tilde{\beta}_j/(1 + \lambda j^4)|^2 ,$$

where, as usual, $\beta_j$ is used to denote the jth Fourier coefficient of $\mu$. Since $\mu \in W_{2,per}^2[0,1]$ assume that the $\beta_j$ decay algebraically, i.e.

$$|\beta_j|^2 \sim |j|^{-(5 + \delta)}$$

for some $\delta > 0$. Then (c.f. similar calculations in Chapter 3)

$$\sum_{|j|>(n-1)/2} |\beta_j|^2 = O(n^{-(4 + \delta)})$$

and the j=0 term in the second sum in $B_n^2(\lambda)$ is

$$|\beta_0 - E\tilde{\beta}_0|^2 = |\sum_{s\neq 0} \beta_{sn}|^2$$

$$\le O(n^{-(5+\delta)}) \quad .$$

Consequently, it suffices to study

$$\sum_{|j|\le(n-1)/2}' |\beta_j - \sum_{\ell=-\infty}^{\infty} \beta_{j+\ell n}/(1+\lambda j^4)|^2$$

$$= \sum_{|j|\le(n-1)/2}' |\beta_j|^2 |1-(1+\lambda j^4)^{-1}|^2$$

$$- 2\mathrm{Re}\Big( \sum_{|j|\le(n-1)/2}' \beta_j (1-(1+\lambda j^4)^{-1}) \sum_{\ell\ne 0} \beta^*_{j+\ell n}/(1+\lambda j^4)\Big)$$

$$+ \sum_{|j|\le(n-1)/2}' (1+\lambda j^4)^{-2} \left| \sum_{\ell\ne 0} \beta_{j+\ell n} \right|^2 \quad ,$$

where $\Sigma_j'$ denotes summation excluding j=0 and Re(z) is the real part of the complex number z. Attention need be focused only on the first and last sum since the modulus of the middle sum can be dominated by the square roots of the moduli of the other two through the use of the Cauchy-Schwarz inequality.

Now observe that

$$\sum_{|j|\le(n-1)/2}' (1+\lambda j^4)^{-2} \left| \sum_{\ell\ne 0} \beta_{j+\ell n} \right|^2$$

$$\le O(n^{-(5+\delta)}) \sum_{|j|\le(n-1)/2}' (1+\lambda j^4)^{-2}$$

$$\le O(n^{-(5+\delta)}) \int_{-\infty}^{\infty} (1+\lambda x^4)^{-2} dx$$

$$= O(n^{-(4+\delta)}) (n\lambda^{1/4})^{-1} \quad .$$

The last inequality arises from the use of an integral estimate to show that

$$\sum_{|j|\le(n-1)/2} (1+\lambda j^4)^{-2} \sim \lambda^{-1/4} \int_{-\infty}^{\infty} (1+y^4)^{-2} dy \quad .$$

Thus, provided $(n\lambda^{1/4}) \to \infty$ as $n\to\infty$, the asymptotic properties of $B_n^2(\lambda)$ will be determined by

$$\sum_{|j|\le(n-1)/2}' |\beta_j|^2 |1 - (1 + \lambda j^4)^{-1}|^2$$

$$= \sum_{|j|\le(n-1)/2}' |\beta_j|^2 |(\lambda^{1/4} j)^4/(1 + (\lambda^{1/4} j)^4)|^2 .$$

Now break this sum into two parts corresponding to $|j| \le \lambda^{-1/4}$ and $\lambda^{-1/4} < |j| \le (n - 1)/2$ from which it may be seen that, asymptotically, $B_n^2(\lambda)$ will have the same order as

$$\lambda^2 \sum_{|j|\le\lambda^{-1/4}}' |\beta_j|^2 j^8 + \sum_{\lambda^{-1/4}<|j|\le(n-1)/2}' |\beta_j|^2 .$$

There are two cases to consider. If $|\beta_j|^2 \sim |j|^{-(5+\delta)}$ for $0 < \delta < 4$, then the first sum in this expression behaves like

$$\lambda^2 \int_0^{\lambda^{-1/4}} x^{-(5+\delta)} x^8 dx \propto \lambda^{(4+\delta)/4}$$

and

$$\sum_{\lambda^{-1/4}<|j|\le(n-1)/2}' |\beta_j|^2 \sim \int_{\lambda^{-1/4}}^{\infty} x^{-(5+\delta)} dx$$

$$\propto \lambda^{(4+\delta)/4} .$$

So the squared bias part of $IR_n(\lambda)$ is $O(\lambda^{(4+\delta)/4})$ in this case. However, if $\Sigma_{j=-\infty}^{\infty} |\beta_j|^2 j^8 < \infty$, the first sum is $O(\lambda^2)$ while the second will decay at a rate faster than $\lambda^2$. Consequently, the squared bias is $O(\lambda^2)$ in this latter instance. The case of $\Sigma_{j=-\infty}^{\infty} |\beta_j|^2 j^8 < \infty$ arises when $\mu$ is very smooth in the sense that $\mu \in W_{2,per}^4[0,1]$.

To summarize we have shown that if $|\beta_j|^2 \sim |j|^{-(5+\delta)}$ for $0 < \delta < 4$ then

$$B_n^2(\lambda) = O(\lambda^{(4+\delta)/4}) .$$

If $\sum_{j=-\infty}^{\infty} |\beta_j|^2 j^8 < \infty$ then

$$B_n^2(\lambda) = O(\lambda^2) .$$

Now consider the variance portion of the risk which can be written as

$$V_n(\lambda) = \int_0^1 \mathrm{Var}\ \mu_{\lambda p}(t)dt$$

$$= \frac{\sigma^2}{n} \sum_{|j| \le (n-1)/2} (1 + \lambda j^4)^{-2} ,$$

through use of the fact that

$$\mathrm{Cov}(\tilde{\beta}_j, \tilde{\beta}_u) = \begin{cases} \sigma^2/n , & j = u , \\ 0 , & \text{otherwise} . \end{cases}$$

Combining this with our previous approximations gives the following theorem.

**Theorem 6.2.** Assume that $|\beta_j|^2 \sim |j|^{-(5+\delta)}$ for $0 < \delta < 4$. Then, if $n \to \infty$, $\lambda \to 0$ and $n\lambda^{1/4} \to \infty$,

$$IR_n(\lambda) \sim C_1 \lambda^{(4+\delta)/4} + C_2(n\lambda^{1/4})^{-1} \tag{29a}$$

for positive constants $C_1$ and $C_2$. The asymptotically optimal risk in this case is $O(n^{-(4+\delta)/(5+\delta)})$. If $\Sigma |\beta_j|^2 j^8 < \infty$ then

$$IR_n(\lambda) \sim C_1 \lambda^2 + C_2(n\lambda^{1/4})^{-1} , \tag{29b}$$

and the asymptotically optimal risk is $O(n^{-8/9})$.

The essential conclusion of Theorem 6.2 is that, if $n \to \infty$ and $\lambda \to 0$ in such a way that $(n\lambda^{1/4}) \to \infty$, $\mu_{\lambda p}$ is a consistent estimator of $\mu$. By viewing $\mu_{\lambda p}$ as being essentially a kernel estimator of order 4 with bandwidth $\lambda^{1/4}$, the conditions imposed on n and $\lambda$ are precisely what one would expect from our study of kernel estimators in Chapter 4. The rate of decay for the risk is also what would have been expected for, if $\mu$ has four continuous derivatives and satisfies $\mu^{(j)}(0) = \mu^{(j)}(1)$, $j=0,\ldots,3$, the optimal risk for a kernel estimator of order 4 (modified for periodicity) will be $O(n^{-8/9})$ (c.f. (4.41)). However, these restrictions on $\mu$ imply that $\Sigma|\beta_j|^2 j^8 < \infty$ in which case the optimal risk for $\mu_{\lambda p}$ is also $O(n^{-8/9})$.

The above analysis can be extended to general m (Exercise 10) and to the development of asymptotic theory for variants of the smoothing spline estimator. As an illustration consider a deconvolution problem where the regression function has the special form

$$\mu(t) = \int_0^1 \eta(s)K(t - s)ds$$

for some $\eta \in W^m_{2,per}[0,1]$ and K a smooth, periodic, square integrable function. If the objective is to estimate $\eta$ instead of (or, in addition to) $\mu$, one might use the MOR estimator obtained by minimizing

$$n^{-1}\sum_{j=1}^{n}(y_j - \int_0^1 f(s)K(t_j - s)ds)^2 + \frac{\lambda}{(2\pi)^{2m}}\int_0^1 f^{(m)}(t)^2 dt \qquad (30)$$

over $f \in W^m_{2,per}[0,1]$.

Assume, as before, that $t_j = (j - 1)/n$, n is odd and take m = 2. Then a similar argument to that used to derive $\mu_{\lambda p}$ shows that, to a high order of approximation, the minimizer of (30) in this case is given by

$$\eta_{\lambda p}(t) = \sum_{|j|\le(n-1)/2}(K_j^*\tilde{\beta}_j/(|K_j|^2 + \lambda j^4))e^{2\pi ijt} \quad ,$$

where $K_j$ is the jth Fourier coefficient of K.

Let $\alpha_j$ denote the jth Fourier coefficient for $\eta$ and assume that $|K_j|^2 \sim |j|^{-2\gamma}$ for some $\gamma > 0$, and that $|\alpha_j|^2 \sim |j|^{-(5+\delta)}$. Then if $n\to\infty$, $\lambda\to 0$ and $n\lambda^{(2\gamma+1)/(4+2\gamma)} \to 0$ one finds that the integrated domain variance is

$$IDV_n(\lambda) = \int_0^1 \text{Var } \eta_{\lambda p}(t)dt$$

$$= \frac{\sigma^2}{n} \sum_{|j|\le(n-1)/2} |K_j|^2/(|K_j|^2 + \lambda j^4)^2$$

$$= O((n\lambda^{(2\gamma+1)/(4+2\gamma)})^{-1})$$

and the integrated domain bias is

$$\mathrm{IDB}_n^2(\lambda) = \int_0^1 (\eta(t) - E\eta_{\lambda p}(t))^2 dt$$

$$\sim \sum_{|j|\le(n-1)/2} |\alpha_j|^2 (\lambda j^{4+2\gamma}/(1 + \lambda j^{4+2\gamma}))^2 .$$

Splitting $\mathrm{IDB}_n^2(\lambda)$ into sums over $|j| \le \lambda^{-1/(4+2\gamma)}$ and over $\lambda^{-1/(4+2\gamma)} < |j| \le (n - 1)/2$ shows that $\mathrm{IDB}_n^2(\lambda)$ behaves asymptotically like

$$\lambda^2 \sum_{|j|\le\lambda^{-1/(4+2\gamma)}} |\alpha_j|^2 |j|^{8+4\gamma} + \sum_{\lambda^{-1/(4+2\gamma)}<|j|\le(n-1)/2} |\alpha_j|^2 .$$

By combining our asymptotic expressions for $\mathrm{IDV}_n(\lambda)$ and $\mathrm{IDB}_n^2(\lambda)$ we find, for example, that if $|\alpha_j|^2 \sim |j|^{-(5+\delta)}$, $0 < \delta < 4 + 4\gamma$, the optimal domain risk for $\eta_{\lambda p}$ is $O(n^{-(4+\delta)/(5+2\gamma+\delta)})$.

To conclude our discussion of periodic smoothing splines let us establish the asymptotic normality of the estimator. To apply the Liapunov Theorem of Section 3.4.4 we take

$$V_{rn} = n^{-1} y_r \sum_{|j|\le(n-1)/2} (1 + \lambda j^4)^{-1} e^{2\pi i j(t-(r-1)/n)} , \quad r=1,\ldots,n,$$

and will choose $\theta = 2 + \alpha$ for some $\alpha > 0$. It has already been established that

$$\sigma_n^2 = \sum_{j=1}^{n} E|V_{jn} - EV_{jn}|^2 = O((n\lambda^{1/4})^{-1}) .$$

Thus, we need only work with

$$C_n(\theta) = \sum_{j=1}^{n} E|V_{jn} - EV_{jn}|^{2+\alpha} .$$

Assume that the random errors $\varepsilon_1,\ldots,\varepsilon_n$ are independent and identically distributed with finite absolute $(2+\alpha)$th moments and observe that

$$\left|\sum_{|j|\le(n-1)/2}(1+\lambda j^4)^{-1}e^{2\pi iju}\right|$$

$$\le \sum_{|j|\le(n-1)/2}(1+\lambda j^4)^{-1}$$

$$\le \lambda^{-1/4}A$$

with $A = \int_{-\infty}^{\infty}(1+y^4)^{-1}dy$ . Thus,

$$C_n(\theta) = n^{-\theta}E|\varepsilon_1|^{\theta}\sum_{r=1}^{n}\left|\sum_{|j|\le(n-1)/2}(1+\lambda j^4)^{-1}e^{2\pi ij(t-(r-1)/n)}\right|^{\theta}$$

$$\le n^{-\theta}E|\varepsilon_1|^{\theta}(A\lambda^{-1/4})^{\alpha}\sum_{r=1}^{n}\left|\sum_{|j|\le(n-1)/2}(1+\lambda j^4)^{-1}e^{2\pi ij(t-(r-1)/n)}\right|^2$$

$$= n^{-(1+\alpha)}E|\varepsilon_1|^{\theta}(A\lambda^{-1/4})^{\alpha}\sum_{|j|\le(n-1)/2}(1+\lambda j^4)^{-2}$$

$$= O((n\lambda^{1/4})^{-(1+\alpha)}) .$$

Therefore,

$$C_n(\theta)/\sigma_n^{\theta} = O((n\lambda^{1/4})^{-\alpha/2})$$

and $[\mu_{\lambda p}(t) - E\mu_{\lambda p}(t)]/\sigma_n$ converges in distribution to a standard normal random variable provided only that $n\to\infty$, $\lambda\to 0$ and $n\lambda^{1/4}\to\infty$.

It is of more interest to know whether $[\mu_{\lambda p}(t) - \mu(t)]/\sigma_n$ will be asymptotically standard normal. For that to happen one needs the additional condition

$$(n\lambda^{1/4})^{1/2}[E\mu_{\lambda p}(t) - \mu(t)] \to 0$$

as $n \to \infty$ and $\lambda \to 0$. Calculations such as those leading to (3.46) reveal that if, for example, $|\beta_j|^2 \sim |j|^{-(5+\delta)}$, $0 < \delta < 4$, this condition will be satisfied if $n\lambda^{(4+\delta)/4} \to 0$. The optimal $\lambda$ for this case is $O(n^{-4/(5+\delta)})$ which does not have this property. This is a consequence of the fact that bias and standard error will be roughly balanced under the optimal $\lambda$. Thus to ensure asymptotic normality $\lambda$ should converge to zero at something faster than the optimal rate. This has various consequences for interval estimation which have already been discussed in Chapters 3 and 4 where similar problems arose for Fourier series and kernel estimators.

### 6.3.2 The General Case

We now relax the assumption that the regression function is periodic and assume only that $\mu \in W_2^m[a,b]$. In this case there are essentially two approaches which have been used to study the asymptotic behavior of the risk. One employs Fourier analysis and parallels the methodology of the previous section. The other relies on the approximation of $E\mu_\lambda$ and the eigenvalues $\delta_{jn}$, $j=1,\ldots,n$, from the Demmler-Reinsch basis by quantities from related continuous (in t) problems which are easier to study. We will discuss the principles behind both approaches but first will detail the connection between spline smoothing and kernel regression estimators.

#### Relationship with Kernel Estimators

Recall that $\mu_\lambda$ is a linear estimator. Consequently, for any t in [a,b] and $n \geq m$ there is a collection of weight functions $\{K_n(t;t_j,\lambda)\}_{j=1}^n$, such that

$$\mu_\lambda(t) = \sum_{j=1}^n y_j K_n(t;t_j,\lambda) .$$

The asymptotic behavior of the weight function sequence for smoothing splines has been studied by Speckman (1981), Cox (1983a) and Silverman (1984b).

Assume that m=2 and the $t_j$ are generated by a positive density p on [a,b], in the sense that

$$\int_a^{t_j} p(t)dt = (2j - 1)/2n. \tag{31}$$

Then, Silverman (1984b) shows that, for n large, $\lambda$ small and $t_j$ not too close to the boundary of [a,b],

$$K_n(t;t_j,\lambda) \doteq K((t - t_j)/h(t_j))/p(t_j)h(t_j) \quad , \tag{32a}$$

where K is the kernel (28) and

$$h(s) = (\lambda/p(s))^{1/4} . \tag{32b}$$

A more precise statement of this result and conditions under which it is valid can be found in Silverman (1984b) along with more refined approximations that are better suited for use near the boundaries. Further refinements and extensions of Silverman's results are provided in Nychka (1986) and Messer (1986).

Silverman's result (32) has several important consequences. First, it indicates that a cubic smoothing spline is very nearly a variable bandwidth kernel estimator (c.f. Section 4.7) with bandwidth determined by h(s). The fact that h(s) involves both $\lambda$ and p(s) means that for points near s the bandwidth is determined by the global smoothing factor $\lambda^{1/4}$ as well as the density of the design points near s. Through dependence on p the smoothing spline estimator adjusts the amount of averaging which takes place to take account of the sparsity or richness of data. In data poor regions, where p(s) is small, the bandwidth h(s) will be larger than in regions of high density, where p(s) is large. This is the type of qualitative behavior one would want from a variable bandwidth kernel estimator and Silverman (1984b) even argues that the fact that h(s) involves the 4th root of p(s) is close to optimal in a certain sense (Exercise 11).

Another consequence of (32) is the fact that $\mu_\lambda$ is closely related to the periodic smoothing spline $\mu_{\lambda p}$ of the previous section. We saw that when $p(t) \equiv 1$ the periodic estimator was essentially a kernel estimator with kernel (28) and bandwidth $\lambda^{1/4}$. Silverman's approximation (32) indicates the same is true for $\mu_\lambda$ if boundary regions are excluded. Thus, one would anticipate that smoothing splines and periodic smoothing splines should behave similarly away from the boundaries. In particular, this means that our view of how periodic smoothing splines smooth the data by damping higher order Fourier coefficients is also applicable, apart from edge effects, in the general case where $\mu$ is not periodic.

There is actually nothing special about m = 2 and analogs of (32) hold for general m. The kernel in each case will be the Fourier transform of $(1 + y^{2m})^{-1}$ (c.f. Exercise 9).

## Analysis of the Risk

Our discussion now turns to the description of the asymptotic properties of the integrated risk for the general smoothing spline estimator $\mu_\lambda$ obtained by minimizing (2). For notational simplicity it will be assumed throughout that [a,b] = [0,1].

One approach to studying the integrated risk is through the use of Fourier analysis. The essential technique is the same as that demonstrated in Section 6.3.1 for the periodic case. It is first necessary to obtain a representation for $\mu_\lambda$ in terms of its Fourier coefficients. To accomplish this it is helpful to adjust criterion (2) to allow for the fact that the Fourier series of a function f in $W_2^m[0,1]$ will converge to $(f(0) + f(1))/2$ at the boundaries. For uniformly spaced $t_j$ and observations $y_0$ and $y_n$ obtained at t=0 and t=1, respectively, this can be accomplished by minimizing

$$n^{-1}\{(y_0 + y_1 - f(0) - f(1))^2/4 + \sum_{j=1}^{n-1} (y_j - f(j/n))^2\}$$

$$+ \lambda\int_0^1 f^{(m)}(t)^2 dt \tag{33}$$

over $f \in W_2^m[0,1]$. The derivation of the estimator parallels that of $\mu_{\lambda p}$ for the periodic case. However, the details are more tedious due to the fact that

$$b_j = \int_0^1 f(t)e^{-2\pi ijt}dt$$

$$= \sum_{r=0}^{m-1} (-1/2\pi ij)^{r+1}(f^{(r)}(1) - f^{(r)}(0)) - (-1/2\pi ij)^m b_j^{(m)},$$

where $b_j^{(m)} = \int_0^1 f^{(m)}(t)e^{-2\pi ijt}dt$. Thus the relationship between the Fourier coefficients for f and its derivatives is not as simple as in the periodic case and this must be taken into account when deriving the estimator.

To analyze the integrated risk corresponding to the minimizer of (33) the first step is to express the risk as the sum of an integrated variance and squared bias component. The asymptotic properties of the integrated variance component can be derived in a similar manner to the periodic case using the uncorrelated nature of the sample Fourier coefficients. The integrated squared bias component can be written as a sum of the squared differences between the Fourier coefficients of $\mu$ and those of the minimizer of (33) which can then be studied under various assumptions on $\mu$. The details become quite complicated however due to the boundary behavior of the estimator and will not be presented here. The interested reader is referred to Rice and Rosenblatt (1983) for a complete analysis using this program for the case of cubic smoothing splines.

There is an alternative approach to the use of Fourier analysis for studying the risk. This method has its roots in early work by Wahba (1975, 1977a) and Craven and Wahba (1979). It has been carried out for general m by Speckman (1981).

Consider the case where the design points are generated by a positive density p through relation (31). The integrated risk in this case is

$$IR_n(\lambda) = \int_0^1 E(\mu(t) - \mu_\lambda(t))^2 p(t)dt$$

$$= B_n^2(\lambda) + V_n(\lambda) ,$$

where

$$B_n^2(\lambda) = \int_0^1 (\mu(t) - E\mu_\lambda(t))^2 p(t)dt$$

and

$$V_n(\lambda) = \int_0^1 \mathrm{Var}(\mu_\lambda(t))p(t)dt.$$

The idea is to replace $B_n^2(\lambda)$ and $V_n(\lambda)$ by other quantities which have the same asymptotic properties but are more mathematically tractable.

Let us focus first on the integrated squared bias component $B_n^2(\lambda)$. Observe that $E\mu_\lambda$ is the minimizer of

$$n^{-1}\sum_{j=1}^{n} (\mu(t_j) - f(t_j))^2 + \lambda\int_0^1 f^{(m)}(t)^2 dt$$

over $f \in W_2^m[0,1]$ (Exercise 13). Thus, for large n and small $\lambda$ it might be expected that $E\mu_\lambda$ could be well approximated by $f_\lambda$, the minimizer of

$$\int_0^1 (\mu(t) - f(t))^2 p(t)dt + \lambda\int_0^1 f^{(m)}(t)^2 dt$$

over $f \in W_2^m[0,1]$. Speckman (1981) uses this motivation to suggest replacing $E\mu_\lambda$ by $f_\lambda$ in $B_n^2(\lambda)$ and instead studying

$$B^2(\lambda) = \int_0^1 (\mu(t) - f_\lambda(t))^2 p(t)dt \quad .$$

He justifies this practice by showing that, for $m \geq 2$, $B_n^2(\lambda) = B^2(\lambda)(1 + o(1))$, where $o(1) \to 0$ as $n\to\infty$, $\lambda\to 0$ and $n^2\lambda \to \infty$.

The function $f_\lambda$ is what is known as a Tikhonov regularizer for $\mu$ (see, e.g., Tikhonov and Arsenin, 1977). Speckman obtains an explicit form for $f_\lambda$ using the Green's function for a particular differential equation. Utreras (1986a) derives an expression for $f_\lambda$ using a limiting version of the Demmler-Reinsch basis that will be introduced subsequently.

A useful asymptotic bound for $B_n^2(\lambda)$ follows immediately upon noting that

$$B^2(\lambda) \le B^2(\lambda) + \lambda \int_0^1 f_\lambda^{(m)}(t)^2 dt$$

$$\le \int_0^1 (\mu(t) - f(t))^2 p(t)dt + \lambda \int_0^1 f^{(m)}(t)^2 dt$$

for any f in $W_2^m[0,1]$. In particular by taking $f = \mu$ we obtain

$$B_n^2(\lambda) \le \lambda J_m(\mu)(1 + o(1)) \quad . \tag{34}$$

Thus, $B_n^2(\lambda) = O(\lambda)$ as $n \to \infty$, $\lambda \to 0$ and $n^2\lambda \to \infty$.

Now consider $V_n(\lambda)$. Under the Demmler-Reinsch basis

$$\mu_\lambda(t) = \sum_{j=1}^{n} (1 + n\lambda\delta_{jn})^{-1} \tilde{\beta}_j x_{jn}(t),$$

where

$$\tilde{\beta}_j = \sum_{r=1}^{n} y_r x_{jn}(t_r) \ ,$$

$x_{1n},\ldots,x_{nn}$ are the Demmler-Reinsch natural spline basis functions and $0 = \delta_{1n} = \ldots = \delta_{mn} < \delta_{m+1,n} \le \ldots \le \delta_{nn}$ are the associated eigenvalues (c.f. Section 5.3.3). Thus,

$$\text{Var}\ \mu_\lambda(t) = \sigma^2 \sum_{j=1}^{n} (1 + n\lambda\delta_{jn})^{-2} x_{jn}^2(t) \ ,$$

since

$$\text{Cov}(\tilde{\beta}_j, \tilde{\beta}_k) = \sigma^2 \sum_{r=1}^{n} x_{jn}(t_r) x_{kn}(t_r)$$

$$= \begin{cases} \sigma^2 , & j = k , \\ 0 , & j \neq k , \end{cases}$$

due to the orthonormal nature of the basis elements. Consequently,

$$V_n(\lambda) = \int_0^1 \mathrm{Var}(\mu_\lambda(t))p(t)dt = \sigma^2 \sum_{j=1}^{n} (1 + n\lambda\delta_{jn})^{-2} \int_0^1 x_{jn}^2(t)p(t)dt.$$

Now, for n large we should have

$$n^{-1} = n^{-1} \sum_{r=1}^{n} x_{jn}^2(t_r)$$

$$\sim \int_0^1 x_{jn}^2(t)p(t)dt,$$

which suggests that

$$V_n(\lambda) \sim \frac{\sigma^2}{n} \sum_{j=1}^{n} (1 + n\lambda\delta_{jn})^{-2} . \tag{35}$$

Speckman (1981) gives a rigorous proof of this conjecture through establishing that if $n \to \infty$, $\lambda \to 0$ and $n\lambda \to \infty$ there is a constant C independent of $\lambda$ such that

$$\left| V_n(\lambda) - \sigma^2 n^{-1} \sum_{j=1}^{n} (1 + n\lambda\delta_{jn})^{-2} \right| \le C/n^2\lambda . \tag{36}$$

Since $n^{-2}\lambda^{-1} = o(n^{-1})$ if $n\lambda \to \infty$, this means that we are justified in replacing $V_n(\lambda)$ by the simpler approximation (35).

To establish (36) Speckman uses the previously mentioned limiting version of the Demmler-Reinsch basis, which plays an important role in the study of the large sample characteristics of the $\delta_{jn}$ as well. The motivation for this approach stems from work by Utreras (1979, 1981b) who recognized that the $x_{jn}$ and $\delta_{jn}$ were essentially characterized as the solutions to the problem of finding a function f and value $\delta$ such that

$$\int_0^1 f^{(m)}(t)g^{(m)}(t)dt = \delta \sum_{j=1}^{n} f(t_j)g(t_j) \tag{37}$$

for every $g \in W_2^m[0,1]$ (Exercise 14). To be somewhat more precise, the problem of finding f and $\delta$ such that (37) is satisfied for all functions g in $W_2^m[a,b]$ has as its only solutions for $\delta$ the values $\delta_{1n},\ldots,\delta_{nn}$. The functions $x_{jn}$ corresponding to the $\delta_{jn}$ form a basis for the set of possible functions solving this problem and if $\delta_{jn}$ is distinct, $x_{jn}$ is the unique solution to (37) corresponding to $\delta_{jn}$.

By letting n grow large the left hand side of (37) is seen to be approximately equal to $n\delta\int_0^1 f(t)g(t)p(t)dt$. This prompts the consideration of the problem: find f and $\gamma$ which satisfy

$$\int_0^1 f^{(m)}(t)g^{(m)}(t)dt = \gamma\int_0^1 f(t)g(t)p(t)dt \tag{38}$$

for all $g \in W_2^m[0,1]$. It can be shown that there are values $0 = \gamma_1 = \ldots = \gamma_m < \gamma_{m+1} \leq \ldots$ which represent the only solutions to this problem and there is a corresponding sequence of functions $\{\xi_j\}$ which provide a basis for the set of solutions for f. The $\xi_j$ may be taken to be orthonormal and therefore satisfy

$$\int_0^1 \xi_j(t)\xi_k(t)p(t)dt = \begin{cases} 0 & , \quad j \neq k \quad , \\ 1 & , \quad j = k \quad , \end{cases}$$

and

$$\int_0^1 \xi_j^{(m)}(t)g^{(m)}(t)dt = \gamma_j\int_0^1 \xi_j(t)g(t)p(t)dt$$

for all functions g in $W_2^m[0,1]$. See, e.g., Weinberger (1974) for a detailed discussion of problems of this type.

By comparing (37) and (38) we are led to the conjecture that $n^{1/2}x_{jn} \to \xi_j$ and $n\delta_{jn} \to \gamma_j$ as $n\to\infty$. This turns out to be the case. The convergence of $n^{1/2}x_{jn}$ to $\xi_j$ in $L_2[0,1]$ is established in Speckman (1981) while Utreras (1979, 1981b) and Nussbaum (1985) have shown that $n\delta_{jn} \to \gamma_j$ for any fixed j. For this latter result to be useful we need to know how the $\gamma_j$ behave as j grows large.

Known results from the theory of differential equations (Naimark, 1967) have the consequence that for $j \geq 1$

$$\gamma_{j+m} = (\pi j)^{2m}(\int_0^1 p(t)^{1/2m}dt)^{-2m}(1 + o(1)) ,$$

where $o(1) \to 0$ as $j \to \infty$. Speckman (1981) employs this property to justify the approximation

$$n\delta_{j+m,n} \doteq (\pi j)^{2m}(\int_0^1 p(t)^{1/2m}dt)^{-2m} \tag{39}$$

which is found to hold uniformly over $j = o(n^{2/(2m+1)})$. Exact bounds for the $\delta_{jn}$ have been obtained by Utreras (1983) for the case of $p(t) \equiv 1$.

Now write

$$V_n(\lambda) \sim \frac{\sigma^2}{n} \sum_{j=1}^{n} (1 + n\lambda\delta_{jn})^{-2}$$

and break the sum into two parts corresponding to those value of $j \leq n^{3/4m}$, say, and $j > n^{3/4m}$. Since $3/4m < 2/(2m + 1)$ for $m \geq 2$ our approximation can be used to see that

$$n^{-1} \sum_{j \leq n^{3/4m}} (1 + n\lambda\delta_{jn})^{-2}$$

$$\doteq n^{-1} \sum_{j \leq n^{3/4m}} (1 + \lambda\gamma_k)^{-2} \sim \frac{\ell_m}{n\lambda^{1/2m}} \pi^{-1} \int_0^1 p(t)^{1/2m}dt ,$$

where

$$\ell_m = \int_{-\infty}^{\infty} (1 + y^{2m})^{-2}dy .$$

To treat the remaining sum use the fact that the $\delta_{jn}$ are non-decreasing along with approximation (39) to see that there is a positive constant $C$ such that for $n$ large and $j > n^{3/4m}$

$$(1 + n\lambda\delta_{jn})^{-2} \leq (1 + C\lambda n^{3/2})^{-2}$$

$$= O(\lambda^{-1} n^{-3}) .$$

Thus,

$$n^{-1} \sum_{j>n^{3/4m}} (1 + n\lambda\delta_{jn})^{-2} = O(\lambda^{-2}n^{-3})$$

which is $o(n^{-1})$ if $n\lambda \to \infty$.

Combining the above results about $V_n(\lambda)$ with those concerning $B_n^2(\lambda)$ shows that, for large n, $IR_n(\lambda)$ is bounded above by

$$\lambda J_m(\mu) + \sigma^2 \frac{\ell_m}{n\lambda^{1/2m}} \pi^{-1} \int_0^1 p(t)^{1/2m} dt \quad .$$

Consequently, by taking $\lambda = O(n^{-2m/(2m+1)})$ the integrated risk can always be made to decay at a rate of $n^{-2m/(2m+1)}$. Recall that this is the best uniform rate of convergence for functions in $W_2^m[0,1]$.

In some cases, however, the bound $\lambda J_m(\mu)$ used for $B_n^2(\lambda)$ can be improved. Whether or not this occurs depends on $\mu$ being smoother than originally stated and the characteristics of the function at zero and one. Recall that $\mu_\lambda$ is a natural spline of order 2m which means it must satisfy the boundary constraints

$$\mu_\lambda^{(j)}(0) = \mu_\lambda^{(j)}(1) = 0, \quad j = m,\ldots,2m - 1.$$

These restrictions limit the flexibility of the estimator at the boundary. It would therefore be expected that if $\mu$ admits more than m derivatives the rate of decay of the squared bias term would depend on whether or not $\mu$ satisfies similar conditions.

Speckman (1981) shows that if $\mu \in W_2^{2m}[0,1]$ satisfies

$$\mu^{(j)}(0) = \mu^{(j)}(1) = 0, \quad j=m,\ldots,2m - 1$$

and $p(t) \equiv 1$ then

$$B_n^2(\lambda) \sim \lambda^2 J_{2m}(\mu) \ .$$

Thus, in this case the optimal $\lambda$ is $O(n^{-2m/(4m+1)})$ and the optimal risk is

$$IR_n(\lambda) \sim (4m + 1)[\sigma^2 \ell_m (\textstyle\int_0^1 p(t)^{1/2m} dt)/4mn\pi]^{4m/(4m+1)}$$

$$\times \ (J_{2m}(\mu))^{1/(4m+1)} .$$

This result is under the unrealistic assumption that the boundary behavior of $\mu$ is matched to that of $\mu_\lambda$. Assume, instead, that we only have

$$\mu^{(j)}(0) = \mu^{(j)}(1) = 0, \quad j = m,\dots,k ,$$

for some k between m - 1 and 2m - 1, with the convention that k = m - 1 means that $\mu^{(m)}(0)$ or $\mu^{(m)}(1)$ is not zero. Utreras (1986a) gives upper bounds for $B^2(\lambda)$ for such instances which have the consequence that $B_n^2(\lambda)$ will be bounded above, for large n, by a constant multiple of $\lambda^{(2k+3)/2m}$.

To exemplify the consequences of Utreras' results assume that m = 2. Then, if $\mu^{(2)}(0)$ or $\mu^{(2)}(1)$ differs from zero, we cannot be assured of $B_n^2(\lambda)$ decaying any faster than $\lambda^{5/4}$ which corresponds to an attainable rate of $n^{-5/6}$ for $IR_n(\lambda)$. If both $\mu^{(2)}(0)$ and $\mu^{(2)}(1)$ vanish, but $\mu^{(3)}(0)$ or $\mu^{(3)}(1)$ is not zero, the rate improves to $n^{-7/8}$. However, neither rate is as good as $n^{-8/9}$ which can be attained when $\mu^{(2)}(0) = \mu^{(2)}(1) = \mu^{(3)}(0) = \mu^{(3)}(1) = 0$. Although we have used only an upper bound for $B_n^2(\lambda)$ to obtain these conclusions it has been shown by Rice and Rosenblatt (1983) and Speckman (1981) that, for m = 2, $n^{-5/6}$ and $n^{-7/8}$ are the best possible rates for $IR_n(\lambda)$ under the two types of boundary conditions which were just considered.

The dependence of $B_n^2(\lambda)$ on the boundary properties of $\mu$ is often viewed as a type of smoothing spline analog of the boundary bias difficulties associated with kernel estimators. While there are similarities, the two problems differ to an extent which merits discussion.

Our intuition suggests that smoothing splines behave roughly like kernel estimators of order 2m (e.g., order 4 for m = 2). Thus, if $\mu \in C^{2m}[0,1]$ we would expect $\mu_\lambda$ to have an optimal risk which is $O(n^{-4m/(4m+1)})$, that being the behavior of the optimal integrated risk for a kernel estimator of order 2m (c.f., Section

4.7). As we have seen this will not generally be the case and the best rate for general $\mu \in C^{2m}[0,1]$ would seem to be $O(n^{-2m/(2m+1)})$. This makes smoothing splines appear inferior to kernel estimators of order 2m. However, there is a flaw in this argument.

It was not assumed that $\mu$ had 2m continuous derivatives but rather that $\mu$ was in $W_2^m[0,1]$. As noted in Chapter 4, the use of kernel estimator of order 2m is guaranteed to improve convergence only when it can be assumed that $\mu \in C^{2m}[0,1]$, since the argument showing the optimal risk is $O(n^{-4m/(4m+1)})$ for such estimators rests on that assumption. If it can only be assumed that $\mu \in C^m[0,1]$, then a kernel estimator of order m would ordinarily be used for which the optimal rate of decay for the risk is $n^{-2m/(2m+1)}$. If $\lambda$ is chosen appropriately, smoothing splines will share this latter rate and, hence, are competitive with the kernel estimator which is actually appropriate for this situation. In fact they are superior to kernel estimators of order m in the sense that no boundary modifications are required for smoothing splines to obtain $n^{-2m/(2m+1)}$ as a rate of decay for their integrated risk. Recall that this cannot be accomplished for kernel estimators without some type of boundary modification.

To summarize, what has been discovered through our study of asymptotic theory for smoothing splines is that they perform competitively (in terms of decay rate for the risk) to the comparable kernel estimators and, if $\mu$ admits more than m derivatives, they may or may not perform like the higher order kernel estimator which is appropriate in that instance. However, this last property seems of little consequence; for if we knew that $\mu$ had more than m derivatives we would use a smoothing spline with an appropriate larger value of m that would then be assured of performing like the higher order kernel estimator that is suitable for this case.

The asymptotic theory associated with the eigenvalues $\delta_{jn}$ for the Demmler-Reinsch basis is useful in other contexts than analysis of the variance component of the integrated risk. In particular, it has proved useful in the study of certain problems associated with generalized cross-validation for smoothing splines.

Recall that the GCV criterion can be written as

$$GCV_n(\lambda) = MSE(\lambda)/(1 - \tau_1(\lambda))^2 \quad ,$$

where $MSE(\lambda)$ is the average of the squared residuals and $\tau_1(\lambda) = n^{-1}\mathrm{tr}H(\lambda)$. Using the Demmler-Reinsch basis we find that

$$\tau_1(\lambda) = n^{-1} \sum_{j=1}^{n} (1 + n\lambda\delta_{jn})^{-1} \ .$$

As previously noted, the difficulty in computing $GCV_n(\lambda)$ lies in the computation of $\tau_1(\lambda)$ which can now be viewed as related to the computation of the $\delta_{jn}$.

For cubic smoothing splines ($m = 2$), Silverman (1984a) uses (39) to motivate replacing the $\delta_{jn}$ in $\tau_1(\lambda)$ by values $d_j$ defined by

$$d_1 = d_2 = 0$$

and

$$d_j = \{\pi(j - 1.5)(\textstyle\int_0^1 p(t)^{1/4}dt)^{-1}\}^4$$

for $j > 2$. This assumes, of course, that p is known. When that is not the case a density estimator may be used to approximate the design density. Through replacing the $\delta_{jn}$ by the $d_j$ Silverman obtains an $O(n)$ algorithm for computing an approximation to the GCV criterion.

Another application of smoothing spline eigenvalue theory is to the proof of results such as (5.40) concerning the asymptotic efficiency of the GCV criterion for selecting $\lambda$. To see how this is done recall that as a result of the GCV Theorem (Theorem 2.1), we have

$$|EGCV_n(\lambda) - \sigma^2 - R_n(\lambda)|/R_n(\lambda) \le g_n(\lambda) \ ,$$

where $R_n(\lambda) = n^{-1} \Sigma_{j=1}^n E(\mu(t_j) - \mu_\lambda(t_j))^2$ and

$$g_n(\lambda) = [2\tau_1(\lambda) + \tau_1^2(\lambda)/\tau_2(\lambda)]/(1 - \tau_1(\lambda))^2$$

with

$$\tau_j(\lambda) = n^{-1}\mathrm{tr}[H(\lambda)^j] , \quad j = 1,2.$$

Using Speckman's approximation (39) it has already been established that $\tau_2(\lambda) = n^{-1}\Sigma_{j=1}^n(1 + n\lambda\delta_{jn})^{-2} = O((n\lambda^{1/2m})^{-1})$. A similar calculation gives

$$\tau_1(\lambda) \sim \frac{k_m}{n\lambda^{1/2m}} \pi^{-1}\int_0^1 p(t)^{1/2m}dt ,$$

where

$$k_m = \int_0^1(1 + y^{2m})^{-1}dy .$$

Thus, if $n \to \infty$, $\lambda \to 0$ in such a manner that $n\lambda^{1/2m} \to \infty$, then $g_n(\lambda) \to 0$ which suggests that the minimizer of $EGCV_n(\lambda)$ and $R_n(\lambda)$ should have similar properties for large n.

Let $\tilde{\lambda}_n$ be the minimizer of $EGCV_n(\lambda)$ and $\lambda_{n,opt}$ the minimizer of $R_n(\lambda)$. An argument that $R_n(\tilde{\lambda}_n)/R_n(\lambda_{n,opt}) \to 1$ can then be constructed as follows. As a result of the GCV Theorem

$$GCV_n(\lambda_{n,opt}) - \sigma^2 \le R_n(\lambda_{n,opt})(1 + g_n(\lambda_{n,opt}))$$

and

$$R_n(\tilde{\lambda}_n)(1 - g_n(\tilde{\lambda}_n)) \le GCV_n(\tilde{\lambda}_n) - \sigma^2 .$$

Combining these inequalities with the fact that $GCV_n(\lambda_{n,opt}) \ge GCV_n(\tilde{\lambda}_n)$ gives

$$R_n(\tilde{\lambda}_n)(1 - g_n(\tilde{\lambda}_n)) \le R_n(\lambda_{n,opt})(1 + g_n(\lambda_{n,opt})) .$$

Thus if $\tilde{\lambda}_n$ converges to zero at a rate which insures that $g_n(\tilde{\lambda}_n)$ converges to zero we may divide both sides by $(1 - g_n(\tilde{\lambda}_n))$ for n sufficiently large to obtain

$$1 \le R_n(\tilde{\lambda}_n)/R_n(\lambda_{n,opt}) \le (1 + g_n(\lambda_{n,opt}))/(1 - g_n(\tilde{\lambda}_n)) .$$

The right hand side converges to one as $n \to \infty$, provided $g_n(\lambda_{n,opt}) \to 0$, thereby establishing the result.

The key assumption in the above is that both $g_n(\tilde{\lambda}_n)$ and $g_n(\lambda_{n,opt})$ tend to zero as $n \to \infty$. This occurs if $\tilde{\lambda}_n \to 0$ and $\lambda_{n,opt} \to 0$ as $n \to \infty$ in such a way that $n\tilde{\lambda}_n^{1/2m} \to \infty$ and $n\lambda_{n,opt}^{1/2m} \to \infty$. Since $R_n(\lambda)$ will converge to $IR_n(\lambda)$ as $n \to \infty$, our experience with the integrated risk makes it intuitively clear that $\lambda_{n,opt}$ should have the desired properties. That the same is true for $\tilde{\lambda}_n$ is not so transparent. However, the conditions are satisfied as has been demonstrated by Craven and Wahba (1979) and Utreras (1985b). (See, also Exercises 16-18).

To conclude our discussion of the asymptotic theory of $\mu_\lambda$ we mention results by Cox (1984b) concerning asymptotic normality of the smoothing splines estimator. Assume that the design points are generated by a positive density p through relation (31) and that the $\varepsilon_j$ are independent and identically distributed with finite tenth moments. Then, subject to certain differentiability conditions on p, results in Cox (1984b) have the implication that, for $m \geq 2$, $(\mu_\lambda(t) - E\mu_\lambda(t))/(Var\mu_\lambda(t))^{1/2}$ has an asymptotic standard normal distribution provided that $\lambda \to 0$ as $n \to \infty$ and $\lambda$ is bounded above by $n^{-4m/5}(\log n)$. Cox actually establishes much more than this and is able to show that $\mu_\lambda(t) - E\mu_\lambda(t)$ can be treated as a normal process on [0,1] under certain conditions. The usefulness of results such as this pertains to the construction of both interval estimates at a point as well as confidence bands for the regression function. Provided $\lambda$ is small enough so that the bias can be assumed small relative to the estimator's variance, Cox's results could be used to construct interval estimates for $\mu$ as well as $E\mu_\lambda$.

### 6.3.3 Asymptotics for Smoothing Spline Variants

The asymptotic properties of some of the generalization of the smoothing spline estimator described in Section 6.2 have also been studied. We now present a brief summary of some of these results.

Wahba (1977a) has examined the large sample properties of MOR estimators in the special case where $\eta$ and $\mu$ satisfy a Fredholm integral equation of the first kind, i.e.,

$$\mu(t) = \int_0^1 K(t,s)\eta(s)ds$$

for some known kernel function $K(\cdot,\cdot)$. Under the assumption of uniformly spaced design points and under restrictions on K and $\eta$ she obtains the asymptotically optimal $\lambda$ and asymptotically optimal form for the (range) risk

$$RR(\lambda) = n^{-1} \sum_{j=1}^{n} E(\mu(t_j) - \tilde{\mu}_\lambda(t_j))^2 \quad ,$$

where

$$\tilde{\mu}_\lambda(t) = \int_0^1 K(t,s)\eta_\lambda(s)ds.$$

She is also able to establish a parallel of (5.40) for this case and demonstrate a consistency property of $\eta_{\tilde{\lambda}}$ with $\tilde{\lambda}$ the minimizer of the expected GCV criterion. More recent results concerning MOR estimators for the approximate solution of integral equations can be found in Nychka and Cox (1984). Cox (1983b) provides asymptotic theory for general MOR estimators and Cox (1984b) gives results concerning the asymptotic distribution theory of such estimators. Rice and Rosenblatt (1983) have studied the integrated risk corresponding to $\eta_\lambda$ in the special case of a deconvolution problem and are able to characterize the influence of boundary behavior on the risk.

The use of smoothing splines for estimation of derivatives, which can be viewed as a special case of MOR estimators, has received a good bit of individual attention. Consequently, somewhat more is known about this problem. Ragozin (1983) has shown that if the design points are uniformly spaced and $\lambda = O(n^{-2m/(2m+1)})$, then for $d < m$

$$\int_0^1 E(\mu^{(d)}(t) - \mu_\lambda^{(d)}(t))^2 dt = O(n^{-2(m-d)/(2m+1)}) \ .$$

Thus smoothing splines compare favorably with the kernel method for derivative estimation (c.f., Section 4.8). Other related results which also describe the influence of boundary behavior in derivative estimation can be found in Rice and Rosenblatt (1983) and Utreras (1986b).

The asymptotic properties of smoothing splines under convexity constraints have been studied by Utreras (1984, 1985a) and Wong (1984). Utreras (1985a) shows that constrained univariate smoothing splines have the same optimal rates of convergence for the risk as in the unconstrained case. Utreras (1984) and Wong (1984) give results applicable to the asymptotic behavior of certain special types of constrained multivariate smoothing splines.

Convergence rates and other properties for multivariate smoothing splines have been studied by Wahba (1979), Cox (1984a) and Utreras (1985b, 1986b). Wahba (1979) originally conjectured that the optimal risk associated with a p-dimensional Laplacian smoothing spline of order m is $O(n^{-2m/(2m+p)})$. This conjecture has essentially been proved by Utreras (1986b) under certain regularity conditions. Cox (1984a) shows that the same convergence rates can be obtained for multivariate smoothing splines on bounded domains. Convergence rates for derivative estimation using multivariate splines are also provided by Cox (1984a) and Utreras (1986b). Utreras (1985b) has even established a version of (5.40) that is applicable to both Laplacian smoothing splines and multivariate smoothing splines on bounded domains.

The asymptotic properties of partial spline estimators have been studied by Heckman (1986) and Rice (1986a). If we think of partial splines as a nonparametric approach to analysis of covariance, then the interest is primarily in estimation of the parametric part of the model. One might then hope that the partial spline estimator $\underset{\sim}{\zeta}_\lambda$ for the parameter vector $\underset{\sim}{\zeta}$ would behave like an estimator from a purely parametric model in the sense that its bias and variance are both $O(n^{-1})$, independently of $\lambda$, provided, e.g., that $n \to \infty$ and $\lambda \to 0$ and $n\lambda^{1/2m} \to \infty$. In some cases this will be true while in others it will not. For example, when $p = 1$, Heckman (1986) shows that $\underline{\zeta}_\lambda$ behaves like a parametric estimator if $\underline{z}$ does not admit a regression on t and certain other conditions are satisfied. She shows that $\underline{\zeta}_\lambda$ is consistent for $\underline{\zeta}$ and $n^{1/2}(\underline{\zeta}_\lambda - \underline{\zeta})$ is asymptotically normal in that case. An alternative set of assumptions, which allows $\underline{z}$ and t to be related, has been used by Rice (1986a). In that case he finds that, although the variance for $\underline{\zeta}_\lambda$ is $O(n^{-1})$, the bias is $O(\lambda)$ and, hence $\underline{\zeta}_\lambda$ converges to $\underline{\zeta}$ at a nonparametric, rather than a parametric, rate. This has the implication that the bias properties of $\underline{\zeta}_\lambda$ must be taken into account when constructing interval estimates unless $\underline{z}$ and t can be assumed to be unrelated.

## 6.4 Applications

This final section is devoted to describing several applications of smoothing spline estimators and their variants which have appeared in the literature. We begin with a description of how constrained Laplacian smoothing splines can be used to effect a solution to problems in discriminant analysis.

### 6.4.1 Estimation of Posterior Probabilities

An illustration of the use of inequality constrained smoothing splines is provided by Villalobos and Wahba (1985, 1987). They use such methods to obtain estimates of posterior probabilities in discriminant analysis.

For simplicity let us restrict attention to the case where we have two populations of individuals, $\pi_1$ and $\pi_2$. For a particular individual we observe the value of a vector of random variables $\underline{T} = (T_1,\ldots,T_p)'$. On the basis of this vector of values we want to decide which population produced the individual in question.

Let $g_1$ and $g_2$ denote the $\underline{T}$ densities for $\pi_1$ and $\pi_2$ and let

$p_i$ = probability of drawing an observation from $\pi_i$.

Then one solution to our classification problem is to assign the individual to the population having the largest posterior probability. A simple calculation shows that the posterior probability of $\pi_1$ given the observed value of $\underline{T}$ is

$$p(\pi_1|\underline{t}) = \text{Prob}(\pi_1|\underline{T} = \underline{t}) = p_1g_1(\underline{t})/(p_1g_1(\underline{t}) + p_2g_2(\underline{t})).$$

The posterior probability for $\pi_2$ is obtained from the fact that $p(\pi_2|\underline{t}) = 1 - p(\pi_1|\underline{t})$. The classification rule that results is: classify the individual as coming from $\pi_1$ if

$$p(\pi_1|\underline{t}) \geq .5$$

and to $\pi_2$ otherwise.

In some cases one has training data available for this problem. This is a collection of n observations $\underline{t}_1,\ldots,\underline{t}_n$ of which $n_1$ are known to have come from $\pi_1$ and the remaining $n_2 = n - n_1$ come from $\pi_2$. If a parametric form can be assumed for $g_1$ and $g_2$ the observations from each population can be used to estimate any unknown parameters and, assuming $p_1$ and $p_2$ are known, construct an estimated classification rule. The classical approach is to assume that both $g_1$ and $g_2$ are p-variate normal densities and then estimate the unknown mean vectors and covariance matrices.

There are numerous other ways to solve the discrimination problem. See, e.g., Johnson and Wichern (1982) and Villalobos and Wahba (1983) for a discussion of parametric and nonparametric methods respectively.

Villalobos and Wahba (1985, 1987) use smoothing spline techniques to estimate $p(\pi_1|\underline{t})$. They estimate the function

$$\mu(\underline{t}) = \hat{p}_1g_1(\underline{t})/(\hat{p}_1g_1(\underline{t}) + \hat{p}_2g_2(\underline{t})),$$

where

$$\hat{p}_i = n_i/n .$$

Note that $\mu(\underline{t})$ is obtained from $p(\pi_1|\underline{t})$ by substituting the observed proportions $\hat{p}_i$ for the $p_i$. Given an estimate of $\mu$ one can estimate $p(\pi_1|\cdot)$ by use of the relation

$$p(\pi_1|\underline{t}) = p_1\mu(\underline{t})/\hat{p}_1/(p_1\mu(\underline{t})/\hat{p}_1 + p_2\mu(\underline{t})/\hat{p}_2) \ . \tag{40}$$

The reason for estimating $\mu$ is as follows. Suppose we knew that our training sample consisted of $n_1$ observations from $\pi_1$ and $n_2$ from $\pi_2$ but did not know which observation came from which population. Now draw one of the training sample elements at random and denote its values by $\underline{t}$. Then, by the same argument used to obtain $p(\pi_1|\underline{t})$,

$$\text{Prob}(\pi_1|T = \underline{t}) = \mu(\underline{t}) \ ,$$

since, in this case, the probability that the observation we drew was from $\pi_i$ is $\hat{p}_i$. If we let

$$y = \begin{cases} 1, & \underline{t} \ \varepsilon \ \pi_1 \ , \\ 0, & \underline{t} \ \varepsilon \ \pi_2 \ , \end{cases}$$

the above reasoning shows that

$$\mu(\underline{t}) = E[y|\underline{T} = \underline{t}] = \text{Prob}(\pi_1|\underline{T} = \underline{t}).$$

Consequently, if $y_1,\ldots,y_n$ are the observed values of y from our training sample we can view them as following the model

$$y_i = \mu(\underline{t}_i) + \varepsilon_i, \quad i = 1,\ldots,n \ ,$$

where the $\varepsilon_i$ are zero mean uncorrelated random variables.

As a result of the above reasoning Villalobos and Wahba (1983, 1985, 1987) proposed using a constrained Laplacian smoothing spline to estimate $\mu$. Specifically, they considered estimating $\mu$ by the minimizer of

$$n^{-1}\sum_{j=1}^{n}(y_j - f(\underline{t}_j))^2 + \lambda J_{m,p}(f) \tag{41a}$$

over $f \in W_2^m(\mathbb{R}^p)$ subject to the constraints

$$0 \le f(\underline{s}_i) \le 1 , \quad i = 1,\ldots,k , \tag{41b}$$

for a fine regular grid of points $\underline{s}_i$ in $\mathbb{R}^p$. These particular constraints are motivated by the fact that, ideally, we want $0 \le f(\underline{s}) \le 1$ for all $\underline{s}$. By imposing condition (41b) for a sufficiently fine grid one can make this appear to be true for at least the subset of $\mathbb{R}^p$ which is of interest.

An illustration of the types of results which can be expected from this procedure is provided by an example with real data taken from Villalobos and Wahba (1985, 1987). They consider the case of $p = 2$ where the two populations consist of psychotic and normal individuals, respectively, and $t_1$ and $t_2$ represent scores related to "size" and "shape" (see Penrose, 1945). The data came from Smith (1947) and consisted of 25 psychotic and 25 normal patients.

Figure 6.2 shows the contours of the estimator of $p(\pi_1|\cdot)$ obtained by Villalobos and Wahba (1985, 1987) using the minimizer of (41) in (40) with $p_1 = p_2 = .5$. A variant of GCV suitable for constrained problems was used to select $\lambda$. Also plotted are contours of the estimate of $p(\pi_1|\cdot)$ which result from assuming that $g_1$ and $g_2$ are normal densities and estimating their parameters in the usual fashion. These latter contours are necessarily ellipses. However, the contours of the nonparametric estimator are not constrained to have this property and for this data are decidedly non-elliptical. This suggests that normal theory methods may not be appropriate, even though the methods fare similarly in terms of reclassification of the training sample values.

### 6.4.2 Estimation of Tumor Size Distributions

An interesting application of MOR estimates has been given by Nychka, Wahba, Goldfarb and Pugh (1983, 1984) (hereafter

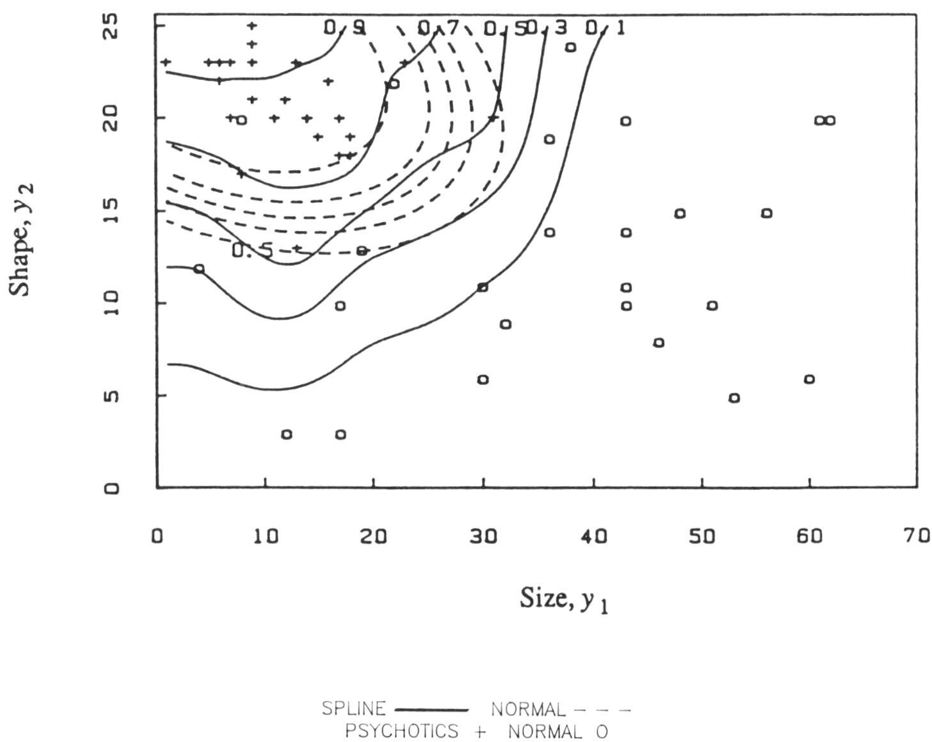


**Figure 6.2** Contours of constrained thin plate spline and bivariate normal fits to size and shape score data. [From Villalobos and Wahba (1986). Reprinted with permission of the American Statistical Association.]

referred to as NWGP). They use an MOR estimator to estimate the three dimensional size distribution of malignant tumors in mice livers.

The data used by NWGP came from a study where mice were injected at fifteen days of age with a carcinogen which induced the formation of malignant tumors in the liver. After a fixed amount of time the mice were sacrificed, at which point their livers could be examined. The way this was done was to stain samples of the liver tissues and then embed them in a paraffin matrix. Thin slices could then be made which can be examined under a microscope. Tumors will appear in cross section and their size and number can be recorded.

Biological theory indicates that the tumors will be uniformly distributed throughout the liver and experience indicates they tend to be approximately spherical. A way to model these properties is to treat the tumors as spheres in a medium with centers distributed according to a Poisson process with constant intensity and equatorial radii having some cumulative distribution function G. The problem to be considered here is estimation of G or, rather, its density $g = G'$.

While it is possible to examine many closely spaced slices and completely reconstruct a tumor, this process is both time consuming and expensive. Thus one will generally want to work with only a moderate number of slices. The problem with this is that, depending on where the slice hits each tumor, different size tumors can show up with similar cross-sectional (not equatorial) radii. Let H denote the distribution function for cross-sectional radii. It can then be shown that g and H are related by

$$H(x) = 1 - \alpha^{-1}\int_x^R (r^2 - x^2)^{1/2} g(r)dr , \tag{42}$$

where R is an upper bound for the largest possible equatorial radius and

$$\alpha = \int_0^R rg(r)dr.$$

Given data on a mouse we can easily construct an estimate of H. For example, the empirical histogram of observed cross-sectional radii can be used for this purpose. However, what is needed is an estimate of g which requires us to try to "invert" (42) in some fashion. One approach to this is to use an MOR estimator of h.

Let $\tilde{H}$ be the empirical distribution function of cross-sectional radii and let $0 = u_0 < u_1 < u_2 < \ldots < u_n = R$ be a partition of $[0,R]$. Then, define

$$y_i = \tilde{H}(u_i) - \tilde{H}(u_{i-1}), \quad i = 1,\ldots,n .$$

If the cross-sectional radii can be viewed as a random sample from H then

$$Ey_i = H(u_i) - H(u_{i-1}) ,$$

$$\text{Var } y_i = n^{-1}\{H(u_i) - H(u_{i-1}) - (H(u_i) - H(u_{i-1}))^2\}$$

and, for $i \neq j$ ,

$$\text{Cov}(y_i,y_j) = -n^{-1}\{H(u_i) - H(u_{i-1})\}\{H(u_j) - H(u_{j-1})\}.$$

Thus the $y_i$ will be only weakly correlated and we can treat them as following the model

$$y_i = L_i g + \varepsilon_i, \quad i = 1,\ldots,n ,$$

where

$$L_i g = H(u_i) - H(u_{i-1})$$

$$= \alpha^{-1}\int_{u_{i-1}}^{R} (r^2 - u_{i-1}^2)^{1/2} g(r)dr$$

$$- \alpha^{-1}\int_{u_i}^{R} (r^2 - u_i^2)^{1/2} g(r)dr$$

and the $\varepsilon_i$ are zero mean uncorrelated random errors. If the fact that the $\varepsilon_i$ in this case will have unequal variances is ignored, g can be estimated by the minimizer of

$$n^{-1} \sum_{j=1}^{n} (y_j - L_j f)^2 + \lambda J_2(f) ,$$

over $f \varepsilon W_2^2[a,b]$, with $\lambda$ selected by GCV.

An MOR estimator of the type discussed above was used by NWGP to estimate g. Their formulation differs slightly from ours due to a correction they imposed on the model which accounts for the fact that tumors with very small radii cannot be seen.

Figure 6.3 shows the histogram for cross-sectional radii of 154 sampled sections for the mouse liver studied by NWGP. They computed an MOR estimate of g using this histogram and the result is shown in Figure 6.4. Notice that there is an indication of a mode at about 280 microns.

For the particular tissue sample used by NWGP on exhaustive slicing was undertaken allowing them to determine the true tumor size distribution. This is also shown in Figure 6.4 along with the estimate. The mode of 280 microns is actually an aspect of the distribution that was picked up by the estimator. However, the estimator does not perform particularly well near the lower limit of observable tumor size of 38.46 microns. NWGP give several proposals for methods to correct this latter difficulty.

### 6.4.3 The Stack Loss Data Revisited

Figure 6.5 is a plot of data which describes successive days of operation of a plant that oxidizes ammonia to nitric acid. The response variable, Y, is ten times the percentage of ingoing ammonia that is lost as unabsorbed nitric oxides and, therefore, is a measure of the plant's inefficiency. The dependent variables, $t_1$ and $t_2$, are the flow of air in the plant and temperature of cooling water, respectively. This data set has a long history of statistical analysis dating back to Brownlee (1965). It has come to be known as the Stack Loss data.

The classical approach to analysis of the Stack Loss data has been to fit the data using polynomial regression: usually with

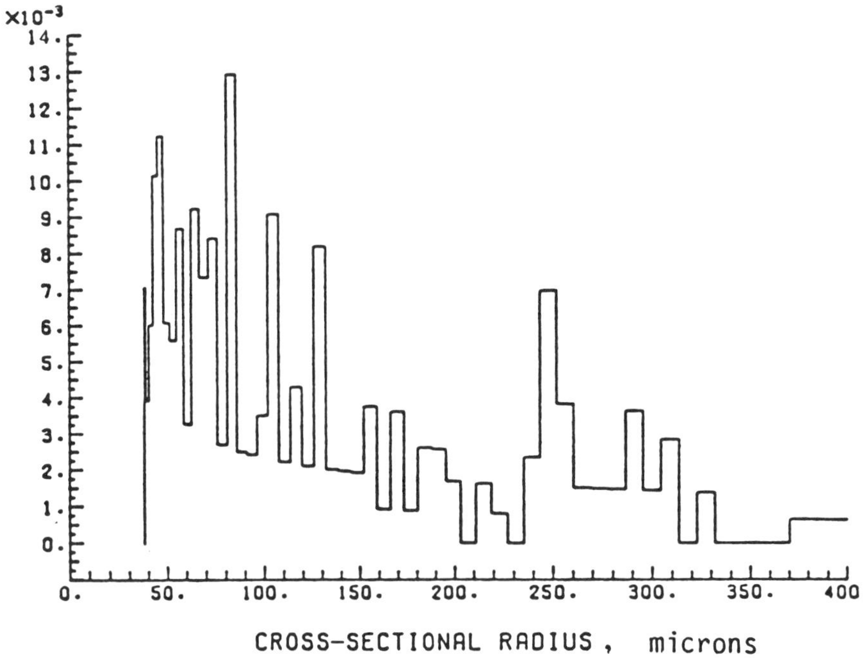


**Figure 6.3** Histogram of cross-sectional radii. [From Nychka et al. (1984). Reprinted with permission of the American Statistical Association.]

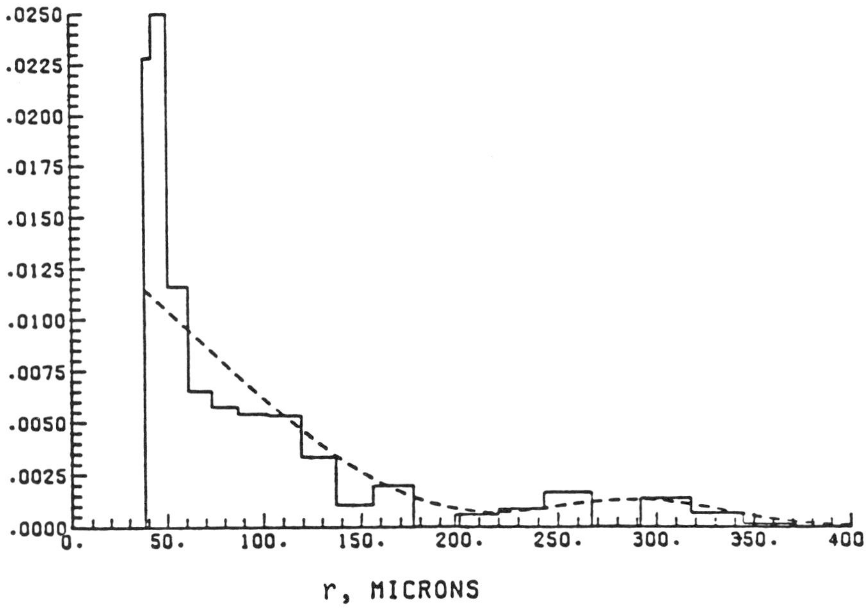


**Figure 6.4** MOR estimate and true histogram for tumor size distribution. [From Nychka et al. (1984). Reprinted with permission of the American Statistical Association.]

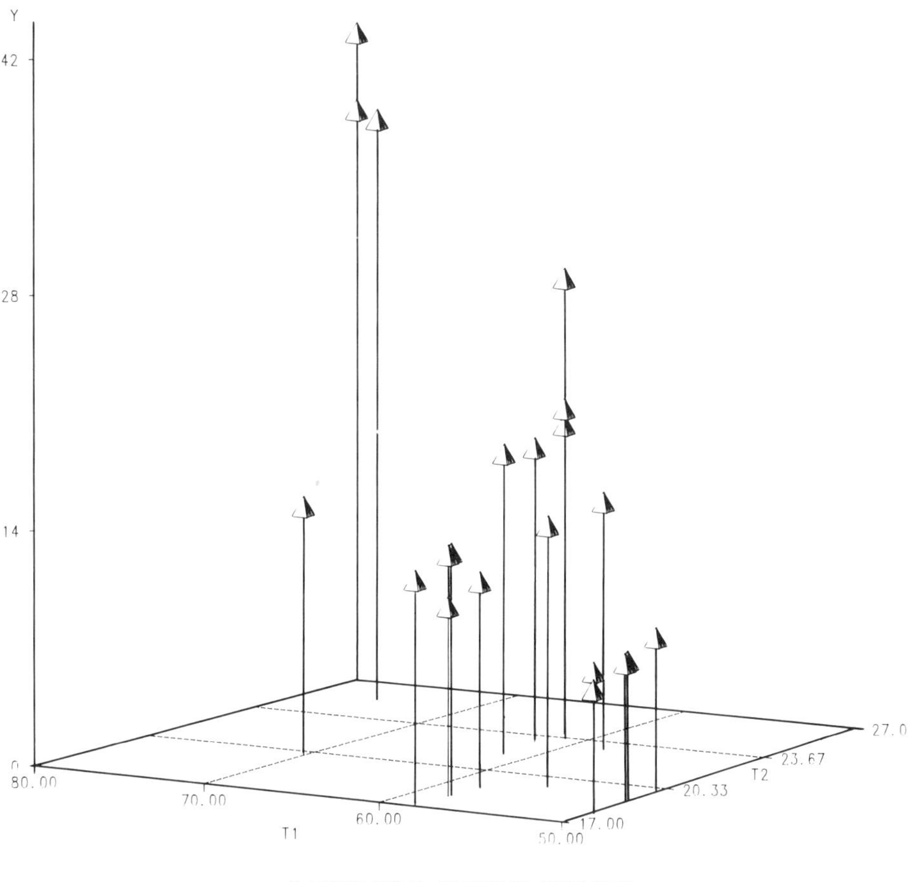


**Figure 6.5** The stack loss data. [From Carmody (1985).]

a polynomial of total order three. The reason this particular data set has received so much attention is that it contains a number of known outliers which are detectable, to varying degrees, under a polynomial fit. Consequently, the Stack Loss data provides a medium for testing diagnostic and robust regression methodology.

One of the most complete analyses of the Stack Loss data was undertaken by Daniel and Wood (1971). They fit the data using a quadratic polynomial and identified the observations numbered 1, 3, 4 and 21 in Table 6.1 as being outliers. The explanation that Daniel and Wood found for this was that each of the four observations was taken when the plant was in a transitional phase while the remaining observations were taken when the plant was in equilibrium. Consequently, the observations for days 1, 3, 4 and 21 can be viewed as following a different type of model than the rest of the data.

The Stack Loss data has also been analyzed by a number of other authors, including Andrews (1974), Dempster and Gasko-Green (1981) and Carrol and Rupert (1985). In addition to observations 1, 3, 4 and 21 some of these analyses have also indicated that day 2 may be an outlier.

A possible criticism of previous analyses of the Stack Loss data is the assumption of a polynomial regression model. There seems to be no reason to assume, *a priori*, that the relationship between Y and $\underline{t} = (t_1, t_2)'$ should adhere to a polynomial form. For this reason Carmody (1985) considered fitting the Stack Loss data using a Laplacian smoothing spline with m = 2 and a value of $\lambda$, $\hat{\lambda}$, selected by GCV. By using this type of estimator non-linearities other than those of a quadratic nature can be taken into account. Also, since $\lambda$ is selected by the data, one would anticipate the choice of a large (possibly infinite) value of $\lambda$ if the relationship between Y and $\underline{t}$ is linear.

Table 6.1 gives leverage values, Studentized residuals and Cook's distance measures (see Section 5.5.5) for Carmody's fit to the Stack Loss data. For this fit the trace of the hat matrix is

6.962 $\doteq$ 7 which can be used to obtain approximate degrees of freedom for comparison of the Studentized residuals and Cook's distance values to values from the Student's t and F-distribution, respectively. We find that points 4 and 21 have significantly large Studentized residuals and points 1, 2, 3, 7 and 9 may also be cause for concern with Studentized residuals of magnitude larger than one. However, comparison of the Cook's distance measures to the rough bound of $F_{7,14,.10} = .378$ reveals that, among these points, only 3, 4, 21 and, to a lesser extent, 2 are actually influential.

**Table 6.1 Diagnostics for the Stack Loss Data**

| Observation Number | $t_1$ | $t_2$ | y | $h_{ii}(\hat{\lambda})$ | $r_{\hat{\lambda}i}$ | $D_{\hat{\lambda}i}$ |
|---|---|---|---|---|---|---|
| 1 | 80 | 27 | 42 | .465 | 1.103 | .152 |
| 2 | 80 | 27 | 37 | .465 | -1.535 | .294 |
| 3 | 75 | 25 | 37 | .653 | 1.574 | .669 |
| 4 | 62 | 24 | 28 | .233 | 2.862 | .358 |
| 5 | 62 | 22 | 18 | .306 | - .003 | .000 |
| 6 | 62 | 23 | 18 | .213 | - .719 | .020 |
| 7 | 62 | 24 | 19 | .233 | -1.105 | .053 |
| 8 | 62 | 24 | 20 | .233 | - .665 | .019 |
| 9 | 58 | 23 | 15 | .494 | -1.098 | .169 |
| 10 | 58 | 18 | 14 | .188 | .586 | .011 |
| 11 | 58 | 18 | 14 | .188 | .586 | .011 |
| 12 | 58 | 17 | 13 | .348 | .656 | .032 |
| 13 | 58 | 18 | 11 | .188 | -.699 | .016 |
| 14 | 58 | 19 | 12 | .215 | -.615 | .015 |
| 15 | 50 | 18 | 8 | .280 | .328 | .006 |
| 16 | 50 | 18 | 7 | .280 | -.126 | .001 |
| 17 | 50 | 19 | 8 | .214 | -.125 | .001 |
| 18 | 50 | 19 | 8 | .214 | -.125 | .001 |
| 19 | 50 | 20 | 9 | .375 | -.170 | .002 |
| 20 | 56 | 20 | 15 | .374 | .652 | .036 |
| 21 | 70 | 20 | 15 | .802 | -1.974 | 2.266 |

**Source: Carmody (1985).**

An examination of the leverage values shows that points 1, 2, 3 and 21 have high leverage. In drawing this conclusion we have used the property that the leverage values for points 1 and 2 are bounded by .5 due to the fact that both observations are taken at the same design point (Exercise 5.18). A plot of $t_1$ versus $t_2$ in Figure 6.6 shows that observations 1, 2, 3 and 21 are all taken at extreme points in the design space which provides the explanation for their high leverage.

To summarize, we have found that observations 2, 3, 4 and 21 can be viewed as outliers in the smoothing spline fit to the Stack Loss data in the sense of having significant influence on the fit. This occurs for points 2 and 21 due to a combination of both large residuals and high leverage. Points 3 and 4 appear to be influential as a result of high leverage and large residuals, respectively. What is interesting about this result is that, apart from observation 1, these are the same observations identified by others using polynomial regression analysis. Although observation 1 has not been identified as being influential on the fit it is on the borderline (along with observation 9). Due to this and the fact that it is known to come from a day when the plant was in a transitional phase we will include it among the other detected outliers and consider it as a candidate for deletion.

Figure 6.7 shows the Laplacian smoothing spline fit to the Stack Loss data when observations 1, 3, 4 and 21 have been deleted. Figure 6.8 shows the Daniel and Wood quadratic polynomial fit to this same set of data. The two fits differ appreciably thereby suggesting that a polynomial fit may not be entirely satisfactory for this data.

If observation 2 is deleted, along with 1, 3, 4 and 21, GCV selects $\hat{\lambda} = \infty$. The conclusion to be drawn from this is that observation 2 induces a nonlinearity into the fit. Since observations 1, 2, 3 and 21 are extreme points in the design space, this suggests that the true response function is at least approximately linear in the interior of the design. Results by Cook (1979) lead to a similar conclusion. He found that the nonlinear portion of the Daniel and Wood quadratic polynomial fit was retained only because of observation 2.

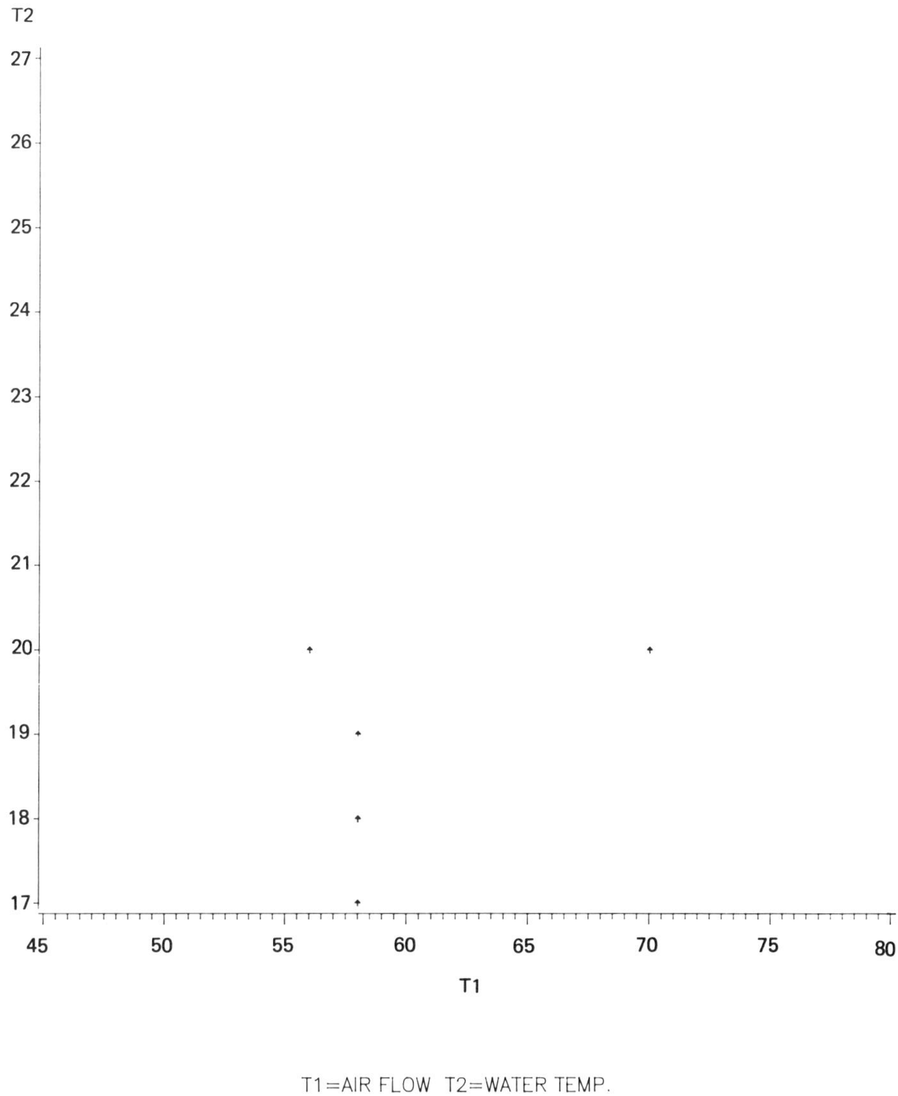


**Figure 6.6** Scatter plot of air flow versus water temperature. [From Carmody (1985).]

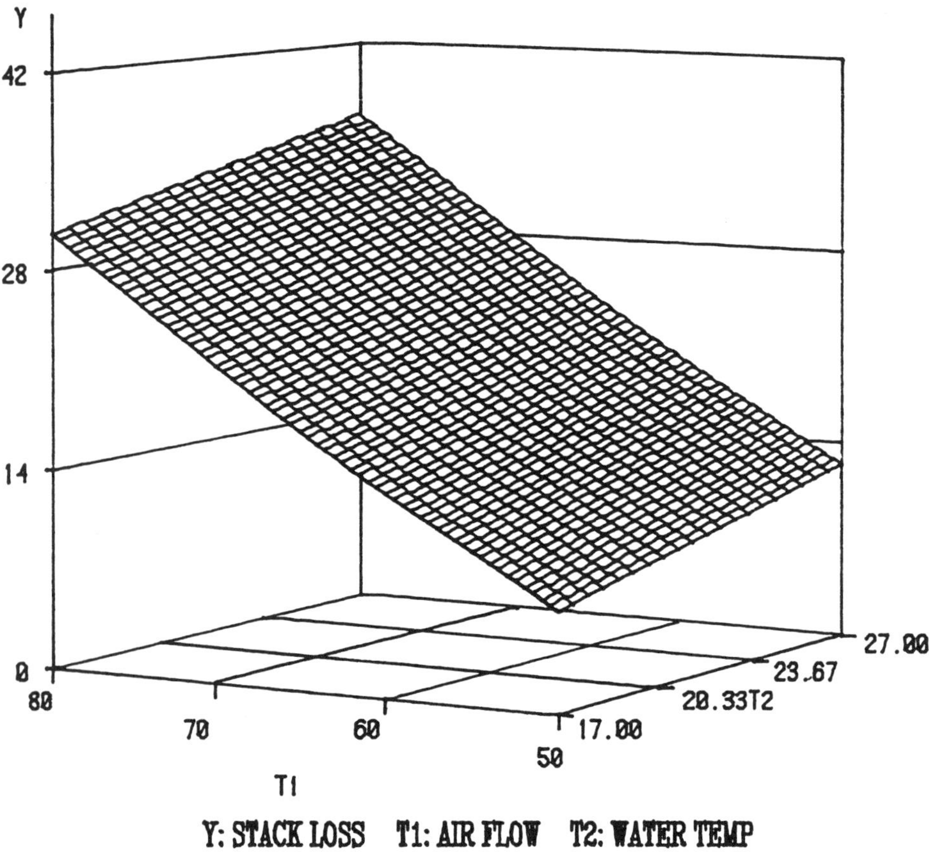


**Figure 6.7** Spline fit to the stack loss. Data points 1, 3, 4, and 21 deleted. [From Carmody (1986).]

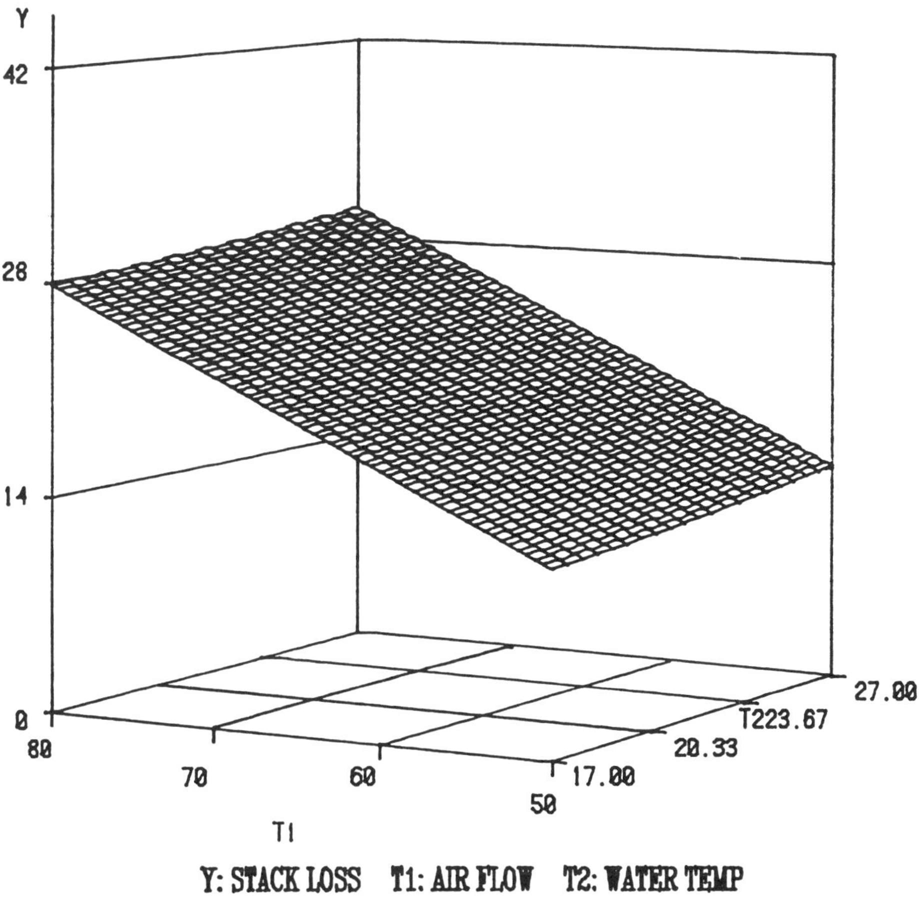


**Figure 6.8** Daniel and Wood fit to the stack loss data. [From Carmody (1986).]

### 6.4.4 Other Applications

There are a number of other examples in the literature of practical uses of spline smoothing methodology. We conclude by mentioning a few of these.

Silverman (1985) has given an example concerning the duration and time of eruption of the Old Faithful geyser where a smoothing spline fit suggests a possible piecewise linear parametric model. Ansley and Wecker (1981) have used a smoothing spline to estimate the elasticity of demand during the German hyperinflation. Largo, Gasser, Prader, Stuetzle and Huber (1978) have employed smoothing splines for growth curve analysis.

MOR estimators have found a number of applications in meterology and physics. We mention only the work of O'Sullivan (1983), O'Sullivan and Wahba (1985) and Merz (1980).

Laplacian and partial smoothing splines have also received considerable attention in the modeling of meteorological data. Examples of their use for this purpose can be found in Wahba and Wendelberger (1980), Hutchison and Bischof (1983), Shiau (1985) and Shiau, Wahba and Johnson (1985).

Laplacian smoothing and interpolating splines have also been utilized to smooth various types of data aggregated by geographical region (see Wahba, 1981a, and references therein). An illustration of the use of a discretized version of a partial spline for modelling the relation between weather and electricity demand is provided by Engle, Granger, Rice and Weiss (1986). Several examples of the use of discretized partial splines for analysis of covariance are given in Green, Jennison and Seheult (1985).

## EXERCISES

1. It can be shown that any linear functional L on $W_2^m[a,b]$ can be written as

$$Lf = \sum_{i=1}^{m} v(s_i)f(s_i) + \int_a^b v^{(m)}(s)f^{(m)}(s)ds$$

for some $v \in W_2^m[a,b]$ and the $s_i$ arbitrary distinct points in $[a,b]$. Use this result along with the fact that

$$[(r-1)!]^{-1}\int_a^b (t-s)_+^{r-1} f^{(r)}(s)ds = f(t) + p(t)$$

for some polynomial p of order r to establish (8).

**2.** Prove Theorem 6.1.

**3.** Show that for an MOR estimator $\eta_\infty = \lim_{\lambda\to\infty}\eta_\lambda$ is a polynomial of order m. Give an explicit expression for $\eta_\infty$ and relate this estimator to polynomial regression. Also show that $\eta_0$ solves the problem of minimizing $J_m(f)$ subject to $L_j f = y_j$, $j = 1,\ldots,n$.

**4.** Derive a closed form expression for the minimizer of (2) subject to (10).

**5.** Given that (2) has a unique minimizer over $C_a$ in (12) show that the minimizing function must be of the form (11).

**6.** Let $\mu_{\lambda(i)}(\cdot;z)$ be the minimizer of

$$n^{-1}\{\sum_{j\neq i}(y_j - f(t_j))^2 + (z - f(t_i))^2\} + \lambda J_m(f)$$

over $C_a$ in (12) and let $\mu_{\lambda(i)}(\cdot)$ denote the minimizer of (2) over $C_a$ when $(t_i,y_i)$ has been deleted from the data.

a) Establish a parallel of Lemma 5.2 for the constrained case by verifying that

$$\mu_{\lambda(i)}(\cdot) = \mu_{\lambda(i)}(\cdot;\mu_{\lambda(i)}(t_i)) .$$

b) Define the influence of $(t_i,y_i)$ on the fit by

$$(\mu_{\lambda(i)}(t_i) - \mu_\lambda(t_i))/(\mu_{\lambda(i)}(t_i) - y_i) .$$

Check that this is just the leverage value $h_{ii}(\lambda)$ in the unconstrained case.

c) In view of b), suggest a form for the GCV criterion for the constrained setting.

[Wahba, 1982]

7. Prove (21).

8. Carry out the details concerning minimization of (24) when n is even and derive the periodic smoothing spline for this case. Give the parallel of approximation (26) for when n is even.

9. Derive the form of the periodic smoothing spline and the parallel of approximation (26) for general m.

10. Working with your approximation from Exercise 9 prove an analog of Theorem 6.2 applicable to general m.

11. Consider a kernel estimator of the form

$$\tilde{\mu}_\lambda(t) = \sum_{j=1}^{n} y_j \int_{s_{j-1}}^{s_j} \lambda^{-1}K(\lambda^{-1}(t - s))ds \; .$$

Find the optimal bandwidth for this estimator for estimating $\mu(t)$ when $\mu \in C^2[0,1]$, the $t_j$ are generated from a positive differentiable density, p, using (4.35) and the $s_j$ are defined by (4.13). Substitute the optimal bandwidth $\lambda_{opt}(t)$ for $\lambda$ in $\tilde{\mu}_\lambda(t)$ and justify approximating

$$\int_{s_{j-1}}^{s_j} \lambda^{-1}K(\lambda^{-1}(t - s))ds$$

by $[n\lambda p(t_j)]^{-1}K(\lambda^{-1}(t - t_j))$ to obtain a new estimator

$$\tilde{\mu}(t) = n^{-1} \sum_{j=1}^{n} y_j[\lambda_{opt}(t)p(t_j)]^{-1}K((t - t_j)/\lambda_{opt}(t)).$$

Using Silverman's approximation (32) compare the qualitative behavior of $\tilde{\mu}$ to that of a smoothing spline paying particular attention to how both estimators depend on the design density p. [Silverman, 1984b]

12. In Section 5.5.5 it is noted that leverage values for smoothing splines have a distance interpretation in that they measure the distance of $\underline{x}_i = (x_1(t_i),\ldots,x_n(t_i))'$ from the mean vector $\bar{\underline{x}} = n^{-1}X'\underline{1}$. This, however, does not relate them directly to the $t_i$. Show that, for large n, the magnitude of $h_{ii}(\lambda)$ can be tied directly to the position of $t_i$ among the design points.

Describe the consequences of this relationship in terms of what situations would tend to result in observations with high leverage. [Hint: Use Silverman's approximation (32).]

**13.** Let $\mu_\lambda$ be the smoothing spline estimator for some fixed $\lambda$ and $E\mu_\lambda$ its expectation. Verify that $E\mu_\lambda$ is the minimizer of

$$n^{-1}\sum_{j=1}^{n}(\mu(t_j) - f(t_j))^2 + \lambda J_m(f)$$

over $f \in W_2^m[a,b]$.

**14.** Consider the problem of finding f in $W_2^m[a,b]$ and $\delta \in \mathbb{R}$ such that

$$\int_a^b f^{(m)}(t)g^{(m)}(t)dt = \delta \sum_{i=1}^{n} f(t_i)g(t_i) .$$

Show that the only solutions for $\delta$ are the eigenvalues $\delta_1,\ldots,\delta_n$ of the matrix $U(U'Q_nU)^{-1}U'$ used in defining the Demmler-Reinsch basis. Further show that the set of functions which solves this problem forms a vector space of dimension n for which the Demmler-Reinsch basis is a basis. Verify that if $\delta_j$ is distinct then the corresponding Demmler-Reinsch basis function $x_j$ is the only function which solves this problem. [Hint: Begin by noting that any solution x, must satisfy

$$\int_a^b x^{(m)}(t)g^{(m)}(t)dt = \sum_{j=1}^{n}\gamma_j g(t_j)$$

for some set of coefficients $\gamma_1,\ldots,\gamma_n$ which are independent of g and satisfy $\Sigma_{j=1}^{n}\gamma_j g(t_j) = 0$ if g is a polynomial of order m. Use this along with Taylor's Theorem to establish that x must be in $NS^{2m}(t_1,\ldots,t_n)$. Now employ the fact that the Demmler-Reinsch basis is a basis for $NS^{2m}(t_1,\ldots,t_n)$ to obtain the desired conclusions.]

**15.** Use (39) to argue that $\tau_1(\lambda) = n^{-1}\mathrm{tr}H(\lambda)$ satisfies

$$\tau_1(\lambda) \sim [n\lambda^{1/2m}]^{-1}\, k_m\pi^{-1}\int_0^1 p(t)^{1/2m}dt \qquad ,$$

where

$$k_m = \int_{-\infty}^{\infty}(1 + y^{2m})^{-1}dy ,$$

if the $t_j$ are generated from p according to (31) and $n\to\infty$, $\lambda\to 0$ and $n\lambda\to\infty$.

**16.** Prove that $\mu_\lambda$ is unbiased, i.e.,

$$E\mu_\lambda(t) = \mu(t),$$

if $\mu$ is a polynomial of order m. Use this to show that both $EGCV_n(\lambda)$ and $R_n(\lambda)$ are minimized at $\lambda=\infty$ when $\mu$ is a polynomial of order m. This proves (5.40) in a special case.
[Craven and Wahba, 1979]

**17.** To see that $\tilde{\lambda}_n$, the minimizer of $EGCV_n(\lambda)$, must tend to zero when $\mu$ is not a polynomial of order m establish the following:

a) $b_n^2(\lambda) = n^{-1}\Sigma_{j=1}^{n}(\mu(t_j) - E\mu_\lambda(t_j))^2$ is a monotone decreasing function of $\lambda$ and, therefore, bounded away from zero if $\lambda$ is bounded away from zero.

b) $EGCV_n(\lambda) \geq \sigma^2$ and $EGCV_n(\tilde{\lambda}_n) \downarrow \sigma^2$ as $n\to\infty$ (you may use the fact that $\lambda_{n,opt}$ satisfies $\lambda_{n,opt} \to 0$ and $n\lambda_{n,opt}^{1/2m} \to \infty$).

c) Conclude from b) that $b_n^2(\tilde{\lambda}_n) \to 0$ and from a) that $\tilde{\lambda}_n \to 0$.
[Craven and Wahba, 1979, Utreras, 1985b]

**18.** To see that $\tilde{\lambda}_n$ in Exercise 17 satisfies $n\tilde{\lambda}_n^{1/2m} \to \infty$ when $p(t) \equiv 1$ one approach can be based on a result of Utreras (1983). He shows that the eigenvalues $\delta_1,\ldots,\delta_n$ from the Demmler-Reinsch basis satisfy

$$n\delta_{j+m} \leq Cj^{2m}, \qquad j = 1,\ldots,n - m ,$$

for some positive constant C independent of j and n. Use this to see that if $1/n\lambda^{1/2m}$ is bounded away from zero then so are $\tau_j(\lambda) = n^{-1}tr[H(\lambda)^j]$, j=1,2. Combine this with part b) of Exercise 17 to reach the desired conclusion.
[For general p see Craven and Wahba, 1979, Speckman, 1981, and Utreras, 1985b]

**19.** Refer to Exercise 5.20. Use your cubic smoothing spline fit obtained there to estimate the derivative of the regression function. Compare your estimate to the true derivative of the regression curve.

# 7

# Least-Squares Splines and Other Estimators

## 7.1 Introduction

In Chapters 3-6 we have studied series, kernel and smoothing spline estimators for nonparametric regression problems. These three estimators represent what are arguably the most popular approaches to nonparametric regression. Since series estimators include polynomial regression estimators they are probably the most widely used of any nonparametric regression technique. However, kernel estimators and smoothing splines seem to be better suited for estimation within the nonparametric framework and their properties as nonparametric estimators are more fully understood.

There are many other types of nonparametric regression estimators which we have not considered. In this final chapter we will introduce several such estimators which have potential value but whose properties and/or practical usefulness have not, as yet, been investigated to the extent of, e.g., kernel or smoothing spline estimators. In the next two sections we consider two types of spline estimators for $\mu$: least-squares splines and Speckman's minimax estimator. These two estimators provide alternative methods of extending the polynomial regression concept to that used in deriving smoothing splines. Then, in Section 7.4, we give a brief discussion of nearest neighbor estimators. These are a type of estimator which bear certain similarities to kernel estimators. The final section, Section 7.5, gives an overview of several promising new nonparametric estimation techniques for analysis of high-dimensional data.

## 7.2 Least-Squares Splines

To begin let us recall our basic model where the observations $(t_i, y_i)$ satisfy

$$y_i = \mu(t_i) + \varepsilon_i \quad , \; i = 1,\ldots,n, \tag{1}$$

with the $\varepsilon_i$ representing zero mean, uncorrelated random errors having common variance $\sigma^2$. Throughout this and the next section it will be assumed that $\mu \in W_2^m[a,b]$. If this is the case then, through the use of Taylor's Theorem, (1) can be rewritten as

$$y_i = \sum_{j=1}^{m} \theta_j t_i^{j-1} + \text{Rem}(t_i) + \varepsilon_i \quad , \tag{2a}$$

where

$$\text{Rem}(t) = [(m-1)!]^{-1} \int_a^b \mu^{(m)}(t)(t-\xi)_+^{m-1} d\xi \; . \tag{2b}$$

If $\text{Rem}(t_1),\ldots,\text{Rem}(t_n)$ are of a uniformly small magnitude, then polynomial regression would seem to provide a reasonable method of analyzing the data. However, if this is not the case, polynomial regression will not be satisfactory and, therefore, some type of protection against this possibility should be built into any estimation scheme utilized for model (2).

One method of estimation which attempts to guard against departures from polynomial models is smoothing splines which were studied in Chapters 5-6. When the smoothing parameter is selected by data driven methods, they accomplish this by, loosely speaking, estimating a bound on the magnitude of Rem(t) in (2b) and then incorporating the information about this bound into the estimator.

Least-squares splines provide another generalization of polynomial regression estimators which also utilize information in the data about Rem(t) to protect against possible inadequacies of a polynomial model. The basic premise behind these estimators is that, by using a quadrature rule, we can approximate the integral in (2b) by a sum of the form

$$\sum_{j=1}^{k} \delta_j (t - \xi_j)_+^{m-1}$$

for some set of coefficients $\delta_1, \ldots, \delta_k$ and points $a < \xi_1 < \ldots < \xi_k < b$. Combining this with the original polynomial approximation leads to an overall approximation of $\mu$ by a function of the form

$$s(t) = \sum_{j=1}^{m} \theta_j t^{j-1} + \sum_{j=1}^{k} \delta_j (t - \xi_j)_+^{m-1} . \tag{3}$$

Recall from Section 5.3 that s in (3) is what is known as a spline of order m with knots at $\xi_1, \ldots, \xi_k$. The set of all such functions, $S^m(\xi_1, \ldots, \xi_k)$, is an m + k dimensional vector space consisting of piecewise polynomials of order m which have m - 2 continuous derivatives and discontinuities in their (m - 1)st derivatives at the $\xi_j$.

Given a choice for $\lambda = \{\xi_1, \ldots, \xi_k\}$, we can then estimate $\mu$ by estimating the coefficients in the approximation (3). One method of accomplishing this is to use least squares. Let

$$\begin{aligned}
x_1(t) &\equiv 1, \\
x_2(t) &= t, \\
&\vdots \\
x_m(t) &= t^{m-1}, \\
x_{m+1}(t) &= (t - \xi_1)_+^{m-1}, \\
&\vdots \\
x_{m+k}(t) &= (t - \xi_k)_+^{m-1}
\end{aligned} \tag{4}$$

and set

$$\underline{\beta} = (\theta_1, \ldots, \theta_m, \delta_1, \ldots, \delta_k)' . \tag{5}$$

The *least-squares spline estimator* of $\mu$ is

$$\mu_\lambda = \sum_{j=1}^{m+k} \beta_{\lambda j} x_j \quad , \tag{6}$$

where $\underline{\beta}_\lambda = (\beta_{\lambda 1}, \ldots, \beta_{\lambda, m+k})$ is any minimizer of

$$\text{MSE}(\underline{\beta};\lambda) = n^{-1} \sum_{i=1}^{n} (y_i - \sum_{j=1}^{m+k} \beta_j x_j(t_i))^2 \tag{7}$$

with respect to $\underline{\beta}$. More specifically, if we define

$$X(\lambda) = \{x_j(t_i)\}_{\substack{i=1,n, \\ j=1,m+k}} \quad , \tag{8}$$

$\underline{\beta}_\lambda$ is found as any solution to the normal equations

$$X(\lambda)'X(\lambda)\underline{\beta} = X(\lambda)'\underline{y} \ , \tag{9}$$

where, as usual,

$$\underline{y} = (y_1, \ldots, y_n)' \ .$$

If $X(\lambda)$ has rank $m + k$, which we will assume to be the case, then

$$\underline{\beta}_\lambda = [X(\lambda)'X(\lambda)]^{-1}X(\lambda)'\underline{y} \ . \tag{10}$$

It follows from (6) and (10) that, given $\lambda = \{\xi_1, \ldots, \xi_k\}$, $\mu_\lambda$ is a linear estimator of $\mu$. The set $\lambda$ is actually a smoothing parameter. Its choice requires the selection of both the number and positioning of the knots for the spline. As the number of knots, k, is increased, this allows for increasingly more flexible and, hence, more wiggly estimators. Placement of a knot in a certain area allows the estimator to adapt more to the data in the region where the knot is located. By locating knots very close to one another the smoothness of the fitted curve can actually be reduced in the sense of a near reduction in the number of continuous derivatives (see de Boor, 1978, pg. 125). When only a few knots are used, the estimator tends to be smoother in the respect that it becomes more like a polynomial of order m. The order, m, is also a smoothing parameter. However, for $m \geq 4$ its selection tends to be of less importance than selecting $\lambda$. Thus, we now turn to the problem of how to select the number and location of the knots for a least-squares spline estimator of $\mu$.

### 7.2.1 Selecting $\lambda$

There are many possible methods for selecting $\lambda = \{\xi_1, \ldots, \xi_k\}$ for the least-squares spline estimator $\mu_\lambda$. For example, since $\mu_\lambda$ is a linear estimator, the methods discussed in Chapter 2 are applicable. We will return to this point shortly.

Perhaps one of the simplest methods for selecting $\lambda$ is through visual inspection of the data. This is the analog of the trial and error approach to selecting, e.g.,the bandwidth for a kernel estimator or the smoothing parameter for a smoothing spline. While this approach is heuristic, it often tends to work quite well in terms of giving a visually pleasing fit to a set of data and has a definite computational advantage over the other methods for selecting $\lambda$ to be discussed subsequently.

The idea to keep in mind when making knot placements through inspection of a scatter plot of the data is that more knots will be needed where $\mu$ seems to change more rapidly. This is because the estimator will need more flexibility in such regions. By combining this premise with knowledge of the pliancy of the individual polynomial segments for a spline, guidelines can be advanced for knot placement. A set of *ad hoc* rules for locating knots which arise from such consideration are the following:

i) For $m = 2$, linear splines, place knots at points where the data exhibit a change in slope.

ii) For $m = 3$, quadratic splines, locate knots near local maxima, minima or inflexion points in the data.

iii) For cubic splines, $m = 4$, arrange the knots so that they are close to inflexion points in the data and not more than one extreme point (maximum or minimum) and one inflexion point occurs between any two knots.

The motivation for the knot placement guidelines for each case is related to the ability of the corresponding spline to adapt to changes in $\mu$ that are reflected in the data. Thus the motivation for guideline i) is intuitively clear. In the quadratic case the guideline for knot placement is a consequence of the fact that a quadratic function has constant curvature and must behave symmetrically about its extreme point. Recognition that a cubic polynomial can accommodate different types of behavior on each side of its maximum or minimum but must behave asymmetrically about its inflexion point provides the reasoning for iii). Guideline iii) was suggested by Wold (1974).

It will also usually be necessary to decide on a choice of m, the order of the spline. If this is also done by examining the scatter plot of the data, instances where linear or quadratic splines are suitable will usually be apparent. The determination of m in other cases may not be so simple. Fortunately, when m = 2 or 3 is not appropriate, there appear to be few cases where cubic splines (m = 4) will not suffice if enough knots are used. Thus knot placement can often be used as a surrogate to compensate for order inadequacies of the estimator.

To illustrate the use of the visual inspection approach to selecting $\lambda$ consider the data on the ratio of weight to height, as a function of age (in months), for boys shown in Figure 7.1. The data is given in Exercise 5.21 and was analyzed in Chapter 5 by the use of smoothing splines. Examination of Figure 7.1 suggests that we might fit this data with a quadratic spline with one knot located around what appears to be an inflexion point at roughly 12 months. The resulting fit obtained from (10) with $\lambda = \{12\}$ is

$$\mu_{\{12\}}(t) = .423 + .055t - .002t^2 + .002(t - 12)_+^2$$

for which the coefficient of determination is .964. A simple calculation shows us that $\mu_{\{12\}}$ can be expressed in piecewise polynomial form as

$$\mu_{\{12\}}(t) = \begin{cases} .423 + .055t - .002t^2 , & t < 12 \\ .731 + .004t , & t \geq 12. \end{cases}$$

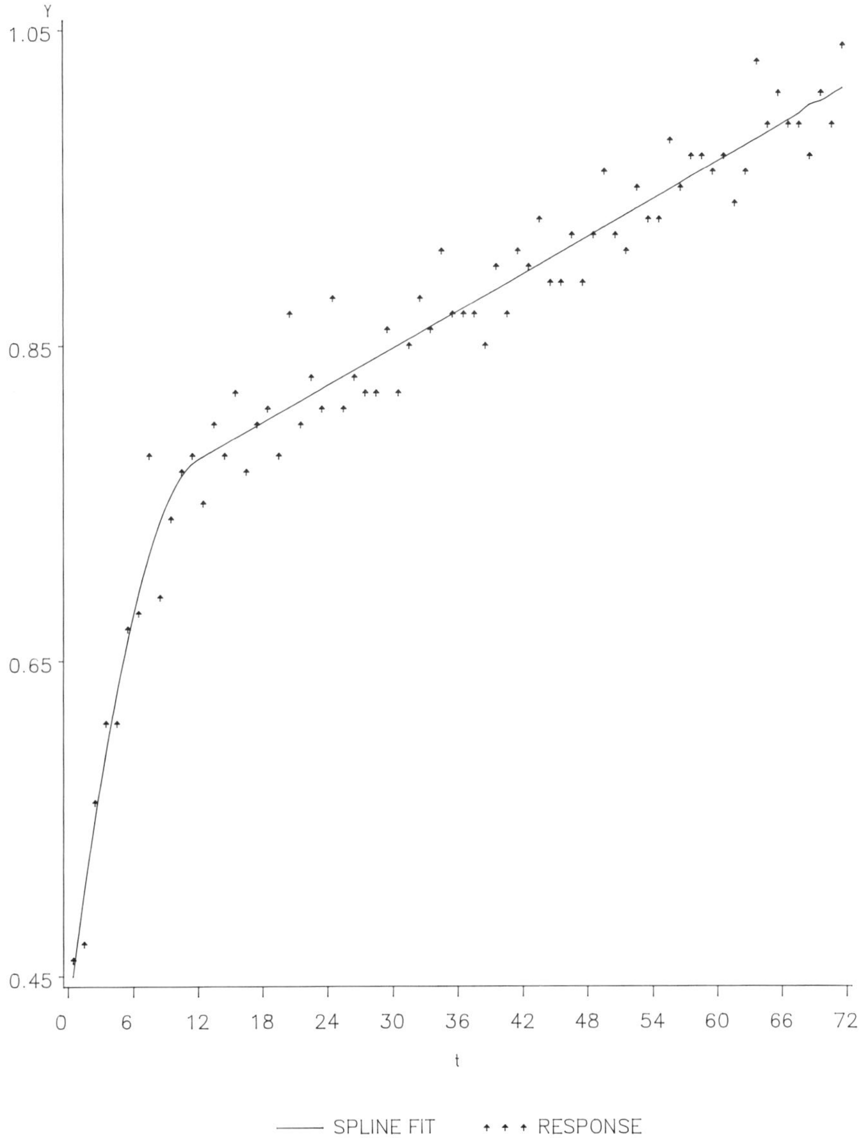


**Figure 7.1** Spline fit to the weight-height ratio data.

Consequently the estimator is linear for $t \geq 12$ and, therefore, is a piecewise polynomial consisting of two different order segments. This provides an illustration of a situation where it would be useful to depart from the basic function (3) and allow for the possibility of different order polynomial segments. We will discuss this further in Section 7.2.3.

Another illustration is provided by the data in Figure 7.2. The response values in this case are the natural logarithms of specific retention volumes of cycloheptene in polyethylene terephthalate which have been plotted as a function of t, the reciprocal times $10^3$ of temperature in degrees kelvin (see Gallant, 1974a). One might attempt to fit this data also using a quadratic spline. An application of guideline (ii) would suggest the use of a single knot near the minimum at about 2.9 or 3.

Given a choice of k and m there is an alternative, more objective, method of selecting $\lambda = \{\xi_1, \ldots, \xi_k\}$ than by looking at scatter plots. One can simply minimize $\mathrm{MSE}(\underline{\beta};\lambda)$ with respect to both $\underline{\beta}$ and the knot locations. Since a spline is nonlinear in the knot locations, this is a nonlinear regression problem which can be solved using the techniques described in Section A.2 of the Appendix. In particular, Gallant and Fuller (1973) have reported some success in the use of the modified Gauss-Newton algorithm for estimating $\underline{\beta}$ and $\xi_1, \ldots, \xi_k$.

To illustrate the type of results which can be obtained from nonlinear least-squares fits of splines, consider again the data in Figure 7.2. A quadratic spline with one knot was fit to this data using the modified Gauss-Newton algorithm (c.f. also Gallant, 1974a). The value of 3 was given as a starting value for the knot. The starting values for the coefficients of the basis functions in (3) were obtained by first fitting a quadratic spline to the data with a knot at 3. The resulting optimal value for the knot was then found to be $\hat{\lambda} = \{\hat{\xi}_1\} = \{2.94\}$. The fitted function is shown in Figure 7.2 along with the data.

In the previous example we selected a starting value for the knot and then used the linear regression estimates of the θ's and

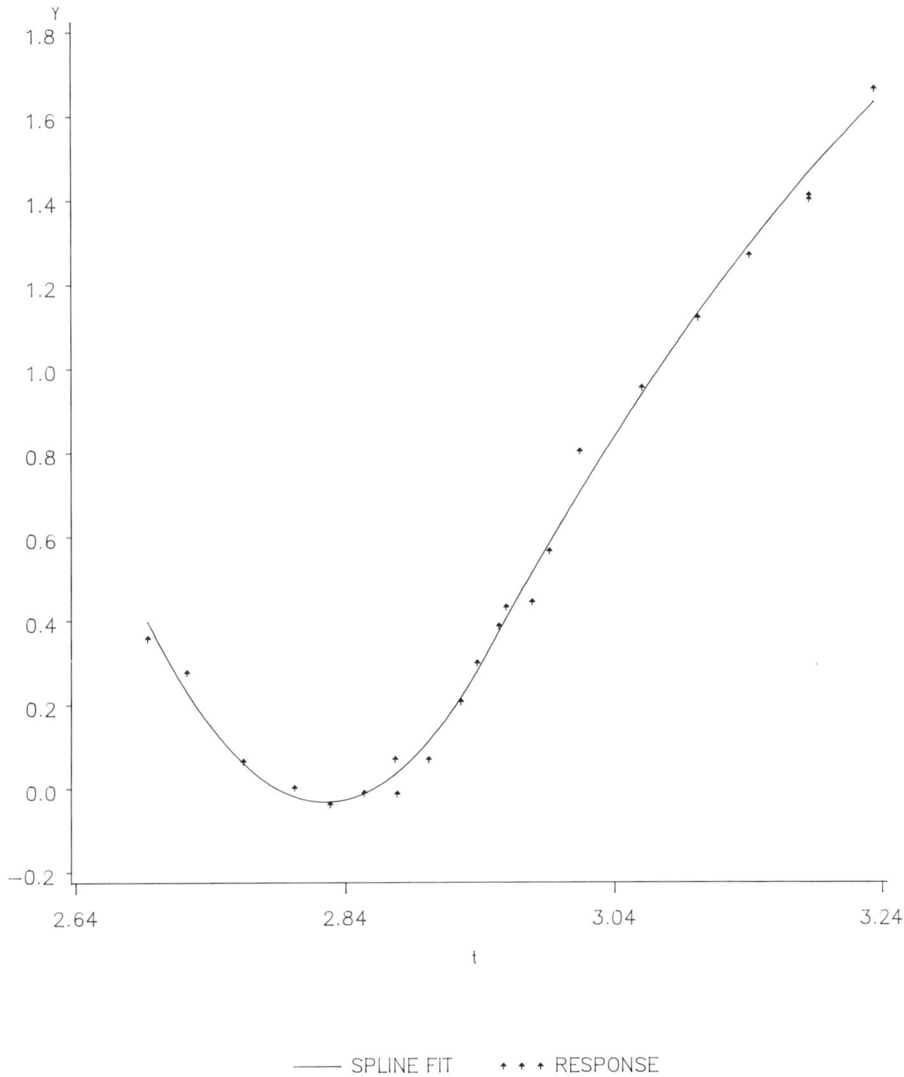


**Figure 7.2** Spline fit to the cycloheptene data.

δ's to initialize the nonlinear least-squares algorithm. This suggests an alternative to the full-blown nonlinear least-squares algorithm used for the example.

Let $\lambda = \{\xi_1, \ldots, \xi_k\}$ be an arbitrary choice for the knot set and let $\underline{\beta}_\lambda$ be any corresponding solution to (9). Then

$$\mathrm{MSE}(\lambda) = \sum_{i=1}^{n} \left(y_i - \sum_{j=1}^{m+k} \beta_{\lambda j} x_j(t_i)\right)^2 \tag{11}$$

is a function of λ alone that can be minimized to obtain an estimate of μ. The latter approach exploits the partially linear and partially nonlinear character of the spline estimator in a more effective fashion than minimizing $\mathrm{MSE}(\underline{\beta};\lambda)$ simultaneously with respect to $\underline{\beta}$ and λ. It has been studied in some detail by Jupp (1972, 1978).

A natural question is when will minimization of MSE(λ) and $\mathrm{MSE}(\underline{\beta};\lambda)$ coincide. The relationship between problems of this type has been investigated by Golub and Pereyra (1973). Applying their work to our problem gives the conclusion that one can expect a local minimum of MSE(λ) to correspond to a local minimum of $\mathrm{MSE}(\underline{\beta};\lambda)$. It will therefore suffice in many cases to search among the critical points of MSE(λ) in computing the estimator.

The minimization of reduced functions such as (11) is called the *variable projection method*. This technique has also been studied by Barham and Drane (1972) and Guttman, Pereyra and Scolnik (1973). The consensus seems to be that, although each iteration may be more expensive, the variable projection approach tends to significantly reduce the number of iterations and thereby results in decreased run times.

Unfortunately, selection of the knots by nonlinear least squares is a rather nasty nonlinear optimization problem. Both MSE(λ) and $\mathrm{MSE}(\underline{\beta};\lambda)$ will not, in general, be convex functionals (Jupp, 1972) and both will have many solutions on the boundary of the feasible region (Jupp, 1978). Thus, the search for a (feasible) solution can be quite tedious and time consuming. In addition, Jupp (1975, 1978) notes that the variable knot problem exhibits a lethargy property which stems from the behavior of $MSE(\lambda)$ and $MSE(\underline{\beta};\lambda)$ on the boundary of the solution space where some of the $\xi_j$ can coincide. Algorithms attempting to find optimal knots will have a tendency to become trapped on the boundary and, in many cases, obtain numerical convergence quite far from the boundary solution. A solution on the boundary will correspond to one or more coincident knots. This can be shown to imply a reduction in the number of continuous derivatives at that point. If a spline is really considered appropriate such a solution would usually not be acceptable. Jupp (1978) has demonstrated that some of these difficulties can be overcome by transforming from the $\xi_j$ to a new parameter set which reduces the tendency of the knots to coalesce.

Still other methods for selecting $\lambda$ can be devised from the techniques in Chapter 2. It is even possible to effect an objective choice for k and m through this approach. To illustrate the idea, suppose it is desired to obtain a data driven choice for both the location and number of knots. Adopting the notation $\lambda_k = \{\xi_1,\ldots,\xi_k\}$ to specifically indicate dependence on k, a GCV criterion corresponding to $\lambda_k$ is

$$GCV(\lambda_k) = MSE(\lambda_k)/(1 - (m + k)/n)^2 \quad , \tag{12}$$

where $MSE(\lambda_k)$ is defined as in (11). Let $\hat{\lambda}_k$ be the minimizer of $GCV(\lambda_k)$ for any fixed k. Then an optimal knot set is $\hat{\lambda}_{\hat{k}}$, where $\hat{k}$ is the minimizer of $GCV(\hat{\lambda}_k)$. Note that, since the denominator of (12) is independent of the knot location, finding $\hat{\lambda}_k$ for any given k is just the variable projection method for estimating the $\xi_j$ that was discussed above. This particular approach to finding an optimal knot set stems from suggestions by Wahba (1977).

### 7.2.2 Computational Considerations

The set of functions defined in (4) form a basis for $S^m(\xi_1,\ldots,\xi_k)$ that is usually called the *truncated power basis.*

As noted by Smith (1979), this basis has some advantages from a statistical viewpoint. However, the basis tends to be ill-conditioned which is often manifested in a nearly singular $X(\lambda)$ matrix in (8). Among other things, this makes the normal equations (9) difficult to solve. Thus, it is desirable to explore the use of other bases for $S^m(\xi_1,\ldots,\xi_k)$ which are better conditioned. One basis with this property is the B-spline basis. We encountered B-splines briefly in Section 5.3.3. A more complete treatment of these functions will be given here.

To develop the concept of B-splines let us first suppose we are interested in splines of order m with some given set of knots satisfying

$$a < \xi_1 < \ldots < \xi_k < b .$$

Then, define 2m additional knots $\xi_{-(m-1)},\ldots,\xi_{-1},\xi_0,\xi_{k+1},\ldots,\xi_{k+m}$ by

$$\xi_{-(m-1)} = \ldots = \xi_0 = a$$

and

$$\xi_{k+1} = \ldots = \xi_{k+m} = b.$$

Also recall the recursive definition for the mth order divided difference of a function g at $\xi_i,\ldots,\xi_{i+m}$ given in Section 5.3.3: namely

$$[\xi_i,\ldots,\xi_{i+m}]g = \frac{[\xi_{i+1},\ldots,\xi_{i+m}]g - [\xi_i,\ldots,\xi_{i+m-1}]g}{\xi_{i+m} - \xi_i}$$

with

$$[\xi_j]g = g(\xi_j) .$$

The B-spline of order m corresponding to the knots $\xi_i,\ldots,\xi_{i+m}$ is then given by

$$N_{i,m}(t) = (\xi_{i+m} - \xi_i)[\xi_i,\ldots,\xi_{i+m}](\cdot - t)_+^{m-1}, \qquad (13)$$

$$i = -(m - 1),\ldots,k.$$

The $(\cdot - t)_+^{m-1}$ notation used in (13) indicates that the divided difference is applied to $(x - t)_+^{m-1}$ as a function of x for fixed t.

It is possible to use the recursive definition of divided differences to obtain a recursion formula for $N_{i,m}$. Adopting the convention that a B-spline of order d corresponding to d + 1 coincident knots is the zero function, $N_{i,m}$ can be defined by

$$N_{i,m}(t) = \frac{t - \xi_i}{\xi_{i+m-1} - \xi_i} N_{i,m-1}(t) + \frac{\xi_{i+m} - t}{\xi_{i+m} - \xi_{i+1}} N_{i+1,m-1}(t) \quad (14)$$

with

$$N_{j,1}(t) = \begin{cases} 1 , & t \in [\xi_j, \xi_{j+1}) , \\ 0 , & \text{otherwise} . \end{cases}$$

A proof of the equivalence of (13) and (14) is given in de Boor (1978). Thus we may regard either (13) or (14) as a definition of the B-splines. However, our attention will be focused on (14) in this chapter. To aid in understanding (14) it will be helpful to consider an example of its use.

Consider the case of linear (m=2) B-splines. In this instance 2m=4 additional knots are defined by $\xi_{-1} = \xi_0 = a$ and $\xi_{k+1} = \xi_{k+2} = b$. Since $N_{-1,1}$ has coincident knots $\xi_{-1}$ and $\xi_0$ (both equal to a), by convention we set $N_{-1,1}(t) \equiv 0$. Similarly, $N_{k+1,1}(t) \equiv 0$. Thus, (14) gives

$$N_{-1,2}(t) = \frac{\xi_1 - t}{\xi_1 - \xi_0} N_{0,1}(t)$$

$$N_{0,2}(t) = \frac{t - \xi_0}{\xi_1 - \xi_0} N_{0,1}(t) + \frac{\xi_2 - t}{\xi_2 - \xi_1} N_{1,1}(t)$$

$$\vdots$$

$$N_{k,2}(t) = \frac{t - \xi_k}{\xi_{k+1} - \xi_k} N_{k,1}(t) .$$

Several important properties of B-splines are established in the Exercises. In particular we note that (Exercise 7)

$$N_{i,m}(t) = 0 \quad \text{if } t \notin [\xi_i, \xi_{i+m}] . \tag{15}$$

This local support characteristic is one of the key reasons for the popularity of B-splines.

Up to this point the term B-spline has been used without justifying that the $N_{i,m}$ are splines, although this fact might have been anticipated from their definition (13) involving the truncated power functions. It can in fact be shown (de Boor, 1978, Chapter 9) that $N_{i,m}$, $i = -(m - 1),\ldots,k$ are linearly independent and are all elements of $S^m(\xi_1,\ldots,\xi_k)$. Since there are $m + k$ B-splines, this has the consequence that $N_{-(m-1),m},\ldots,N_{k,m}$ provide a basis for $S^m(\xi_1,\ldots,\xi_k)$. Thus there are coefficients $\gamma_1,\ldots,\gamma_{m+k}$ such that any element $s$ of $S^m(\xi_1,\ldots,\xi_k)$ can be written as

$$s = \sum_{j=1}^{m+k} \gamma_j N_{j-m,m} \tag{16}$$

Since any spline can be expressed as a linear combination of B-splines, estimation of $\mu$ can be conducted using either B-splines or the truncated power basis. Thus, estimation of the coefficients $\gamma_1,\ldots,\gamma_{m+k}$ in (16) or estimation of $\theta_0,\ldots,\theta_{m-1},\delta_1,\ldots,\delta_k$ in (3) will give identical results (apart from problems of numerical accuracy) in the sense that the fitted splines will have identical values for all points in $[a,b]$. To estimate the B-spline coefficients, define the matrix

$$N(\lambda) = \{N_{j,m}(t_i)\}_{\substack{i=1,n \\ j=-(m-1),k}} .$$

Then, substituting the alternative form (16) into (7) and minimizing with respect to $\underline{\gamma}$ gives

$$\underline{\gamma}_\lambda = (\gamma_{\lambda 1},\ldots,\gamma_{\lambda,m+k})' = [N(\lambda)'N(\lambda)]^{-1}N(\lambda)'\underline{y} . \tag{17}$$

The resulting estimator of $\mu$ is $\mu_\lambda = \Sigma_{j=1}^{m+k}\gamma_{\lambda j}N_{j-m,m}$.

Using (15) we can now see the utility of the B-spline basis for estimation with least-squares splines. The matrix $N(\lambda)'N(\lambda)$ has typical element

$$\sum_{r=1}^{n} N_{i,m}(t_r)N_{j,m}(t_r) .$$

This sum vanishes if $|i - j| \geq m$ as a result of (15). Thus $N(\lambda)'N(\lambda)$ is $2m - 1$ banded and $\underline{\gamma}_\lambda$ can be obtained by solving the normal equations $N(\lambda)'N(\lambda)\underline{\gamma} = N(\lambda)'\underline{y}$ in $O(m^2(m + k))$ arithmetic operations. This is to be compared to the $O((m + k)^3)$ operations required for computing the coefficients for the truncated power basis. Notice that m will usually be small while k can frequently be quite large. Thus, B-splines have a definite computational advantage over the truncated power basis.

The truncated power basis will be suitable for estimation in many statistical problems. However, when handling large complicated data sets where many knots are used it will be advisable to employ B-splines. A difficulty with B-splines is that they are only defined recursively and, as a result, cannot be evaluated directly. However, computer code for both the evaluation of B-splines and computation of the $\gamma_{\lambda j}$ (for specified $\lambda$) can be found in de Boor (1978), so this is not a serious problem.

### 7.2.3 Extensions

There are several extensions of the basic least-squares spline estimator which are often useful in practice. Some of these will be explored in this section.

A particularly useful generalization of least-squares splines can be motivated by examination of the data on the natural logarithm of specific retention volume of methylene chloride in polyethylene terephthalate (in CC. per gm.) plotted as a function of the reciprocal of temperature (times $10^3$) in Figure 7.3. The data appears to be roughly quadratic in nature on each side of a cusp at around t = 2.87. However, as seen from the figure, a quadratic spline (with one knot at $\xi_1 = 2.87$) is a dreadfully poor estimator. The problem is that it is too smooth to satisfactorily resolve the sharp change in behavior of the regression function near 2.87.

The previous example suggests that it may be useful in some cases to fit "splines" with reduced continuity constraints at the knots. This can be easily accomplished by appending new basis elements to the truncated power basis which involve lower powers at any given knot. To be precise, suppose it is desired to have $\nu_j - 1$ continuous derivatives at $\xi_j$, $j=1,\ldots,k$. Then a modification to our original approximation to $\mu$ which incorporates these characteristics is

$$s(t;\underline{\nu}) = \sum_{j=1}^{m} \theta_j t^{j-1} + \sum_{j=1}^{k} \sum_{r=\nu_j}^{m-1} \delta_{jr}(t - \xi_j)_+^r \ . \tag{17}$$

Approximations of this type can also be motivated as in Section 7.1 through the use of integration by parts on Rem(t) and the approximation of each term in the resulting sum by a quadrature formula. Given a choice for $\underline{\nu} = (\nu_1,\ldots,\nu_k)'$, an estimator for $\mu$ can be obtained essentially as before by minimizing $n^{-1}\Sigma_{i=1}^n(y_i - s(t_i;\underline{\nu}))^2$ with respect to $\theta_1,\ldots,\theta_m$ and $\delta_{ij}$, $i = 1,\ldots,k$, $j = \nu_i,\ldots,m-1$.

Examination of (17) reveals that this function agrees with the one in (3) when $\nu_j = m - 1$, $j = 1,\ldots,k$, and, therefore, generalizes our previous approximation to $\mu$. If no continuous derivatives are desired but the approximating curve is to be continuous this corresponds to $\nu_j = 1$ for all j. A discontinuous piecewise polynomial approximation results if all $\nu_j$ are taken to be zero. The case of $\nu_j = 2$, $j = 1,\ldots,k$, has received attention through work by Fuller (1969, 1976), Gallant (1974a) and Gallant and Fuller (1973). They call curves of this latter type grafted polynomials.

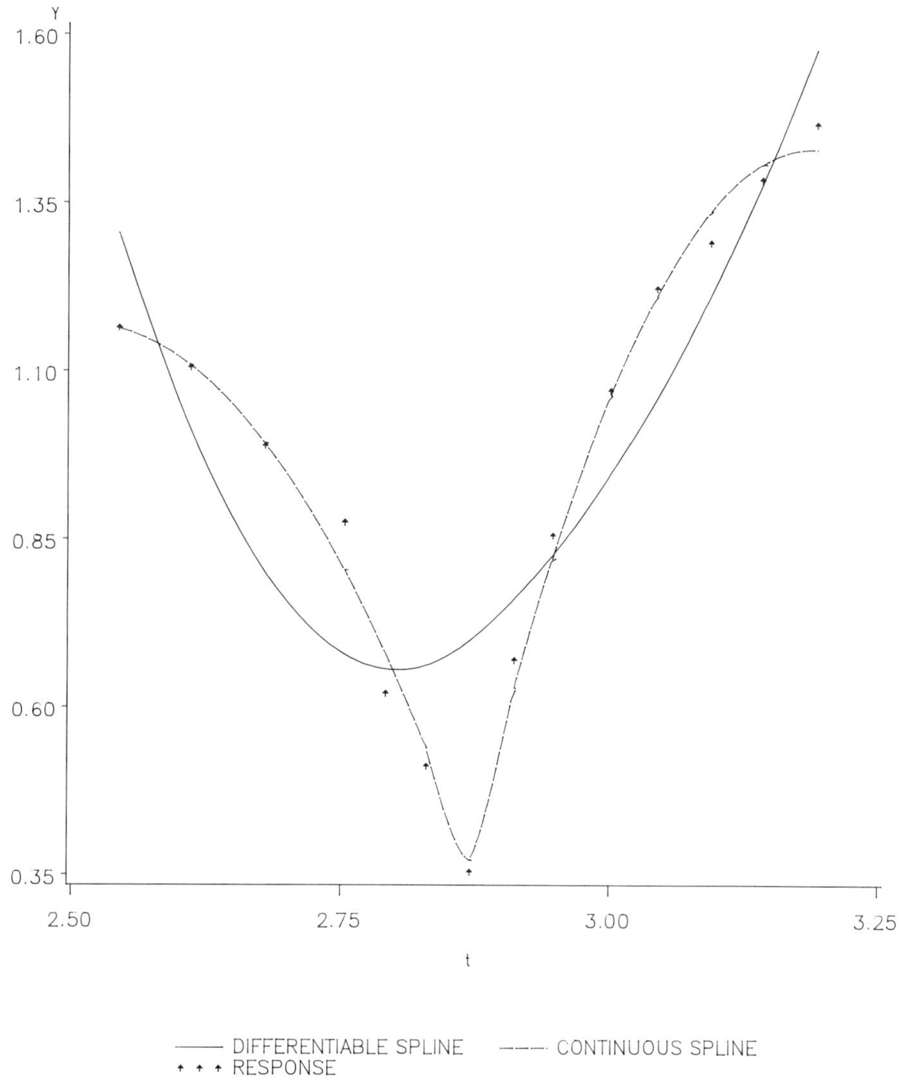


**Figure 7.3** Spline fits to the methylene chloride data.

Functions such as (17) are also frequently called splines. They can be expressed in terms of piecewise polynomials and analogs of B-splines where $m - \nu_j$ knots are "stacked" at $\xi_j$. Such alternative representations will not be explored here. The interested reader is referred to de Boor (1978).

To demonstrate the utility of using reduced continuity constraints return to the methelyene chloride data in Figure 7.3. In addition to our original quadratic spline fit we have also fitted a quadratic piecewise polynomial of the form (17) with one knot at $\xi_1 = 2.87$ and $\nu_1 = 1$. The resulting function is continuous, but not differentiable, and has the form

$$s(t;1) = \sum_{j=1}^{3} \theta_j t^{j-1} + \delta_{11}(t - \xi_1)_+ + \delta_{12}(t - \xi_1)_+^2 \quad . \tag{18}$$

As can be seen from the figure, this function does a much more satisfactory job of resolving the sharp point in the data. In addition it fits the data better than the quadratic spline having an $R^2$ of .98 as compared to .79 for the quadratic spline.

It is, of course, still possible to select the $\xi_j$ (and even $\underline{\nu}$, m and k) using data driven methods for estimation of $\mu$ using functions of the form (17). For example, Gallant (1974a) has found a nonlinear least-squares estimator of $\xi_1$ for fitting the methelyene chloride data by a function of the form (18) is $\hat{\xi}_1 = 2.86$.

Another possible extension of least-squares spline estimators is suggested by our analysis of the weight-height ratio data in Section 7.2.1. We found in that example that the fitted curve was a spline with different order segments.

Estimation of $\mu$ using splines with different order segments has been considered by Fuller (1969), Gallant (1974a), Gallant and Fuller (1973) and Smith (1979). Such functions can be constructed by appropriately constraining the coefficients in (17). To illustrate the idea, suppose we wish to fit a function with knots at $\xi_1$ and $\xi_2$ that consists of three segments which are linear, quadratic and linear, respectively. The general form of (17) in the quadratic case with $\nu_1 = \nu_2 = 1$ is

$$\theta_0 + \theta_1 t + \theta_2 t^2 + \delta_{11} (t - \xi_1)_+ + \delta_{12}(t - \xi_1)_+^2$$
$$+ \delta_{21}(t - \xi_2)_+ + \delta_{22}(t - \xi_2)_+^2 \quad .$$

In order for the first and last segment to be linear it suffices to require that $\theta_2 = 0$ and $\delta_{12} = -\delta_{22}$. We can then fit a curve with the desired properties by deleting the $\theta_2$ term from our original formulation and estimating the remaining coefficients subject to the constraint $\delta_{12} = -\delta_{22}$.

The basic method for constructing splines with different order segments is essentially that illustrated in the example, although the details can be quite tedious in certain cases. The fitting of curves of this nature will generally require the use of restricted least squares. An illustration of this can be found in Gerig and Gallant (1975).

Gallant and Fuller (1973) fit the weight-height ratio data using a spline with one knot that consisted of a quadratic and then a linear segment. Both the basis function coefficients and the knot were estimated via nonlinear least-squares. The resulting optimal knot was found to be $\hat{\xi}_1 = 11.83$.

It is also possible to generalize the concept of least-squares splines to higher dimensions. As an illustration, suppose we have two predictor variables t and z. Assume it is desired to fit a bivariate spline of order m in t with knots at $\xi_1, \ldots, \xi_k$ and of order p in z with knots at $\eta_1, \ldots, \eta_q$. One generalization of least-squares splines to this setting can be based on the so called *tensor product splines* which are linear combinations of all $(m + k)(p + q)$ products of the functions $1, t, \ldots, t^{m-1}, (t - \xi_1)_+^{m-1}, \ldots, (t - \xi_k)_+^{m-1}$ in the t variable with the functions $1, z, \ldots, z^{p-1}$, $(z - \eta_1)_+^{p-1}, \ldots, (z - \eta_q)_+^{p-1}$ in the z variable. For example, if $m = 2$, $k = 1$, $p = 2$ and $q = 2$ we would fit a bilinear spline function of the form

$$\sum_{i=1}^{2} \sum_{j=1}^{2} \theta_{ij} z^{i-1} t^{j-1} + \sum_{i=1}^{2} \alpha_{i1} z^{i-1} (t - \xi_1)_+$$

$$+ \sum_{i=1}^{2} \sum_{j=1}^{2} \gamma_{ij} t^{j-1} (z - \eta_i)_+$$

$$+ \sum_{i=1}^{2} \beta_{i1} (t - \xi_1)_+ (z - \eta_i)_+ \quad .$$

Some of the complications which can arise are obvious from even this simple example. The number of coefficients to be estimated can be quite large and the resulting normal equations can be poorly conditioned. To make matters worse there is, at present, no computationally expedient local support representation for multivariate splines in more than two variables. In addition, the selection of knots, even in the bivariate case, would appear to be a sticky problem.

The use of multivariate least-squares spline estimators is a topic which has received only cursory attention, with work by Fuller (1969) a notable exception. Further research is required by both the mathematical and statistical communities before the use of multivariate least-squares splines can become a standard method of curve fitting.

In Chapter 6 we discussed partial splines which were an extension of smoothing splines to deal with the case of a partially parametric mean value function. Least-squares splines can be adapted to deal with problems of this nature as well. If, for example, $Ey_i = \underline{z}_i'\underline{\zeta} + \mu(t_i)$ with $\mu \in W_2^m[a,b]$ then an estimator of $\underline{\zeta}$ and $\mu$ can be obtained by minimizing

$$\sum_{i=1}^{n} (y_i - \underline{z}_i'\underline{\zeta} - s(t_i))^2$$

with respect to $\underline{\zeta}$ and the $\theta$'s and $\delta$'s in (3). The knot locations, etc. can also be estimated using the obvious parallel of the techniques discussed in Section 7.2.1. For some discussion of estimators of this type see Chen (1986).

Another extension of least-squares splines concerns the use of estimation criteria other than (7). As noted in Sections 4.11 and 6.2.5 a least-squares criterion may not be suitable for data sets containing a number of bad data values. This led to the development of M-type kernel estimators and smoothing splines. A parallel of that development for the present setting would suggest choosing the θ's and δ's in (3) to minimize some criterion of the form

$$n^{-1} \sum_{i=1}^{n} \rho(y_i - s(t_i)) ,$$

where ρ is some symmetric convex function. In particular $\rho(x) = x^2$ gives us our least-squares splines back again. The objective, however, is to instead use other ρ functions which are less sensitive to the influence of large (in magnitude) residuals. Examples of this approach to estimation of μ are given in Lenth (1977) who finds that it can provide an extremely flexible curve fitting procedure.

### 7.2.4 Asymptotic Analysis

In this section we present a brief description of the asymptotic behavior of least-squares splines as estimators of μ. We will deal only with the simple case of functions of the form (3) and assume that μ is to be estimated over some finite interval which, without loss of generality, can be taken as [0,1].

Let $\mu_\lambda$ be defined as in (6). The performance of $\mu_\lambda$ as an estimator of μ will be measured by the integrated risk

$$IR_n(\lambda) = \int_0^1 E(\mu(t) - \mu_\lambda(t))^2 dt . \tag{19}$$

To study the large sample properties of $IR_n(\lambda)$ some simplifying assumptions are needed. First, it will be assumed that the $t_j$ are generated by a positive continuous density p through the relation

$$\int_0^{t_j} p(t)dt = j/(n+1), \quad j = 1,\ldots,n. \tag{20}$$

A similar assumption will be made concerning the knots. Specifically it is assumed that there is a nonnegative continuous density q such that

$$\int_0^{\xi_j} q(\xi)d\xi = j/(k+1) \quad . \tag{21}$$

This latter assumption means that, given k, $\lambda = \{\xi_1,\ldots,\xi_k\}$ is determined immediately by relation (21) and, hence, the actual computation of $\mu_\lambda$ would not involve nonlinear least-squares. However, we will eventually give (asymptotically) optimal choices for both q and k. The asymptotically optimal q can be interpreted as giving an indication of how optimal knots will be located in large samples.

It is also necessary to place some restrictions on the growth of k as a function of n. A natural condition to impose is that $k/n \to 0$ which insures that the number of knots grows at a slower rate than the sample size.

Under the above assumptions, results by Agarwal and Studden (1980) allow us to state the following theorem.

**Theorem 7.1 (Agarwal and Studden, 1980).** Let m be the order of $\mu_\lambda$ and assume that $\mu \in C^m[0,1]$. If the design points and knots are generated by relations (20) and (21), respectively, and $k/n \to 0$ then

$$IR_n(\lambda) \sim \frac{k\sigma^2}{n} \int_0^1 \frac{q(t)}{p(t)}\,dt + \frac{\alpha}{k^{2m}} \int_0^1 \frac{\mu^{(m)}(t)^2}{q(t)^{2m}}\,dt, \tag{22}$$

where $\alpha$ is a constant depending on m but not on $\mu$, p, q or k. The choices of p, q and k which minimize this asymptotic expression are

$$p_{opt}(t) = \left|\mu^{(m)}(t)\right|^{2/(4m+1)} \Big/ \int_0^1 \left|\mu^{(m)}(s)\right|^{2/(4m+1)} ds, \tag{23}$$

$$q_{opt}(t) = \left|\mu^{(m)}(t)\right|^{4/(4m+1)} / \int_0^1 \left|\mu^{(m)}(s)\right|^{4/(4m+1)} ds \quad (24)$$

and

$$k_{opt} = \left\{ \frac{2\alpha mn}{\sigma^2} \int_0^1 \left| \mu^{(m)}(s)\right|^{2/(4m+1)} ds\right\}^{1/(2m+1)}$$

$$\times \left[ \int_0^1 \left|\mu^{(m)}(s)\right|^{4/(4m+1)} ds\right]^{-1} \quad (25)$$

Theorem 7.1 indicates how the integrated risk will behave asymptotically for a given design and knot selection scheme. It also indicates how the design points and knots should be located to insure the best asymptotic behavior. The optimal choices for p, q and k all, unfortunately, involve the unknown function $\mu$ or, rather, its mth derivative. For this reason they will not generally be computable, since it is unlikely that $\mu^{(m)}$ would be known if $\mu$ were unknown. However, they do have some interesting implications which will now be explored.

The optimal knot density, $q_{opt}$, is proportional to $|\mu^{(m)}|^{4/(4m+1)}$ which indicates that knots should be more heavily concentrated where $|\mu^{(m)}|$ is large. Thus, for example, in the cubic case knots should be located in regions where $\mu^{(4)}$ has large magnitude. This would seem difficult to judge from data. However, the overall recommendation is still consistent with the premise that more knots should be located in areas where $\mu$ appears to be wiggly than in regions where it seems to be smooth.

The optimal number of knots $k_{opt}$ is $O(n^{1/(2m+1)})$. In particular, when $m = 4$, $k_{opt} = O(n^{1/9})$. This gives the recommendation that the number of knots should be kept to a minimum and, in any case, should be much smaller than n.

Finally, observe that $p_{opt} \propto (q_{opt})^{1/2}$. Since the square root will tend to suppress the peaks and fatten the tails of $q_{opt}$ this indicates that design points should be more dispersed than knots.

By replacing p, q and k in (22) by the optimal quantities (23) -(25) it is found that the optimal integrated risk is $O(n^{-2m/(2m+1)})$. This corresponds to the optimal rate of decay for the integrated risk discussed in Section 1.2 and is the same rate shared by other competing estimators such as kernel estimators and smoothing splines.

It is of interest to note that least-squares splines attain the optimal rate $n^{-2m/(2m+1)}$ using splines of order m and require no boundary modifications. This is in contrast to smoothing splines where a spline of order 2m is required to attain this rate, and kernel estimators, which require special modifications at the boundary of [0,1]. This property has been viewed by some as an advantage that least-squares splines have over kernel and smoothing spline estimators.

In the case where the process under study can be sampled repeatedly, Agarwal (1978) and Agarwal and Studden (1978 a,b, 1980) have developed an adaptive method of estimating the asymptotically optimal quantities (23)-(25). Their approach will not be pursued here. The interested reader is referred to their papers for the details.

### 7.2.5 Other Inference Problems

In this section we give a brief discussion of the problems of interval estimation and diagnostic analysis for least-squares splines. The problem with currently available theory in this area is that it is applicable only to the case where $\mu$ is itself a spline of order m with k knots and m and k are known. Thus, one must be careful about routine application of the following results in nonparametric regression analysis.

Assume that the data follow a model of the form

$$y_i = s(t_i;\underline{\xi}) + \varepsilon_i \ , \quad i = 1,\ldots,n \qquad , \tag{25a}$$

where the $\varepsilon_i$ are zero mean, uncorrelated errors with common variance $\sigma^2$ and

$$s(t;\underline{\xi}) = \sum_{j=1}^{m} \theta_j t^{j-1} + \sum_{j=1}^{k} \delta_j (t - \xi_j)_+^{m-1}. \tag{25b}$$

If $\underline{\xi} = (\xi_1,\ldots,\xi_k)'$ is known, methods for diagnostic and inferential analysis for this model can be derived from those in Section A.1 of the Appendix since, in this case, (25) is just a special case of a linear model. Thus, assuming normal errors, interval estimation can be conducted using (A.16) and one can even test hypotheses concerning the θ's and δ's. Of particular interest are tests for hypotheses such as $\delta_j = 0$. If $\delta_j = 0$ then s is a single polynomial segment over $[\xi_{j-1}, \xi_{j+1}]$ and the knot $\xi_j$ is said not to be active. Some authors (Poirier, 1973 and Buse and Lim, 1977) propose placing a knot, $\xi_j$, in an area where some type of structural change in the process under study may have occurred. They then view rejection of $\delta_j = 0$ as providing confirmation of this belief.

If the knots are unknown in (25) then methods for inference in nonlinear regression models such as those discussed in Section A.2 of the appendix can be used. Let $\underline{\eta} = (\theta_1,\ldots,\theta_m,\delta_1,\ldots,\delta_k,\xi_1,\ldots,\xi_k)'$ be the vector of m + 2k parameters in model (25) and set

$$\underline{\Delta s}(t;\underline{\eta}) = \left(\frac{\partial s(t;\underline{\eta})}{\partial \eta_1},\ldots,\frac{\partial s(t;\underline{\eta})}{\partial \eta_{m+2k}}\right)' .$$

Also define

$$\Delta S(\underline{\eta})' = [\underline{\Delta s}(t_1;\underline{\eta}),\ldots,\underline{\Delta s}(t_n;\underline{\eta})] .$$

Then, using (A.31), an approximate 100(1 - α)% confidence interval for the value of the true mean value function at t is

$$s(t;\underline{\hat{\eta}}) \pm Z_{\alpha/2}[\hat{\sigma}^2 \underline{\Delta s}(t;\underline{\hat{\eta}})(\Delta S(\underline{\hat{\eta}})'\Delta S(\underline{\hat{\eta}}))^{-1}\underline{\Delta s}(t;\underline{\hat{\eta}})]^{1/2} , \tag{26a}$$

where $\underline{\hat{\eta}}$ is any least-squares estimator of $\underline{\eta}$ and $\hat{\sigma}^2$ is defined as in (A.30) as

$$\hat{\sigma}^2 = \sum_{i=1}^{n} (y_i - s(t_i;\underline{\hat{\eta}}))^2/(n - m - 2k). \tag{26b}$$

Such intervals will be asymptotically valid under model (25) provided certain regularity conditions are met (see Gallant, 1974a,b), the key restriction being that $m \geq 3$. An example of the intervals (26) is shown in Figure 7.4 for the data in Figure 7.2.

It is also possible to develop tests for hypotheses about some of the parameters in model (25) using asymptotic properties of nonlinear least-squares estimators. A discussion of how this can be done is given in Section A.2.2 of the Appendix. Unfortunately, this approach does not appear to be valid for testing hypotheses of the form $\delta_j = 0$ due to the violation of certain regularity conditions. The reasons this approach fails and an alternative successful procedure are discussed by Gallant (1977). Techniques for diagnostic analysis which are suitable for the estimator $s(\cdot;\hat{\underline{\eta}})$ are discussed in Section A.2.3 of the Appendix.

In concluding this section we again caution that the inference results that have been presented apply only to the case where model (25) holds. Their applicability to model (1) when $\mu$ is not a spline is unclear, at present. Note, however, that we can always write $\mu(t) = s(t;\underline{\eta}) + (\mu(t) - s(t;\underline{\eta}))$ and, if the remainder term $\mu(t) - s(t;\underline{\eta})$ is sufficiently small, intervals such as (26) and other inferential methods of this nature may be satisfactory. On the other hand, when this remainder function does not behave properly, the inferential methods discussed in this section will essentially ignore a bias in the estimator that causes their utility to be suspect.

## 7.3 Speckman's Minimax Estimator

In Section 5.2 it was noted that the smoothing spline estimator of $\mu \in W_2^m[a,b]$ was optimal in the sense that it was a minimax estimator of $\mu(t)$ for any given t. More precisely, if L denotes the set of all linear estimators, the smoothing spline estimator attains the bound

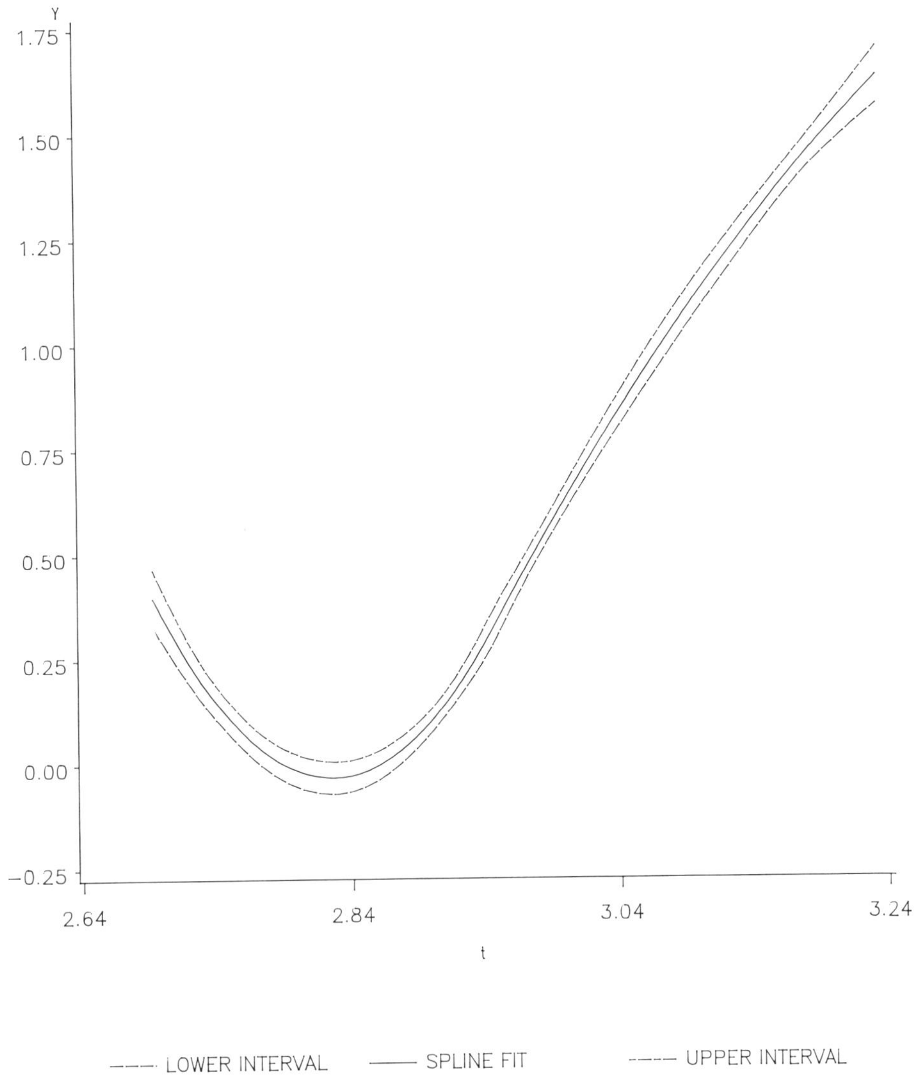


**Figure 7.4** Spline fit and interval estimates for the cycloheptene data.

$$\min_{\hat{\mu}\varepsilon L} \max_{J_m(\mu)\le\rho} E(\mu(t) - \hat{\mu}(t))^2 \quad ,$$

where, recall,

$$J_m(\mu) = \int_a^b \mu^{(m)}(t)^2 dt \ .$$

While this is a good quality for an estimator, it is strictly local in that it applies to estimation at only a single point t.

An alternative method of developing an estimator of $\mu$ might seek to derive an estimator which is globally, rather than locally, minimax. Suppose that our data follow model (1) with $\mu \ \varepsilon \ W_2^m[a,b]$ and define a global performance measure for an estimator $\hat{\mu}$ of $\mu$ by

$$R_n(\hat{\mu},\mu) = n^{-1} E \sum_{i=1}^{n} (\mu(t_i) - \hat{\mu}(t_i))^2 \ . \tag{27}$$

Then a globally minimax linear estimator of $\mu$ is any estimator $\tilde{\mu}\varepsilon L$ which satisfies

$$\max_{J_m(\mu)\le\rho} R_n(\tilde{\mu},\mu) = \min_{\hat{\mu}\varepsilon L} \max_{J_m(\mu)\le\rho} R_n(\hat{\mu},\mu) \ . \tag{28}$$

An estimator with this property has been derived by Speckman (1985).

To describe Speckman's estimator recall again the form of the Demmler-Reinsch basis discussed in Section 5.3.3. This was a basis for the natural splines of order 2m with knots at $t_1,\ldots,t_n$ (i.e., $NS^{2m}(t_1,\ldots,t_n)$). The basis functions $x_1,\ldots,x_n$ satisfy

$$\sum_{i=1}^{n} x_r(t_i)x_k(t_i) = \begin{cases} 1 \ , & r = k \ , \\ 0 \ , & r \ne k \ , \end{cases} \tag{29a}$$

and

$$\int_a^b x_r^{(m)}(t)x_k^{(m)}(t)dt = \begin{cases} \delta_r \ , & r = k \ , \\ 0 \ , & r \ne k \ , \end{cases} \tag{29b}$$

where $0 = \delta_1 = \ldots = \delta_m < \delta_{m+1} \leq \ldots \leq \delta_n$ are an associated set of eigenvalues for a particular matrix deriving from spline interpolation. In particular, $x_1, \ldots, x_m$ form a basis for the polynomials of order m.

Speckman's minimax estimator of μ is defined by

$$\mu_\lambda(t) = \sum_{j=1}^{n} (1 - (\lambda\delta_j)^{1/2})_+ \tilde{\beta}_j x_j(t) , \tag{30a}$$

where

$$\tilde{\beta}_j = \sum_{i=1}^{n} y_i x_j(t_i) , \tag{30b}$$

and $\lambda \geq 0$ is a parameter which must be specified by the user or selected from the data. Note that $\mu_\lambda$ is in $NS^{2m}(t_1, \ldots, t_n)$ but it is not a smoothing spline (c.f. (5.35)). However, like a smoothing spline estimator, it is a generalization of a polynomial regression estimator. To see this, merely let $\lambda \to \infty$ in (30) in which case all terms of the form $(1 - (\lambda\delta_j)^{1/2})_+$ will vanish except those with $\delta_j = 0$. The limiting form of the estimator is $\Sigma_{j=1}^{m} \tilde{\beta}_j x_j$ and is recognized as the polynomial regression estimator of order m for μ.

The parameter λ in (30) is a smoothing parameter. As with smoothing splines, larger values for λ correspond to smoother estimators while smaller values give rise to estimators which are more wiggly. The choice $\lambda = 0$ results in interpolation of the data.

Estimator (30) is a linear estimator, so the vector of fitted values can be written as

$$\underline{\mu}_\lambda = H(\lambda)\underline{y} \tag{31a}$$

for an $n \times n$ matrix $H(\lambda)$. A quick calculation shows that

$$H(\lambda) = XD(\lambda)X' , \tag{31b}$$

where

$$X = \{x_j(t_i)\}_{i,j=1,n} \tag{31c}$$

and

$$D(\lambda) = \mathrm{diag}((1 - (\lambda\delta_1)^{1/2})_+, \ldots, (1 - (\lambda\delta_n)^{1/2})_+). \quad (31d)$$

Thus, $\mu_\lambda$ fits into the framework in Chapter 2 and methods described there can be used for selecting a value of $\lambda$.

To verify the minimax property of $\mu_\lambda$ we follow Speckman (1985) and first note that corresponding to each estimator $\hat{\mu}$ in the set of all linear estimators L there is an $n \times n$ matrix H such that

$$\underline{\hat{\mu}} = (\hat{\mu}(t_1), \ldots, \hat{\mu}(t_n))' = H\underline{y}.$$

Thus, (27) becomes

$$R_n(\hat{\mu}, \mu) = n^{-1}\{\underline{\mu}'(I - H)'(I - H)\underline{\mu} + \sigma^2 \mathrm{tr} H'H\}.$$

Now, let $s_\mu$ be the unique interpolating spline to $\mu$ at $t_1, \ldots, t_n$ and observe that, since $R_n(\hat{\mu}, \mu)$ involves $\mu$ only through its values at $t_1, \ldots, t_n$,

$$R_n(\hat{\mu}, \mu) = R_n(\hat{\mu}, s_\mu) .$$

Also, as a result of Corollary 5.1, $J_m(s_\mu) \le J_m(\mu)$. This has the consequence that for any $\hat{\mu} \in L$ the maximum of $R_n(\hat{\mu}, \cdot)$ will occur in

$$S(\rho) = \{s \in NS^{2m}(t_1, \ldots, t_n): J_m(s) \le \rho\}.$$

Any function $s \in S(\rho)$ can be expressed as $s = \Sigma_{j=1}^n \beta_j x_j$ for coefficients $\beta_1, \ldots, \beta_n$ and $x_1, \ldots, x_n$ the Demmler-Reinsch basis functions. Under this representation

$$\underline{s} = (s(t_1), \ldots, s(t_n))' = X\underline{\beta} ,$$

with $\underline{\beta} = (\beta_1, \ldots, \beta_n)'$ and X defined in (31c). Using (29) we see that $X'X = XX' = I$ and

$$J_m(s) = \underline{\beta}'\mathrm{diag}(\delta_1, \ldots, \delta_n)\underline{\beta}.$$

Thus, if we set $\Delta = \mathrm{diag}(\delta_1, \ldots, \delta_n)$, it follows that

$$\max_{J_m(\mu)\le\rho} R_n(\hat{\mu},\mu) = \max_{s\in S(\rho)} R_n(\hat{\mu},s)$$

$$= n^{-1} \max_{\underline{\beta}'\Delta\underline{\beta}\le\rho} \{\underline{\beta}'(I - X'HX)'(I - X'HX)\underline{\beta} + \sigma^2 tr[(X'HX)'(X'HX)]\}. \quad (32)$$

The implication of equation (32) is that we can trade the problem of finding an estimator which satisfies (28) for the equivalent problem of finding a matrix H which minimizes the right hand side of (32). The solution to this latter problem can be obtained by the use of the following lemma.

**Lemma 7.1 (Kuks and Olman, 1971, Speckman, 1985).** Let D be a diagonal matrix with diagonal elements $0 = d_1 = \ldots = d_m < d_{m+1} \le \ldots \le d_n < \infty$. The solution to the minimax problem

$$\min_{B} \max_{\underline{\beta}'D\underline{\beta}\le\rho} \{|(I - B)\underline{\beta}|^2 + \sigma^2 trB'B\}$$

is $B = diag((1 - (\lambda d_1)^{1/2})_+,\ldots,(1 - (\lambda d_n)^{1/2})_+)$ for some $\lambda > 0$.

By making the identifications $D = \Delta$ and $B = X'HX$, it follows immediately from Lemma 7.1 that the optimal matrix H in (32) is $H(\lambda) = XB(\lambda)X'$, where $B(\lambda) = D(\lambda)$ in (31b). Comparison with (31) indicates that $\mu_\lambda$ does, in fact, satisfy (28) and is, therefore, a global minimax estimator of $\mu$.

Speckman (1985) has investigated some of the large sample properties of $\mu_\lambda$ in (30). He shows that if $\lambda$ is selected appropriately $\mu_\lambda$ attains the best uniform rate of $n^{-2m/(2m+1)}$ for the risk. More precisely he shows that for a particular sequence of smoothing parameters with $\lambda_n = O(n^{-1/(2m+1)})$

$$\max_{J_m(\mu)\le\rho} R_n(\mu,\mu_{\lambda_n}) = C(\alpha,m)n^{-2m/(2m+1)} \quad ,$$

where $C(\alpha,m)$ is defined in (1.5).

Speckman has also studied the asymptotic properties of the adaptive estimator which results when $\lambda$ is selected by GCV. Fix $\mu$

and set $R_n(\lambda) = R_n(\mu_\lambda, \mu)$. Then Speckman (1985) shows that, if the $\varepsilon_j$ are normally distributed and $J_m(\mu) = \rho > 0$, there is a sequence $\{\hat{\lambda}_n\}$ of suitably restricted minimizers of the GCV criterion for which

$$R_n(\hat{\lambda}_n)/\inf_{0 \le \lambda \le \infty} R_n(\lambda) \xrightarrow{P} 1$$

as $n \to \infty$. This result is similar in nature to (5.40) for smoothing splines. However, it is stronger in the respect that $\hat{\lambda}_n$ is a minimizer of the GCV criterion rather than its expectation and is therefore determined from the data. Other related work on large sample properties of minimax estimators can be found in Nussbaum (1985).

An example of a minimax spline estimator is shown in Figure 7.5. The data for this example is the voltage drop data discussed in Section 3.4.5. The estimated curve was kindly provided by Professor Paul Speckman who used generalized cross-validation to select a value for the smoothing parameter.

## 7.4 Nearest Neighbor Estimators

Another popular type of linear estimator is what is known as a nearest neighbor estimator. In many cases these estimators bear similarities to the kernel estimators that were studied in Chapter 4. However, as we shall see, they can take on many other forms.

For simplicity we will work only with scalar valued t, although the more general case of t a p-vector can be handled similarly. It will also be assumed throughout that the $t_j$ in model (1) are distinct. Let $x_1, \ldots, x_q$ denote q functions of t and, for any positive integer $\lambda$, let $I_\lambda(t)$ denote the set of indices $\{i_1, \ldots, i_\lambda\}$ corresponding to the $\lambda$ closest design points, among $t_1, \ldots, t_n$, to the point t. Then, one possible definition of a *nearest neighbor estimator* of $\mu(t)$ is

$$\mu_\lambda(t) = \sum_{j=1}^{q} \beta_{\lambda j}(t) x_j(t) , \tag{33a}$$

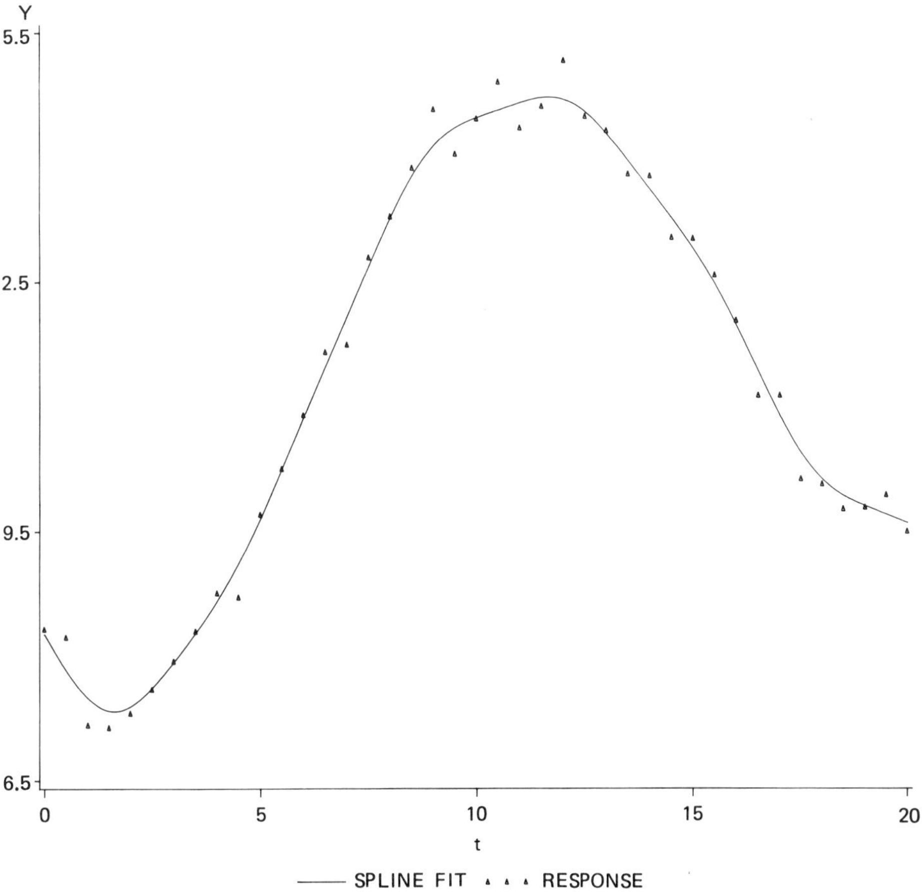


**Figure 7.5** Minimax spline fit to the voltage drop data.

where $\beta_{\lambda 1}(t),\ldots,\beta_{\lambda q}(t)$ are the minimizers of

$$\sum_{i\in I_\lambda(t)} W_{\lambda i}(t)(y_i - \sum_{j=1}^{q} \beta_j x_j(t_i))^2 \tag{33b}$$

and the $W_{\lambda i}(t)$ are associated weights.

To illustrate the types of estimators which derive from this definition consider the special case where $q = 1$ and $x_1(t) \equiv 1$. One might then define the weights in this case by

$$W_{\lambda i}(t) = K_\lambda(\nu_i(t)) ,$$

where

$$\nu_i(t) = 1 + \#\{j: 1 \le j \le n,\ j \ne i,\ |t_j - t| < |t_i - t|\}$$

and $K_\lambda$ is a kernel function that vanishes except on the integers between 1 and $\lambda$. The estimator of $\mu(t)$ which results from this choice is

$$\mu_\lambda(t) = \sum_{i\in I_\lambda(t)} K_\lambda(\nu_i(t))y_i / \sum_{i\in I_\lambda(t)} K_\lambda(\nu_i(t)) . \tag{34}$$

Some typical choices for $K_\lambda$ in (34) include

$$K_\lambda(i) = \lambda^{-1} , \tag{35}$$

$$K_\lambda(i) = 2(\lambda - i + 1)/\lambda(\lambda+1) \tag{36}$$

and

$$K_\lambda(i) = 6(\lambda^2 - (i - 1)^2)/\lambda(\lambda+1)(4\lambda - 1). \tag{37}$$

Estimators of the form (34) are quite similar to kernel estimators. In particular, the choices (35), (36) and (37) for K give estimators that are much like kernel estimators based on the uniform, triangular and quadratic kernels, respectively. Thus they may be motivated from the same viewpoint as kernel estimators in that they utilize a local average of observations near t to estimate $\mu(t)$.

A generalization of estimator (34) that is of particular interest can be derived from definition (33) by taking $x_1(t) \equiv 1,\ldots,x_q(t) = t^{q-1}$. In that case, for $\lambda \geq q$, $\mu_\lambda(t)$ will be a local polynomial fit to the data near t. Numerical examples of these types of nearest neighbor estimators can be found in Stone (1975) and Cleveland (1979). Cleveland (1979) terms this variety of nearest neighbor estimator a locally weighted regression estimator. He shows that a robustified version of these estimators provides a useful tool for scatter plot smoothing. Locally weighted regression estimators are also closely related to kernel estimators with the order of the polynomial playing a parallel role to the order of the kernel. For a detailed explanation of their relationship see Müller (1987).

Nearest neighbor estimators were first studied by Royall (1966) and Cover (1968). Their consistency properties under very weak conditions have been established by Stone (1977), Devroye (1978b) and Cheng (1984b). Li (1984) has shown that nearest neighbor estimators are consistent when $\lambda$ is selected by cross validation.

## 7.5 Additive Nonparametric Regression

In this final section we describe several methods for nonparametric regression that are appropriate for high-dimensional data sets. These methods are related by the fact that they all use some type of additive approximation to describe the underlying regression model.

Let Y and $T_1,\ldots,T_p$ be random variables and let $(t_{i1},\ldots,t_{ip},y_i) = (\underline{t}_i',y_i)$, $i = 1,\ldots,n$, represent a random sample of size n on the random vector $(T_1,\ldots,T_p,Y)$. The problem we consider, initially, is the estimation of the conditional expectation or regression function $\mu(t_1,\ldots,t_p) = E[Y|T_1 = t_1,\ldots,T_p=t_p]$.

There are a number of methods for estimating $\mu$ in general, some of which we have studied in previous chapters. For example, one could use a Laplacian smoothing spline or multivariate kernel estimator for this purpose. However, problems arise with such estimators when p, the number of independent variables, is large. One difficulty stems from the sparcity of data in the multivariate setting. To illustrate what can happen, consider a multivariate kernel estimator for $\mu$ such as that of Section 4.2. The variance of this estimator will be inversely proportional to $n\lambda^p$, where $\lambda$ is a common bandwidth. This quantity represents, roughly, the amount of data that is being averaged in computing the estimator and grows small rapidly for fixed n and $\lambda < 1$. This has the consequence that the variance of a kernel estimator will tend to be unacceptably large for high-dimensional problems unless the sample size is extremely large. Similar conclusions can be drawn for most other linear estimators. The problem of rapidly increasing variance for increasing dimensionality is sometimes referred to as the *curse of dimensionality*.

Another problem with general multivariate estimators of $\mu$, such as kernel and smoothing spline estimators, is their interpretability. The information they impart about the relationship between the dependent and independent variables is often difficult to visualize or comprehend. For example, since one cannot plot the regression surface for $p > 2$, one must instead rely on plots of various projections of the surface into 2 or 3 dimensions for information about how the estimate of $\mu$ behaves. This is a tedious process at best and can result in certain features of the estimate being overlooked.

An alternative to the use of a multivariate smoother for estimation of $\mu$, that overcomes some of the difficulties described above, can be based on estimation of an additive approximation to $\mu$. The idea is to assume that $\mu$ admits an additive approximation of the form

$$\mu(t_1,\ldots,t_p) \doteq \alpha + \sum_{j=1}^{p} \mu_j(t_j) \tag{38}$$

with $E\mu_j(T_j) = 0$, $j = 1,\ldots,p$. Stone (1985) suggests one possible additive approximation where the $\mu_j$ are defined to minimize

$$E[(\mu(T_1,\ldots,T_p) - \alpha - \sum_{j=1}^{p} \mu_j(T_j))^2] \tag{39}$$

subject to the zero mean constraint. He then shows that the $\mu_j$ defined in this fashion are unique under mild restrictions. Given that one is willing to employ an additive approximation to $\mu$, the $\mu_j$ can be estimated by various methods, some of which will be described subsequently, and an estimator of the regression function can then be obtained.

The benefits of using an additive approximation to $\mu$ are at least twofold. First, since each of the $\mu_j$ may be estimated using modifications of univariate estimators, the curse of dimensionality has been circumvented. Secondly, estimators of the $\mu_j$ provide easily interpretable information about how Y changes with the corresponding independent variable $T_j$. One can, for example, plot an estimator versus $T_j$ to obtain insight into the relationship between the independent and dependent variable. Because the estimate of the regression function is additive it is not necessary to study projections of the fitted surface in higher dimensions.

While the use of an additive approximation may pay conceptual dividends, there are also risks involved. There is always the danger that estimation methods based on an additive approximation may overlook interactions among the T variables that might have been detected by a nonadditive multivariate method. Some difficulties of this sort can be avoided by, e.g., including products of independent variables, which may be suspected of interacting, in the original list of predictor variables. However, interactions need not be embodied in variable products or any other easily described functional form. There is even some question as to what interaction means in this setting. (See Section 2 of the

Rejoinder in Hastie and Tibshirani, 1986 for some discussion of this point.)

There are a variety of possible methods for estimating the $\mu_j$ in (38). One approach suggested by Stone (1985) is to use an estimate of the form

$$\mu_\lambda(t_1,\ldots,t_p) = \bar{y} + \sum_{j=1}^{p} \mu_{\lambda j}(t_j) ,$$

where $\bar{y} = n^{-1}\Sigma_{i=1}^{n} y_i$ and $\mu_{\lambda j}$ is a spline of order m having $\lambda$ uniformly spaced knots over the range of $T_j$. The basis function coefficients for the $\mu_{\lambda j}$ are chosen to minimize mean squared error subject to the constraints $\Sigma_{i=1}^{n}\mu_{\lambda j}(t_{ij}) = 0$. Stone (1985) shows that the $\mu_{\lambda j}$ are consistent estimators of the $\mu_j$ from (38)-(39) in a certain sense, provided the number of knots grows large at an appropriate rate.

Another approach to constructing an additive estimator can be based on the *backfitting algorithm* of Friedman and Stuetzle (1981). Suppose that we have some type of univariate smoothing technique such as a nearest neighbor, kernel or even a smoothing spline estimator. Let $\underline{z}$ be a vector of n values and let $\underline{s}(\underline{z})$ represent the smoothed set of values obtained by applying our smoother to the values in the $\underline{z}$ vector. The backfitting algorithm then proceeds as follows. Upon defining the residual vector $\underline{r}_1 = \underline{y} - \bar{y}\underline{1}$ an estimator of $\underline{\mu}_1 = (\mu_1(t_{11}),\ldots,\mu_1(t_{n1}))'$ is taken to be $\hat{\underline{\mu}}_1^{(1)} = \underline{s}(\underline{r}_1)$. A new residual vector $\underline{r}_2 = \underline{y} - \bar{y}\underline{1} - \hat{\underline{\mu}}_1^{(1)}$ is then defined and $\underline{\mu}_2 = (\mu_2(t_{12}),\ldots,\mu_2(t_{n2}))'$ is estimated by $\hat{\underline{\mu}}_2^{(1)} = \underline{s}(\underline{r}_2)$. The process continues until we come to $\underline{r}_p = \underline{y} - \bar{y}\underline{1} - \Sigma_{j=1}^{p-1}\hat{\underline{\mu}}_j^{(1)}$ and $\hat{\underline{\mu}}_p^{(1)} = \underline{s}(\underline{r}_p)$. To allow for re-adjustments of earlier smooths one usually iterates on this basic method. Thus on the $\ell$th iteration ($\ell \geq 2$) $\underline{r}_j$ is redefined to be

$$\underline{r}_j = \underline{y} - \bar{y}\,\underline{1} - \sum_{k=1}^{j-1} \hat{\underline{\mu}}_k^{(\ell)} - \sum_{k=j+1}^{p} \hat{\underline{\mu}}_k^{(\ell-1)}$$

and $\underline{\mu}_j^{(\ell-1)}$ is replaced by

$$\hat{\underline{\mu}}_j^{(\ell)} = \underline{s}(\underline{r}_j) ,$$

successively, for $j = 1,\ldots,p$. This iterative procedure is continued until some criterion such as mean squared error fails to decrease appreciably. The backfitting algorithm can be used with essentially any type of smoother and even different smoothers can be used for different T variables. Usually the smoothers being used will involve some type of smoothing parameter which may be chosen by cross-validation or some other data driven method.

A number of generalizations of the additive nonparametric regression approach described above have been proposed. One such generalization is *projection pursuit* regression as described in Friedman and Stuetzle (1981). This procedure essentially estimates an additive approximation to the regression function involving linear combinations of the variables rather than each variable separately. Thus one considers an approximation to $\mu$ of the form

$$\mu(t_1,\ldots,t_p) \doteq \alpha + \sum_{j=1}^{q} \mu_j\Big( \sum_{k=1}^{p} \alpha_{jk} t_k \Big). \tag{40}$$

Usually q is assumed to be less than p so there is a variable reduction aspect, similar to that of principle components regression, associated with the method. One would hope that the use of linear combinations of the variables in (40) would add flexibility to the approximation and possibly enhance its ability to reflect variable interactions (c.f. Exercise 12).

To choose q and estimate the vectors $\underline{\alpha}_j = (\alpha_{j1},\ldots,\alpha_{jp})$, $j=1,\ldots,q$, in (40), Friedman and Stuetzle (1981) suggest the following approach. Using a modification of our previous notation, let $\underline{s}_{\underline{a}}(\underline{z})$ be the vector of smoothed values obtained by applying some arbitrary smoother to the "data" $(\underline{a}'\underline{t}_1, z_1),\ldots,(\underline{a}'\underline{t}_n, z_n)$. Now define a residual vector by

$$\underline{r} = \underline{y} - \bar{y}\,\underline{1}$$

and set

$$\hat{\underline{\mu}}_1(\underline{a}_1) = \underline{s}_{\underline{a}_1}(\underline{r}).$$

An estimate of $\underline{\alpha}_1$ in (40) is then provided by the minimizer $\hat{\underline{\alpha}}_1$ of the mean squared error

$$\text{MSE}(\underline{a}_1) = (\underline{r} - \hat{\underline{\mu}}_1(\underline{a}_1))'(\underline{r} - \hat{\underline{\mu}}_1(\underline{a}_1))$$

with respect to $\underline{a}_1$. We now redefine $\underline{r}$ by subtracting off the smooth that was just obtained giving

$$\underline{r} = \underline{y} - \bar{y}\,\underline{1} - \hat{\underline{\mu}}_1(\hat{\underline{\alpha}}_1)\ .$$

A new vector of smoothed values can then be defined by $\hat{\underline{\mu}}_2(\underline{a}_2) = \underline{s}_{\underline{a}_2}(\underline{r})$ for some vector $\underline{a}_2$. To obtain an estimate of $\underline{\alpha}_2$ one now uses the minimizer $\hat{\underline{\alpha}}_2$ of

$$\text{MSE}(\underline{a}_2) = (\underline{r} - \hat{\underline{\mu}}_2(\underline{a}_2))'(\underline{r} - \hat{\underline{\mu}}_2(\underline{a}_2))$$

with respect to $\underline{a}_2$. This procedure is continued until p linear combinations have been found or an integer q is reached for which MSE $(\hat{\underline{\alpha}}_{q+1})$ is not appreciably smaller than $\text{MSE}(\hat{\underline{\alpha}}_q)$. Once suitable linear combinations $\hat{\underline{\alpha}}_1'\underline{t},\ldots,\hat{\underline{\alpha}}_q'\underline{t}$ have been found, a variant of the backfitting algorithm discussed above can be used with these new variables to make appropriate adjustments in the original smooths.

In using projection pursuit techniques the linear combinations $\hat{\underline{\alpha}}_j'\underline{t}$, $j = 1,\ldots,q$, are treated as new variables which are viewed as representing some characteristic of interest. By using the coefficients $\hat{\underline{\alpha}}_j$ that are applied to the p original variables one can try to diagnose what the new variable $\hat{\underline{\alpha}}_j'\underline{t}$ measures and thereby assign it a name. This again bears remarkable similarities to what transpires in principle components analysis. Due to the additive nature of the approximation, plots of the values contained in $\hat{\underline{\mu}}_j(\hat{\underline{\alpha}}_j)$ against their corresponding ordinates $\hat{\underline{\alpha}}_j'\underline{t}_1,\ldots,\hat{\underline{\alpha}}_j'\underline{t}_n$ will generally suffice to describe the relationship between the dependent variable values and the new variable $\hat{\underline{\alpha}}_j'\underline{t}$.

Another way to extend the additive approximation concept is to allow both the Y and the T variables to be transformed. Thus one seeks a transformation of Y, $\nu(Y)$, such that $E\nu(Y) = 0$ and

$$E[\nu(Y)|T_1=t_1,\ldots,T_p=t_p] \doteq \sum_{j=1}^{p} \mu_j(t_j) .$$

This is the approach taken by Brieman and Friedman (1985). They define an optimal set of transformations $\nu$, $\mu_1,\ldots,\mu_p$ to be ones which minimize

$$E[(\nu(Y) - \sum_{j=1}^{p} \mu_j(T_j))^2]/E\nu(Y)^2 \tag{41}$$

subject to the constraints $E\nu(Y) = E\mu_1(T_1) = \ldots = E\mu_p(T_p) = 0$. They show that optimal transformations exist and give an algorithm based on alternately computing conditional expectations with respect to Y, $T_1,\ldots,T_p$ that is assured of finding an optimal set of transformations. For this reason their procedure has been termed *ACE* which stands for *alternating conditional expectations*.

The basic ACE algorithm stems from the following considerations. Suppose we fix $E\nu(Y)^2 = 1$ and $E\nu(Y) = \ldots = E\mu_p(T_p) = 0$. Then, the minimizer of (41) with respect to $\nu(Y)$ is

$$\tilde{\nu}(Y) = E\left[\sum_{j=1}^{p} \mu_j(T_j)|Y\right] \Big/ \left\{E\left(E\left[\sum_{j=1}^{p} \mu_j(T_j)|Y\right]\right)^2\right\}^{1/2} .$$

Substituting this choice for $\nu(Y)$ into (41), we find the minimizer of

$$E[(\tilde{\nu}(Y) - \sum_{j=1}^{p} \mu_j(T_j))^2]$$

with respect to $\mu_1$ is

$$\tilde{\mu}_1(T_1) = E[\tilde{\nu}(Y) - \sum_{j=2}^{p} \mu_j(T_j)|T_1] .$$

Proceeding in this fashion one can now replace $\nu$ and $\mu_1$ with $\tilde{\nu}$ and $\tilde{\mu}_1$ in (41) and obtain the conditional expectation which minimizes this quantity with respect to $\mu_2$, etc. Given initial choices $\nu(Y) = Y/\{EY^2\}^{1/2}$, $\mu_j(T_j) = 0$, $j=1,\ldots,p$, this process can be repeated again and again until convergence occurs to an optimal set of transformations. To apply the algorithm to data the actual conditional expectations are replaced by estimates based on some variety of smoother such as a kernel or nearest neighbor estimator.

The ACE algorithm can be viewed as finding a transformation of Y which is best approximated by an additive model in the $T_j$ in the sense of minimizing (41). When applied to data it can be particularly useful from the point of view of validating a candidate parametric model which involves proposed transformations of the predictor and dependent variables. See Breiman and Friedman (1985) for illustrations of this property.

Yet another extension of the additive approximation technique has been studied by Hastie and Tibshirani (1986). They consider the use of additive approximations in the generalized linear models setting (c.f., Section 6.2.6). Recall that within the generalized linear models framework the conditional expectation of Y given $T_1,\ldots,T_p$ is related to the predictor variables through a link function g with

$$g(E[Y|T_1 = t_1,\ldots,T_p = t_p]) = \eta(t_1,\ldots,t_p)$$

being the focus of modeling efforts. Hastie and Tibshirani (1986) propose additive nonparametric estimators for $\eta$ based on the approximation

$$\eta(t_1,\ldots,t_p) \doteq \alpha + \sum_{j=1}^{p} \eta_j(t_j)$$

for smooth functions $\eta_1,\ldots,\eta_p$. They use likelihood considerations to suggest the use of smoothers applied to certain pseudo-data along with backfitting to construct estimates of the $\eta_j$. The interested reader is referred to their paper for further details.

**Exercises**

**1.** Refer to the data in Exercise 4.25. Discuss fitting this data by a least-squares spline using the heuristic rules given in Section 7.2.1. What order should be used and where should the knot(s) be placed. Fit the spline suggested by these rules and discuss your results.

**2.** Repeat your analysis in Exercise 1 for the voltage drop data in Exercise 3.14.

**3.** Obtain an optimal fit via nonlinear least squares for the data in Exercise 4.25 using a cubic spline with one knot. Compare the optimal knot location to the actual inflexion point of the curve.

**4.** Obtain an optimal fit via nonlinear least squares for the voltage drop data in Exercise 4.25 using a cubic spline with two knots.

**5.** Another approach to selecting both k and the $\xi_i$ for least-squares splines has been suggested by Smith (1983). Her approach utilizes a pool of knots which consists of a sufficiently fine partition of $[a,b]$ to insure that no interesting characteristics of $\mu$ are overlooked. If $\underline{\alpha} = (\alpha_1, \ldots, \alpha_K)'$ is a knot pool, then Smith proposes fitting a curve of the form $\Sigma_{j=1}^{m} \theta_j t^{j-1} + \Sigma_{j=1}^{K} \beta_j (t - \alpha_j)_+^{m-1}$ with variable selection techniques used to delete those terms $(\cdot - \alpha_j)_+^{m-1}$ which do not contribute to the overall fit. When all nonsignificant truncated power terms have been deleted, the final fit will be a spline of order m with $k \leq K$ knots at points $\xi_1, \ldots, \xi_k$ with $\xi_i = \alpha_j$ for some j.

a). Contrast the computational aspects of Smith's approach to nonlinear least-squares for knot selection with GCV used to determine k.

b). The data listed below represents a property of titanium as a function of heat. An optimal knot set for a cubic spline fit to this data with k = 5 is known (Jupp 1972, 1978) to have $\lambda$ = {835.967, 876.4016, 898.1462, 916.3145, 973.9075}. Fit this data by Smith's approach using forward selection, cubic splines and a knot pool consisting of ten equally spaced points over [785,

1055]. (Note: You must retain the variables 1, t, $t^2$, $t^3$ in all fits and consider deleting only terms of the form $(t - \alpha_j)_+^3$ on each step.) Also fit the data using 5 uniformly spaced points on [595, 1075]. Contrast the three fits. Discuss, in particular, the relative performance of optimal and uniformly placed knots.

| $t_i$ | $y_i$ | $t_i$ | $y_i$ | $t_i$ | $y_i$ | $t_i$ | $y_i$ |
|---|---|---|---|---|---|---|---|
| 595 | .644 | 725 | .668 | 845 | .812 | 975 | .615 |
| 605 | .622 | 735 | .676 | 855 | .907 | 985 | .607 |
| 615 | .638 | 745 | .676 | 865 | 1.044 | 995 | .606 |
| 625 | .649 | 755 | .686 | 875 | 1.336 | 1005 | .609 |
| 635 | .652 | 765 | .679 | 885 | 1.881 | 1015 | .603 |
| 645 | .639 | 775 | .678 | 895 | 2.169 | 1025 | .601 |
| 655 | .646 | 785 | .683 | 905 | 2.075 | 1035 | .603 |
| 665 | .657 | 795 | .694 | 915 | 1.598 | 1045 | .601 |
| 675 | .652 | 805 | .699 | 925 | 1.211 | 1055 | .611 |
| 685 | .655 | 815 | .710 | 935 | .916 | 1065 | .601 |
| 695 | .644 | 825 | .730 | 945 | .746 | 1075 | .608 |
| 705 | .663 | 835 | .763 | 955 | .672 | | |
| 715 | .663 | 845 | .812 | 965 | .627 | | |

**Source: de Boor and Rice (1968).**

**6.** Draw graphs of the linear B-splines in Section 7.2.2. Derive expressions for the quadratic B-splines and draw graphs of these functions as well.

**7.** Show that for B-splines of order m

a) $N_{i,m}(t) = 0$ for $t \notin [t_i, t_{i+m}]$ ,

b) $N_{i,m}(t) \geq 0$ and

c) $\sum_{i=-(m-1)}^{k} N_{i,m}(t) = 1$ for $t \in (a,b)$.

**8.** Another basis for splines is provided by piecewise polynomials. Any function of the form (3) can be expressed as

$$s(t) = \sum_{j=1}^{m} \beta_{jr} t^{j-1}/(j-1)!, \quad t \in [\xi_{r-1}, \xi_r) ,$$

where

$$\underline{\beta} = (\beta_{11},\ldots,\beta_{m1},\ldots,\beta_{1,k+1},\ldots,\beta_{m,k+1})'$$

must satisfy

$$R\underline{\beta} = \underline{0}$$

with

$$R' = [R_0',\ldots,R_{m-2}'] \quad ,$$

$$R_i = \begin{bmatrix} \underline{r}_{1i}' & -\underline{r}_{1i}' & & & \\ & \underline{r}_{2i}' & -\underline{r}_{2i}' & 0 & \\ & & \ddots & & \\ 0 & & & \underline{r}_{ki}' & -\underline{r}_{ki}' \end{bmatrix}$$

and

$$\underline{r}_{\ell i}' = (0,\ldots,0,\ 1,\ldots,\xi_\ell^{m-1-i}/(m-1-i)!).$$

Show that the least-squares estimator of the polynomial coefficients is

$$\hat{\underline{\beta}} = (X'X)^{-1}X'\underline{y} - (X'X)^{-1}R'[R(X'X)^{-1}R']^{-1}R(X'X)^{-1}X'\underline{y},$$

where

$$X = \begin{bmatrix} X_1 & & 0 \\ & X_2 & \\ & & \ddots \\ 0 & & X_{k+1} \end{bmatrix}$$

with

$$X_r = \{t_i^{j-1}/(j-1)!\}_{\substack{i=1,n_r \\ j=1,m}}$$

and $n_r$ denotes the number of data values with t ordinates in the interval $[\xi_{r-1},\xi_r)$.

**9.** Discuss the utility of the three bases for least-squares splines (truncated power basis, B-splines, and piecewise polynomials in Exercise 8) for estimation of $\mu$. What advantages or disadvantages do each of the bases have for estimation purposes?

**10.** Discuss how to construct a continuous piecewise polynomial consisting of three segments (i.e., with two knots): the first piece being cubic, the second quadratic and the third linear.

**11.** Consider the following regression curves

$$\mu(t) = \begin{cases} t(1 - t) \\ \exp\{-t\} \\ (\cos 2\pi t)^2 \end{cases}$$

Graph the optimal densities (23) and (24) when m = 2 for these regression curves and discuss their implications concerning knot and design point placement.

**12.** Show that an additive approximation of the form (40) is capable of detecting a product interaction of two variables. [Hint: Take p = 2 and define $\underline{\alpha}_1, \underline{\alpha}_2$, $\mu_1$ and $\mu_2$ appropriately to exactly recover the regression function $\mu(t_1, t_2) = t_1 t_2$.] [Friedman and Stuetzle, 1981].

# Appendix: Linear and Nonlinear Regression

This appendix provides a brief summary of selected topics concerning least-squares estimation from linear and nonlinear parametric regression models. The objective is to provide a sketch of that portion of the theory that is relevant to the developments in the text. Thus its intent is to provide a quick refresher for those who are already knowledgeable of the subject rather than an in depth treatment of the area. References are given for those who wish to obtain a more complete account.

## A.1 Linear Models

Consider the regression model described in Chapter 1. In that case observations $(t_1,y_1),\dots,(t_n,y_n)$ are obtained following the model

$$y_i = \mu(t_i) + \varepsilon_i , \qquad i = 1,\dots,n, \tag{1a}$$

where $\mu$ is an unknown regression function and the $\varepsilon_i$ are zero mean, uncorrelated random variables which have common variance $\sigma^2$. A linear model results when it is possible to write

$$\mu = \Sigma_{j=1}^{p} \beta_j x_j \tag{1b}$$

for known functions $x_1,\dots,x_p$ and unknown regression coefficients $\underline{\beta} = (\beta_1,\dots,\beta_p)'$. Model (1) fits into the classical linear models framework. We will now discuss techniques for inference about $\mu$ using least-squares estimators which arise from this fact.

### A.1.1 Parameter Estimation

The classical method of estimating the regression coefficients in model (1) is to use least-squares estimators. The least-squares estimators of $\beta_1, \ldots, \beta_p$ are obtained by minimizing

$$\mathrm{MSE}(\underline{\beta}) = n^{-1} \sum_{j=1}^{n} \left(y_j - \sum_{i=1}^{p} \beta_i x_i(t_j)\right)^2 \tag{2}$$

with respect to $\underline{\beta}$. This approach can be motivated from several perspectives. One motivation is provided by observing that

$$\underline{\mu} = (\mu(t_1), \ldots, \mu(t_n))' = X\underline{\beta} \quad ,$$

where

$$X = \{x_j(t_i)\}_{\substack{i=1,n \\ j=1,p}} \quad .$$

Thus, from a geometric viewpoint, minimization of (2) provides the estimator of $\underline{\mu}$ which is closest to $\underline{y}$ in the sense of Euclidean distance.

It is easy to see (e.g., differentiate (2) with respect to $\underline{\beta}$) that the minimizer of (2), $\hat{\underline{\beta}}$, must satisfy the *normal equations*

$$X'X\hat{\underline{\beta}} = X'\underline{y} \ . \tag{3}$$

If X has rank p, which we will always assume to be the case, then (3) can be rewritten as

$$\hat{\underline{\beta}} = (\hat{\beta}_1, \ldots, \hat{\beta}_p)' = (X'X)^{-1}X'\underline{y} \ . \tag{4a}$$

Thus, $\mu(t)$ is estimated by

$$\hat{\mu}(t) = \sum_{i=1}^{p} \hat{\beta}_i x_i(t) \ . \tag{4b}$$

In particular (4b) has the implication that the vector of fitted values

$$\hat{\underline{\mu}} = (\hat{\mu}(t_1), \ldots, \hat{\mu}(t_n))' = H\underline{y} \ , \tag{5a}$$

where

$$H = X(X'X)^{-1}X' \ . \tag{5b}$$

The matrix H in (5b) is known as the *hat matrix* since it transforms the response vector $\underline{y}$ to the vector of fitted values $\hat{\underline{\mu}}$. It is also referred to as the *influence matrix* in some cases since its elements provide information about the influence of the observations on the fit.

The estimator (4) obtained by minimizing (2) is an example of a least-squares estimator. In the context of linear models, least-squares estimators have a variety of important properties. First note that

$$E\hat{\underline{\beta}} = (X'X)^{-1}X'E\underline{y} = (X'X)^{-1}X'X\underline{\beta} = \underline{\beta},$$

so that $\hat{\underline{\beta}}$ is an unbiased estimator of $\underline{\beta}$. This has the immediate implication that $\hat{\mu}(t)$ is unbiased for $\mu(t)$ as well.

The variance-covariance matrix for $\hat{\underline{\beta}}$ is found to be

$$\text{Var } \hat{\underline{\beta}} = \sigma^2(X'X)^{-1} \tag{6}$$

as a consequence of the fact that Var $\underline{\varepsilon} = \sigma^2 I$. Writing

$$\hat{\mu}(t) = \underline{x}(t)'\hat{\underline{\beta}} \tag{7}$$

with $\underline{x}(t) = (x_1(t),\ldots,x_p(t))'$, it follows from (6) that the covariance between $\hat{\mu}(t)$ and $\hat{\mu}(s)$ is

$$\text{Cov}(\hat{\mu}(t), \hat{\mu}(s)) = \sigma^2\underline{x}(t)'(X'X)^{-1}\underline{x}(s) \ . \tag{8}$$

In particular, the variance-covariance matrix for $\hat{\underline{\mu}}$ is found to be

$$\text{Var } \hat{\underline{\mu}} = \sigma^2 H \ . \tag{9}$$

Note that $\hat{\mu}(t)$ in (7) is a linear estimator in the sense that it can be written as $\Sigma_{j=1}^{n}\ell_j y_j$ for constants $\ell_j$ not dependent on the $y_j$. A classical result from regression analysis states that the variance of $\hat{\mu}(t)$ is the smallest among all estimators in the class of unbiased linear estimators of $\mu(t)$. Under further restrictions on the distribution of the errors, other optimality properties for least-squares estimators can be established. The reader is referred to Searle (1971), Graybill (1976) or Kshirsagar (1983) for further details.

The other model parameter which requires estimation is $\sigma^2$, the error variance. Now, if the $\varepsilon_j$ were known, $n^{-1}\Sigma_{j=1}^{n}\varepsilon_j^2$ would be an unbiased estimator of $\sigma^2$. The $\varepsilon_j$ are, of course, unobservable but they can be estimated by $e_j = y_j - \hat{\mu}(t_j)$. This suggests using $MSE(\underline{\hat{\beta}})$ to estimate $\sigma^2$. Actually $MSE(\underline{\hat{\beta}})$ is a biased estimator of $\sigma^2$. However, an unbiased estimator is found to be

$$\hat{\sigma}^2 = nMSE(\underline{\hat{\beta}})/(n - p) \tag{10}$$

through application of the following lemma.

**Lemma A.1.** Let $\underline{z}$ be $n \times 1$ with mean $\underline{\theta}$ and variance-covariance matrix $\Sigma$. If A is a symmetric $n \times n$ matrix

$$E\underline{z}'A\underline{z} = \underline{\theta}'A\underline{\theta} + tr(A\Sigma) , \tag{11}$$

where $tr(A\Sigma)$ is the trace of $A\Sigma$. If, in addition, $\underline{z}$ has an n-variate normal distribution then

$$Var\ \underline{z}'A\underline{z} = 2tr[(A\Sigma)^2] + 4\underline{\theta}'A\Sigma A\underline{\theta} . \tag{12}$$

*Proof.* See Searle (1971, Chapter 2).

By applying Lemma A1 with $\underline{z} = \underline{y}$, $\underline{\theta} = X\underline{\beta}$ and $\Sigma = \sigma^2 I$ one can verify that $\hat{\sigma}^2$ is an unbiased estimator of $\sigma^2$. Its variance follows from (12) when the errors are normally distributed. For a discussion of other optimality properties of $\hat{\sigma}^2$ see, e.g., Graybill (1976).

Once the parameters of the model have been estimated it may be desired to determine how well the estimated regression curve fits (or conforms) to the data. An indication of this is provided by the *coefficient of determination*

$$R^2 = 1 - nMSE(\underline{\hat{\beta}})/\sum_{j=1}^{n}(y_j - \bar{y})^2 ,$$

where $\bar{y} = n^{-1}\Sigma_{j=1}^{n} y_j$. $R^2$ is the proportion of observed Y variation explained by the fitted regression function.

### A.1.2 Hypothesis Tests/Interval Estimates

Further inference about $\mu$ or $\underline{\beta}$, beyond that of point estimation, will usually take the form of hypothesis tests and/or interval estimates. We will restrict most of our attention here to interval estimates and tests concerning the $\beta_i$ and the value of $\mu$ at some specified point.

Suppose $\underline{\varepsilon}$ is n-variate normal and we wish to test a general linear hypothesis of the form $L\underline{\beta} = \underline{\ell}$, where L is an $r \times p$ matrix of rank $r \leq p$ and $\underline{\ell}$ is some specified vector. Then, under the hypothesis,

$$\frac{(L\hat{\underline{\beta}} - \underline{\ell})'(L(X'X)^{-1}L')^{-1}(L\hat{\underline{\beta}} - \underline{\ell})}{r\hat{\sigma}^2} \tag{13}$$

has an F-distribution with r numerator and $n - p$ denominator degrees of freedom. This statistic can therefore be utilized for tests of a general nature about $\underline{\beta}$.

By choosing L appropriately, (13) can be used to obtain test statistics for a variety of hypotheses which are of interest. For example, by choosing $r = 1$, L to be the row vector of all zeros except for a one as its ith component and $\ell = \beta_{i0}$ one obtains the statistic $(\hat{\beta}_i - \beta_{i0})^2/\hat{\sigma}^2 c_{ii}$, where $c_{ii}$ is the ith diagonal element of

$$C = (X'X)^{-1},$$

for testing the hypothesis $\beta_i = \beta_{i0}$. This statistic has an F distribution with one numerator and $n - p$ denominator degrees of freedom under the hypothesis. It is, therefore, equivalent to the statistic

$$Z = (\hat{\beta}_i - \beta_{i0})/[\hat{\sigma}^2 c_{ii}]^{1/2} \tag{14}$$

which has a Student's t-distribution when the hypothesis is valid. Consequently, the hypothesis $\beta_i = \beta_{i0}$ will be rejected if $|Z| \geq Z_{n-p;\alpha/2}$, the upper $\alpha/2$ percentage point of a Student's t-distribution with n - p degrees of freedom.

By taking r = 1, $L = (x_1(t),\ldots,x_p(t))$ and $\ell = \mu_0$, (13) can be used to derive a test for the hypothesis that $\mu(t) = \mu_0$, where $\mu_0$ is some specified constant. The resulting statistic is equivalent to

$$Z = [\hat{\mu}(t) - \mu_0]/[\hat{\sigma}^2\underline{x}(t)'C\underline{x}(t)]^{1/2} .$$

The hypothesis is rejected if $|Z| \geq Z_{n-p;\alpha/2}$.

By inverting the test statistics for hypotheses about $\beta_i$ and $\mu(t)$ given above, confidence intervals for $\beta_i$ and $\mu(t)$ can be derived. Specifically, we find that $100(1 - \alpha)\%$ confidence intervals are

$$\hat{\beta}_i \pm Z_{n-p;\alpha/2}[\hat{\sigma}^2 c_{ii}]^{1/2} \tag{15}$$

and

$$\hat{\mu}(t) \pm Z_{n-p;\alpha/2}[\hat{\sigma}^2\underline{x}(t)'C\underline{x}(t)]^{1/2} \quad . \tag{16}$$

In the particular case that $t = t_i$, (16) reduces to

$$\hat{\mu}(t_i) \pm Z_{n-p;\alpha/2}[\hat{\sigma}^2 h_{ii}]^{1/2} , \tag{17}$$

where $h_{ii}$ is the ith diagonal element of H in (5b).

### A.1.3 Diagnostic Analysis

There is a large and still growing literature concerning diagnostic methods for linear regression analysis. The basic objective of such methods is to detect problems in the fit to the data which may indicate the presence of "bad data" (data values which do not seem to conform to the model) and/or model misspecifications. By isolating such trouble spots in the fit it is often possible to undertake remedial action leading to better parameter estimates and more reliable tests and interval estimates.

Excellent elementary accounts of diagnostic methods for linear regression can be found in Gunst and Mason (1980) and Draper and Smith (1981). A somewhat more advanced treatment of this theory is provided by Cook and Weisberg (1982). We refer the reader to these texts for detailed discussions of the theory and practice of diagnostic methods for linear regression beyond what will be given here.

A focal point of diagnostic analysis is the vector of residuals $\underline{e} = (e_1, \ldots, e_n)'$ defined by

$$\underline{e} = \underline{y} - X\hat{\underline{\beta}} . \tag{18}$$

A statistically large residual indicates a poor fit and thereby suggests the corresponding data point merits further inspection. By statistically large we mean large relative to its standard error.

Since $E\hat{\underline{\mu}} = \underline{\mu}$ it follows that $E\underline{e} = \underline{0}$. The variance-covariance matrix for $\underline{e}$ is seen to be

$$\text{Var } \underline{e} = \sigma^2(I - H). \tag{19}$$

Thus, standardized measures associated with the residuals are provided by the *studentized residuals*

$$r_i = e_i/\hat{\sigma}(1 - h_{ii})^{1/2} , \quad i = 1, \ldots, n, \tag{20}$$

where $h_{ii}$ is the ith diagonal element of H. The ith residual is often deemed to be significantly large in magnitude if $|r_i|$ exceeds $Z_{n-p;\alpha/2}$ for some suitably small value of $\alpha$: e.g., $\alpha = .05$.

A common practice in residual analysis is to plot residuals or studentized residuals against t or the $x_i$. The plot should exhibit no patterns or trends. Various tests such as the runs test (see, e.g., Gunst and Mason, 1980) can be used as aids in revealing the presence of patterns in the residuals. If a trend or pattern is detected this may indicate the failure to include an appropriate function in the model, the presence of correlated errors as well as other problems.

The residuals provide information about response values which are outliers in the sense of departing significantly from the fitted curve. It is also possible to have outliers in the values of the predictor functions $x_1, \ldots, x_p$. Information of this nature is contained in the diagonal elements of H, $h_{ii}$, $i = 1, \ldots, n$, which are known as *leverage values.* If $x_1(t) \equiv 1$ it can be shown that

$$h_{ii} = (\underline{x}(t_i) - \bar{\underline{x}})'(X'X)^{-1}(\underline{x}(t_i) - \bar{\underline{x}}) + n^{-1} , \qquad (21)$$

where $\bar{\underline{x}} = n^{-1}\Sigma_{j=1}^{n}\underline{x}(t_j)$. Thus, $h_{ii}$ provides a measure of the departure of the ith observation for the predictor functions from the vector of averages, $\bar{\underline{x}}$.

It can be easily shown that $0 \le h_{ii} \le 1$, so a leverage value near one indicates a vector $\underline{x}(t_i)$ which is an outlier in the sense of being far from $\bar{\underline{x}}$. Points such as this are called high leverage points and indicate places which are particularly sensitive to a discrepant response. This latter property stems from the fact that $\hat{\mu}(t_i) = \Sigma_{j=1}^{n}h_{ij}y_j$ so $h_{ii}$ indicates how much influence $y_i$ has on its own fit.

A somewhat more refined assessment of what is meant by high leverage, rather than just $h_{ii}$ close to one, has been proposed by Velleman and Welsch (1981). They suggest that a data point should be considered as having high leverage if $h_{ii} > 3\ \Sigma_j h_{jj}/n = 3\ p/n$.

As suggested by our discussion of leverage, a response value can be discrepant, in the sense of having a large residual, while not having much influence on the fit if it is coupled with a small leverage value. From a practical viewpoint, if an observation does not significantly influence the fit then it makes little difference whether or not it is an outlier. This line of reasoning has prompted the development of various influence measures. An example of one such measure is Cook's distance measure defined by

$$D_i = (\hat{\underline{\beta}} - \hat{\underline{\beta}}_{(i)})'(X'X)(\hat{\underline{\beta}} - \hat{\underline{\beta}}_{(i)})/p\hat{\sigma}^2 \quad , \tag{22}$$

where $\hat{\underline{\beta}}_{(i)}$ is the estimator $\hat{\underline{\beta}}$ computed without the point $(t_i,y_i)$. It follows from (5) and (22) that $D_i$ can be viewed as an assessment of the overall change in the fit due to deleting the ith observation from the data. Another useful form for $D_i$ is

$$D_i = h_{ii}r_i^2/p(1 - h_{ii}) \ . \tag{23}$$

Cook (1977) suggests that an observation be judged influential if it exceeds $F_{p,n-p;.10}$, the lower 10th percentage point of an F-distribution with p numerator and n - p denominator degrees of freedom.

## A.2 Nonlinear Models

Let us now alter our assumptions and suppose that model (1a) holds except that now $\mu$ follows a nonlinear model. More specifically, assume there is a known functional form $\mu(\cdot;\underline{\beta})$ which is nonlinear in at least one of the parameters and $\mu(\cdot) = \mu(\cdot;\underline{\beta})$. Examples of such functions are given in Chapter 1. We will now discuss the problem of inference via least-squares estimation for such nonlinear models.

### A.2.1 Parameter Estimation

The regression function is now assumed to depend nonlinearly on the parameters $\underline{\beta}$. However, to estimate $\underline{\beta}$ and, hence, $\mu$, we might still proceed as in the linear models case and use a least-squares estimator. A least-squares estimator of $\underline{\beta}$ will be any minimizer, $\hat{\underline{\beta}}$, of

$$\mathrm{MSE}(\underline{\beta}) = n^{-1} \sum_{j=1}^{n} (y_j - \mu(t_j;\underline{\beta}))^2 \tag{24}$$

over all vectors $\underline{\beta}$ in the set of allowable solutions, or parameter space, B. Unlike the case of a linear model, however, there will generally be no closed form expression for $\hat{\underline{\beta}}$. In fact, the minimizer may not even be unique.

Assuming that $\mu(\cdot;\underline{\beta})$ is differentiable with respect to the $\beta_i$ we may differentiate (24) to see that $\hat{\underline{\beta}}$ must satisfy

$$\sum_{j=1}^{n} \hat{x}_{ij}\mu(t_j;\hat{\underline{\beta}}) = \sum_{j=1}^{n} \hat{x}_{ij}y_j \quad , \ i = 1,\ldots,p \ , \tag{25a}$$

where

$$\hat{x}_{ij} = \left. \frac{\partial\mu(t_j;\underline{\beta})}{\partial\beta_i} \right|_{\underline{\beta} = \hat{\underline{\beta}}} \quad . \tag{25b}$$

Equations (25) are the nonlinear analog of the normal equations (3).

There are a variety of algorithms which are designed to find a solution to equations (25). One of the best known of these is the Gauss-Newton method which will now be described.

Let $\underline{\beta}_0$ be some initial guess as to the value of $\hat{\underline{\beta}}$. Then, there is a vector $\delta\underline{\beta}_0 = (\delta\beta_{10},\ldots,\delta\beta_{p0})'$ such that $\hat{\underline{\beta}} = \underline{\beta}_0 + \delta\underline{\beta}_0$. A first order Taylor expansion of $\mu(\cdot;\underline{\beta})$ about $\underline{\beta}_0$ allows us to write

$$\mu(t_j;\hat{\underline{\beta}}) \doteq \mu_{j0} + \sum_{i=1}^{p} x_{ij0}\delta\beta_{i0} \tag{26a}$$

with $\mu_{j0} = \mu(t_j;\underline{\beta}_0)$ and

$$x_{ij0} = \left. \frac{\partial\mu(t_j;\underline{\beta})}{\partial\beta_i} \right|_{\underline{\beta} = \underline{\beta}_0} \quad . \tag{26b}$$

Substituting the approximation (26) into (24) we see that

$$\mathrm{MSE}(\hat{\underline{\beta}}) \doteq \widetilde{\mathrm{MSE}}(\delta\underline{\beta}_0) = n^{-1} \sum_{j=1}^{n} (y_j - \mu_{j0} - \sum_{i=1}^{p} x_{ij0}\delta\beta_{i0})^2 \quad . \tag{27}$$

The vector $\delta\underline{\beta}_0 = \hat{\underline{\beta}} - \underline{\beta}_0$ is still unknown. However, since $MSE(\hat{\underline{\beta}})$ is the minimal mean square error it should correspond to the minimum value of $\widetilde{MSE}(\delta\underline{\beta}_0)$. Arguing as in Section A.1.1 we see that this minimum occurs for any vector $\delta\hat{\underline{\beta}}_0$ satisfying

$$X(\underline{\beta}_0)'X(\underline{\beta}_0)\delta\hat{\underline{\beta}}_0 = X(\underline{\beta}_0)'\underline{e}(\underline{\beta}_0) \quad , \tag{28a}$$

where

$$X(\underline{\beta}) = \left\{\frac{\partial\mu(t_i;\underline{\beta})}{\partial\beta_j}\right\}_{\substack{i=1,n\\ j=1,p}} \tag{28b}$$

and

$$\underline{e}(\underline{\beta}) = (y_1 - \mu(t_1;\underline{\beta}),\ldots,y_n - \mu(t_n;\underline{\beta}))' \ . \tag{28c}$$

Assuming that $X(\underline{\beta}_0)$ has full column rank, $\widetilde{MSE}(\delta\underline{\beta}_0)$ has a unique minimizer and $\underline{\beta}_0 + \delta\hat{\underline{\beta}}_0$ gives an approximation to the minimizer $\hat{\underline{\beta}}$ of $MSE(\underline{\beta})$.

Since (27) is only an approximation it follows that $\underline{\beta}_0 + \delta\hat{\underline{\beta}}_0$ is only an approximation to $\hat{\underline{\beta}}$ and that $MSE(\underline{\beta}_0 + \delta\hat{\underline{\beta}}_0)$ will not be the minimal mean squared error in general. To remedy this problem we can repeat the process and set $\underline{\beta}_1 = \underline{\beta}_0 + \delta\hat{\underline{\beta}}_0$, write $\hat{\underline{\beta}} = \underline{\beta}_1 + \delta\underline{\beta}_1$ and obtain $\underline{\beta}_2 = \underline{\beta}_1 + \delta\hat{\underline{\beta}}_1$ with $\delta\hat{\underline{\beta}}_1 = (X(\underline{\beta}_1)'X(\underline{\beta}_1))^{-1}X(\underline{\beta}_1)'\underline{e}(\underline{\beta}_1)$. This can be done again and again until convergence is obtained in some sense. For example, if $\underline{\beta}_j$ and $\underline{\beta}_{j+1}$ are the approximations to $\hat{\underline{\beta}}$ on the jth and (j + 1)st step, one may decide to iterate until

$$(\underline{\beta}_{j+1} - \underline{\beta}_j)'(\underline{\beta}_{j+1} - \underline{\beta}_j)/\underline{\beta}_j'\underline{\beta}_j < \alpha_1$$

and

$$\left|MSE(\underline{\beta}_{j+1}) - MSE(\underline{\beta}_j)\right|/MSE(\underline{\beta}_j) < \alpha_2$$

for suitably small values of $\alpha_1$ and $\alpha_2$.

The Gauss-Newton method works quite well in many cases but can run into difficulties in regions where a first order Taylor expansion is not satisfactory to describe the disparity between $\mu(t_j;\hat{\underline{\beta}})$ and $\mu(t_j;\underline{\beta}_k)$, $j = 1,\ldots,n$, for some $k \geq 0$. This can lead to a poor choice for $\underline{\beta}_{k+1}$ and can even cause the algorithm to diverge.

One way around the difficulties associated with the Gauss-Newton method is to insure that the mean squared error decreases at each step. Two methods which accomplish this are the modified Gauss-Newton algorithm due to Hartley (1961), and the Marquardt algorithm (Marquardt, 1963). We will discuss only the former of these two methods here. In depth discussions of a variety of nonlinear least-squares algorithms, including the modified Gauss-Newton and Marquardt algorithms, can be found in Kennedy and Gentle (1980).

The principal difference between the Gauss-Newton and modified Gauss-Newton methods is that the modified Gauss-Newton algorithm uses $\underline{\beta}_{j+1} = \underline{\beta}_j + \gamma_j \delta\hat{\underline{\beta}}_j$, for some $0 \le \gamma_j \le 1$, as the approximation to $\hat{\underline{\beta}}$ on the jth step rather than $\underline{\beta}_j + \delta\hat{\underline{\beta}}_j$. A value is chosen for $\gamma_j$ which insures that $\underline{\beta}_{j+1} \in B$ and that $\text{MSE}(\underline{\beta}_j + \gamma_j \delta\hat{\underline{\beta}}_j) \le \text{MSE}(\underline{\beta}_j)$. This is always possible since $\gamma_j = 0$ is an admissible value.

Under certain conditions the modified Gauss-Newton algorithm is guaranteed to converge to $\hat{\underline{\beta}}$. See Hartley (1961) and Gallant and Fuller (1973) for discussion of conditions under which this is true and some practical guidelines for selection of the $\gamma_j$.

We have glossed over one potential problem in the above analysis: namely, how does one select the initializing vector $\underline{\beta}_0$. This is a crucial step and can drastically affect the convergence properties of the algorithm. General methods for selecting $\underline{\beta}_0$ have been proposed by Hartley and Booker (1965). Gallant (1975a) suggests choosing a subset, $(\tilde{t}_1, \tilde{y}_1), \ldots, (\tilde{t}_p, \tilde{y}_p)$, of the n data values and then solving the system of nonlinear equations

$$\tilde{y}_j = \mu(\tilde{t}_j; \underline{\beta}), \quad j = 1, \ldots, p ,$$

for $\underline{\beta}$. The resulting solution can then be used for $\underline{\beta}_0$.

Once $\hat{\underline{\beta}}$ has actually been computed other parameters of interest are easily estimated. The regression function is estimated by

$$\hat{\mu}(\cdot) = \hat{\mu}(\cdot;\hat{\underline{\beta}}) \tag{29}$$

and an estimator of $\sigma^2$ is

$$\hat{\sigma}^2 = n\text{MSE}(\underline{\beta})/(n - p) \ . \tag{30}$$

A large sample estimate of the variance-covariance matrix for $\hat{\underline{\beta}}$ is $\hat{\sigma}^2(X(\hat{\underline{\beta}})'X(\hat{\underline{\beta}}))^{-1}$ (Gallant, 1975 a,b) and an estimator for the variance of $\hat{\mu}(t)$ is

$$\hat{\sigma}^2\underline{x}(t;\hat{\underline{\beta}})'(X(\hat{\underline{\beta}})'X(\hat{\underline{\beta}}))^{-1}\underline{x}(t;\hat{\underline{\beta}}) \tag{31a}$$

with

$$\underline{x}(t;\underline{\beta}) = \left(\frac{\partial\mu(t;\underline{\beta})}{\partial\beta_1}, \ldots, \frac{\partial\mu(t;\underline{\beta})}{\partial\beta_p}\right)' . \tag{31b}$$

### A.2.2 Hypothesis Tests/Interval Estimates

When $\underline{\varepsilon}$ can be assumed to have an n-variate normal distribution the construction of tests and interval estimates for the parameters of nonlinear regression models closely parallels what is done in the linear case. For most cases of interest, tests and interval estimates such as those in Section A.1.2 can be used with the convention that $\hat{\underline{\beta}}$ and $\hat{\sigma}^2$ now denote the minimizer of (24) and the estimator of $\sigma^2$ defined in (30) and that X'X and $\underline{x}(t)$ are replaced by $X(\hat{\underline{\beta}})'X(\hat{\underline{\beta}})$ and $\underline{x}(t;\hat{\underline{\beta}})$. Gallant (1975a) gives an excellent intuitive account of why this should lead to tests which are asymptotically correct. More technical presentations are provided in Gallant (1974b, 1975b).

There are other possible test statistics which include likelihood ratio tests. The asymptotic properties of some of these have been investigated by Gallant (1974b, 1975b).

Halperin (1963), Hartley (1964) and El-Shaarawi and Shah (1980) show that it is possible to construct exact tests for some hypotheses in nonlinear regression. Suppose, for example, we wish to test the hypothesis that $\underline{\beta} = \underline{\beta}_0$. Under this hypothesis $\underline{e}(\underline{\beta}_0) =$

$\underline{\varepsilon}$. Thus, if we employ the model $\underline{e}(\underline{\beta}_0) = Z\underline{\delta} + \underline{\varepsilon}$, with Z an $n \times q$ matrix of suitably chosen regressor variables, the hypothesis $\underline{\beta} = \underline{\beta}_0$ becomes equivalent to $\underline{\delta} = \underline{0}$. The validity of the latter statement can be tested by realizing that this is a linear model so that results in Section A.1.2 are applicable. A test for $\underline{\delta} = \underline{0}$ can therefore be obtained from statistic (13), once we make the identifications $\underline{y} = \underline{e}(\underline{\beta}_0)$, $X = Z$, $L = I$ and $p = q$.

One problem with the Halperin/Hartley type tests is the ambiguity associated with specification of the matrix Z. This will have a direct bearing on the power of the test against alternatives. One natural choice is $Z = X(\underline{\beta}_0)$. Some general guidelines and illustrations concerning the selection of Z can be found in Gallant (1974b, 1975a, 1977).

### A.2.3 Diagnostic Analysis

The development of diagnostic methods for nonlinear regression has lagged somewhat behind that of the linear case. However, the close connection between the methods used for linear and nonlinear regression makes it clear that there are diagnostic measures for nonlinear regression which can be patterned after the linear case. The development of such diagnostics is explored in Cook and Weisberg (1982, Chapter 5). We summarize a few of their proposals.

The residual vector for nonlinear regression is $\underline{e}(\hat{\underline{\beta}})$. As in the linear case its elements should be standardized to give more meaningful interpretations. Thus, we seek a parallel of the studentized residuals (20).

The role of the hat matrix in the nonlinear case is occupied by

$$H(\hat{\underline{\beta}}) = X(\hat{\underline{\beta}})(X(\hat{\underline{\beta}})'X(\hat{\underline{\beta}}))^{-1}X(\hat{\underline{\beta}})' \quad . \tag{32}$$

Thus, a possible definition of studentized residuals is

$$r_i = e_i(\hat{\underline{\beta}})/\hat{\sigma}(1 - h_{ii}(\hat{\underline{\beta}}))^{1/2} \quad , \quad i = 1,\ldots,n, \tag{33}$$

where $\hat{\sigma}$ is obtained from (30) and $h_{ii}(\hat{\underline{\beta}})$ is the ith diagonal element of (32). A difficulty with this definition is that the residuals from nonlinear regression are not an entirely satisfactory measure of the fit since, unlike the linear case, they can depend heavily on the true parameter value. For this reason, Cook and Tsai (1985) have proposed the use of what they term projected residuals for residual analysis in nonlinear regression.

By computing the minimizer of (24) with the ith observation deleted one can obtain a "delete i estimator" $\hat{\underline{\beta}}_{(i)}$ of $\underline{\beta}$. This can be used to obtain a direct parallel of the Cook's distance measure (22): namely

$$D_i = (\hat{\underline{\beta}} - \hat{\underline{\beta}}_{(i)})'X(\hat{\underline{\beta}})'X(\hat{\underline{\beta}})(\hat{\underline{\beta}} - \hat{\underline{\beta}}_{(i)})/p\hat{\sigma}^2 . \tag{34}$$

A convenient approximation to $\hat{\underline{\beta}}_{(i)}$ is

$$\hat{\underline{\beta}}_{(i)} \doteq \hat{\underline{\beta}} - [X(\hat{\underline{\beta}})'X(\hat{\underline{\beta}})]^{-1}\underline{x}(t_i;\hat{\underline{\beta}})e_i(\hat{\underline{\beta}})/(1 - h_{ii}(\hat{\underline{\beta}})). \tag{35}$$

If this approximation is used in place of $\hat{\underline{\beta}}$ in (34) one finds that

$$D_i = r_i^2 h_{ii}(\hat{\underline{\beta}})/p(1 - h_{ii}(\hat{\underline{\beta}})) \tag{36}$$

which is analogous to (23). There are other possible influence measures which may be more appropriate in certain instances. Descriptions of some of these are given in Cook and Weisberg (1982, Chapter 5).

# References


Agarwal, G. G. (1978). Asymptotic design and estimation in polynomial spline regression. Mimeo. Series No. 78-11, Dept of Statist., Purdue Univ.

Agarwal, G. G. and Studden, W. J. (1978a). An algorithm for selection of design and knots in response curve estimation by spline functions. Mimeo. Series No. 78-15. Dept. of Statist., Purdue Univ.

____ (1978b). Asymptotic design and estimation using linear splines. *Commun. Statist. B 7*, 309-319.

____ (1980). Asymptotic integrated mean square error using least squares and bias minimizing splines. *Ann. Statist. 8*, 1307-1325.

Akaike, H. (1970). Statistical predictor identification. *Ann. Inst. Statist. Math 22*, 203-217.

____ (1973). Information theory and an extension of the maximum likelihood principle. In *2nd International Symposium on Information Theory* (B.N. Petrov and F. Csàki, eds.), 267-281. Budapest: Akademia Kiadó.

____ (1974). A new look at the statistical model identification. *I.E.E.E. Trans. Auto. Control 19*, 716-723.

____ (1978). A Bayesian analysis of the minimum AIC procedure. *Ann. Inst. Statist. Math. A 30*, 9-14.

Allen, D. M. (1974). The relationship between variable selection and data augmentation and a method of prediction. *Technometrics 16*, 125-127.

Anderssen, R. S. and Bloomfield, P. (1974a). A time series approach to numerical differentiation. *Technometrics 16*, 69-75.

____ (1974b). Numerical differentiation procedures for non-exact data. *Numer. Math. 22*, 157-182.

Anderssen, R. S., Bloomfield, P. and McNeil, D. R. (1974). Spline functions in data analysis. Tech. Rep. No. 69, Ser. 2, Dept. of Statist., Princeton Univ.

Andrews, D. F. (1974). A robust method for multiple linear regression. *Technometrics 16*, 523-531.

Anselone, P. M. and Laurent, P. J. (1968). A general method for the construction of interpolating or smoothing spline functions. *Numer. Math. 12*, 66-82.

Ansley, C. F. and Kohn, R. (1986). Spline smoothing with repeated values. *J. Statist. Comput. Simul.*, to appear.

____ (1987). Efficient generalized cross-validation for state space models. *Biometrika*, *74*. 139-148.

Ansley, C. F. and Wecker, W. E. (1981). Extensions and examples of the signal extraction approach to regression. *Proc. ASA-CENSUS-NBER Conf. Applied Time Series Analysis of Economic Data*, 181-192. Washington, D.C.: U.S. Bureau of the Census.

Atilgan, T. (1984). Determination of the dimension of an approximation for fitting a given data set. Manuscript.

Atkinson, A. C. (1978). Posterior probabilities for choosing a regression model. *Biometrika 65*, 39-48.

_____ (1980). A note on the generalized information criterion for choice of a model. *Biometrika 67*, 413-418.

_____ (1981). Likelihood ratios, posterior odds and information criteria. *J. of Econometrics 16*, 15-20.

Barham, R. H. and Drane, W. (1972). An algorithm for least squares estimation of nonlinear parameters when some of the parameters are linear. *Technometrics 14*, 757-766.

Barry, D. (1986). Nonparametric Bayesian regression. *Ann. Statist. 14*, 934-953.

Bates, D. M. and Wahba, G. (1983). A truncated singular value decomposition and other methods for generalized cross-validation. Tech. Rep. No. 715, Dept. of Statist., Univ. of Wisconsin-Madison.

Bates, D. M., Lindstrom, M. J., Wahba, G. and Yandell, B. S. (1987). GCVPACK-Routines for generalized cross validation. *Commun. Statist.-Simula. 16*, 263-297.

Benedetti, J. (1974). *Kernel Estimation of Regression Functions.* Doctoral Dissertation, University of Washington, Seattle.

_____ (1975). Kernel estimation of regression functions. *Proc. Computer Science and Statistics: 8th Annual Symposium on the Interface*, 405-412.

_____ (1977). On the nonparametric estimation of regression functions. *J. Roy. Statist. Soc. B. 39*, 248-253.

Bierens, H. J. (1983). Uniform consistency of kernel estimators of a regression function under generalized conditions. *J. Amer. Statist. Assoc. 78*, 699-707.

Blight, B.J.N. and Ott, L. (1975). A Bayesian approach to model inadequacy for polynomial regression. *Biometrika 62*, 79-88.

de Boor, C. (1978). *A Practical Guide to Splines.* New York: Springer-Verlag.

de Boor, C. and Rice, J. (1968). Least squares cubic spline approximation I - fixed knots. CSD Tech. Rep. No. 20.

Bowman, A. W. (1980). A note on the consistency of the kernel method for the analysis of categorical data. *Biometrika 67*, 682-684.

Breiman, L. and Freedman, D. (1983). How many variables should be entered in a regression equation? *J. Amer. Statist. Assoc. 78*, 131-136.

Breiman, L. and Friedman, J. H. (1985). Estimating optimal transformations for multiple regression and correlation. *J. Amer. Statist. Assoc. 80*, 580-619.

Brown, P. J. and Rundell, P.W.K. (1985). Kernel estimates for categorical data. *Technometrics 27*, 293-299.

Brownlee, K. A. (1965). *Statistical Theory and Methodology in Science and Engineering*. New York: John Wiley.

Buse, A. and Lim, L. (1977). Cubic splines as a special case of restricted least squares. *J. Amer. Statist. Assoc. 72*, 64-68.

Carmody, T. J. (1985). *Diagnostics for Multivariate Smoothing Splines*. Doctoral Dissertation, Dept. of Statist., Southern Methodist Univ.

_____ Diagnostics for multivariate smoothing splines. Manuscript.

Carroll, R. J. and Ruppert, D. (1985). Transformation in regression: a robust analysis. *Technometrics 27*, 1-12.

Chen, H. (1986). Data smoothing in analysis of covariance. Manuscript.

Cheng, K. F. and Lin, P. E. (1981). Nonparametric estimation of a regression function. *Z. Wahrsch. verw. Gebiete 57*, 223-233.

Cheng, P. E. (1984). Strong consistency of nearest neighbor regression function estimators. *J. Mult. Anal. 15*, 63-72.

Clark, R. M. (1977). Non-parametric estimation of a smooth regression function. *J. Roy. Statist. Soc. B 39*, 107-113.

_____ (1979). Calibration, cross-validation and carbon-14, I. *J. Roy. Statist. Soc. A, 142*, 47-62.

_____ (1980). Calibration, cross-validation and carbon-14, II. *J. Roy. Statist. Soc. A 143*, 177-194.

Cleveland, W. S. (1979). Robust locally weighted regression and smoothing scatterplots. *J. Amer. Statist. Assoc. 74*, 829-836.

Cogburn, R. and Davis, H. T. (1974). Periodic splines and spectral estimation. *Ann. Statist. 2*, 1108-1126.

Collomb, G. (1981). Estimation non-parametric de la regression: revue bibliographique. *Internat. Statist. Rev. 49*, 75-93.

Conte, S.D. and de Boor, C. (1980). *Elementary Numerical Analysis: An Algorithmic Approach*. New York: McGraw-Hill.

Cook, R. D. (1977). Detection of influential observations in linear regression. *Technometrics 19*, 15-18.

_____ (1979). Influential observations in linear regression. *J. Amer. Statist. Assoc. 74*, 169-174.

Cook, R. D. and Tsai, C-L. (1985). Residuals in nonlinear regression. *Biometrika 72*, 23-29.

Cook, R. D. and Weisberg, S. (1982). *Residuals and Influence in Regression*. New York: Chapman and Hall.

Cover, T. M. (1968). Estimation by the nearest neighbor rule. *IEEE Trans. Inform. Theory IT-14*, 50-55.

Cox, D. D. (1981). Asymptotics for M-type smoothing splines. Tech. Rep. No. 654, Dept. of Statist., Univ. of Wisconsin, Madison.

_____ (1983a). Approximation of method of regularization estimators. Tech. Rep. No. 723, Dept. of Statist., Univ. of Wisconsin-Madison.

_____ (1983b). Asymptotics for M-type smoothing splines. *Ann. Statist. 11*, 530-551.

_____ (1984a). Multivariate smoothing spline functions. *SIAM J. Numer. Anal. 21*, 789-813.

_____ (1984b). Gaussian approximation of smoothing splines. Tech. Rep. No. 743, Dept. of Statist., Univ. of Wisconsin-Madison.

_____ (1986). An analysis of Bayesian inference for non-parametric regression. Manuscript.

Cox, D. D. and O'Sullivan, F. (1985). Analysis of penalized likelihood type estimators with application to the generalized smoothing problem in Sobolev spaces. Tech. Rep. No. 51, Dept. of Statist., Univ. of California-Berkeley.

Craven, P. and Wahba, G. (1979). Smoothing noisy data with spline functions: estimating the correct degree of smoothing by the method of generalized cross-validation. *Numer. Math. 31*, 377-403.

Daniel, C. and Wood, F. S. (1971). *Fitting Equations to Data.* New York: John Wiley.

Davis, P. J. (1975). *Interpolation and Approximation.* New York: Dover Publications, Inc.

DeGroot, M. H. (1970). *Optimal Statistical Decisions.* New York: McGraw Hill.

Demmler, A. and Reinsch, C. (1975). Oscillation matrices with spline smoothing. *Numer. Math. 24*, 375-382.

Dempster, A. P. and Gasko-Green, M. (1981). New tools for residual analysis. *Ann. Statist. 9*, 945-959.

Devroye, L. P. (1978a). The uniform convergence of the Nadaraya-Watson regression function estimate. *Canada. J. Statist. 6*, 179-191.

_____ (1978b). The uniform convergence of nearest neighbor regression function estimators and their application in optimization. *IEEE Trans. Inform. Theory IT-24*, 142-151.

_____ (1981). On the almost everywhere convergence of nonparametric regression function estimates. *Ann. Statist. 9*, 1310-1319.

Devroye, L. and Györfi, L. (1985). *Nonparametric Density Estimation: The $L_1$ View.* New York: John Wiley.

Devroye, L. P. and Wagner, T. J. (1980). Distribution-free consistency results in nonparametric discrimination and regression function estimation. *Ann. Statist. 8*, 231-239.

Dobson, A. J. (1983). *An Introduction to Statistical Modeling.* London: Chapman and Hall.

Dongarra, J. J., Moler, C. B., Bunch, J. R. and Stewart, G. W. (1979). *Linpack Users' Guide.* Philadelphia: SIAM.

Draper, N. R. and Smith, H. (1981). *Applied Regression Analysis*, 2nd Edn. New York: John Wiley.

Duchon, J. (1977). Splines minimizing rotation-invariant semi-norms in Sobolev spaces. In *Constructive Theory of Functions of Several Variables* (W. Schempp and K. Zeller, eds.), 85-100. New York: Springer-Verlag.

Dym, H. and McKean, H. P. (1972). *Fourier Series and Integrals.* New York: Academic Press.

Dyn, N. and Wahba, G. (1982). On the estimation of functions of several variables from aggregated data. *SIAM J. Math. Anal. 13*, 134-152.

Dyn, N., Wahba, G. and Wong, W. H. (1979). Comment to "Smooth pycnophylactic interpolation for geographical regions" by W. R. Tobler. *J. Amer. Statist. Assoc. 74*, 530-536.

Efron, B. (1982). *The Jackknife, the Bootstrap and Other Resampling Plans*. SIAM monograph no. 38, CBMS-NSF. Philadelphia: SIAM.

El-Shaarawi, A. and Shah, K. R. (1980). Interval estimation in non linear models. *Sankhya B 42*, 229-232.

Engle, R. F., Granger, C.W.J., Rice, J. and Weiss, A. (1986). Semiparametric estimates of the relation between weather and electricity sales. *J. Amer. Statist. Assoc. 81*, 310-320.

Epanechnikov, V. A. (1969). Non-parametric estimation of a multivariate probability density. *Theor. Probab. Appl.14*, 153-158.

Eppright, E. S., Fox, H. M., Fryer, B. A., Lamkin, G. H., Vivian, V. M. and Fuller, E. S. (1972). Nutrition of infants and preschool children in the north central region of the United States of America. *World Review of Nutrition and Dietetics 14*, 269-332.

Eubank, R. L. (1984a). The hat matrix for smoothing splines. *Statist. and Prob. Letters 2*, 9-14.

_____ (1984b). Approximate regression models and splines. *Commun. Statist. A 13*, 433-484.

_____ (1985). Diagnostics for smoothing splines. *J. R. Statist. Soc. B 47*, 332-341.

_____ (1986). A note on smoothness priors and nonlinear regression. *J. Amer. Statist. Assoc. 81*, 514-517.

Eubank, R. L. and Gunst, R. F. (1986). Diagnostics for penalized least-squares estimators. *Statist. and Prob. Letters, 4*, 265-272.

Farrell, R. H. (1972). On the best obtainable asymptotic rates of convergence in estimation of a density function at a point. *Ann. Math. Statist. 43*, 170-180.

Franklin, J. N. (1968). *Matrix Theory*. Englewood Cliffs, N. J.: Prentice-Hall.

Frenkel, J. A. (1977). The forward exchange rate, expectations, and the demand for money: the German hyperinflation. *Amer. Econ. Rev. 67*, 653-670.

Friedman, J. H. and Stuetzle, W. (1981). Projection pursuit regression. *J. Amer. Statist. Assoc. 76*, 817-823.

Fuller, W. A. (1969). Grafted polynomials as approximating functions. *Aus. J. Agr. Econ. 13*, 35-46.

_____ (1976). *Introduction to Statistical Time Series*. New York: John Wiley.

Gallant, R. (1974a). The theory of nonlinear regression as it relates to segmented polynomial regressions with estimated join points. Mimeo. Series No. 925, Inst. of Statist., North Carolina State Univ.-Raleigh.

_____ (1974b). Testing a subset of the parameters of a nonlinear regression model. Mimeo. Series No. 943, Inst. of Statist., North Carolina State Univ.-Raleigh.

_____ (1975a). Nonlinear regression. *Amer. Statist.* *29*, 73-81.

_____ (1975b). Inference for nonlinear models. Mimeo. Series No. 875, Inst. of Statist., North Carolina State Univ.-Raleigh.

_____ (1977). Testing a nonlinear regression specification: a nonregular case. *J. Amer. Statist. Assoc.* *72*, 523-530.

Gallant, R. and Fuller, W. (1973). Fitting segmented polynomial regression models whose join points have to be estimated. *J. Amer. Statist. Assoc.* *68*, 144-147.

Gamber, H. (1979a). Choice of an optimal shape parameter when smoothing noisy data. *Commun. Statist. A* *8*, 1425-1435.

_____ (1979b). Confidence regions for periodic functions. *Commun. Statist. A* *8*, 1437-1446.

Gasser, Th. and Müller, H. G. (1979). Kernel estimation of regression functions. In *Smoothing Techniques for Curve Estimation* (Th. Gasser and M. Rosenblatt, eds.), 23-68. Heidelberg: Springer.

_____ (1984). Estimating regression functions and their derivatives by the kernel method. *Scand. J. Statist.* *11*, 171-185.

Gasser, Th., Müller, H. G., Köhler, W., Molinari, L. and Prader, A. (1984). Nonparametric regression analysis of growth curves. *Ann. Statist.* *12*, 210-229.

Gasser, Th., Müller, H. G. and Mammitzsch, V. (1985). Kernels for nonparametric curve estimation. *J. Roy. Statist. Soc. B* *47*, 238-252.

Gasser, Th., Sroka, L. and Jennen-Steinmetz, C. (1986). Residual variance and residual pattern in nonlinear regression. *Biometrika* *73*, 625-633.

Geisser, S. (1975). The predictive sample reuse method with applications. *J. Amer. Statist. Assoc.* *70*, 320-328.

Georgiev, A. A. (1984). Kernel estimates of functions and their derivatives with applications. *Statist. and Prob. Letters* *2*, 45-50.

Gerig, T. M. and Gallant, A. R. (1975). Computing methods for linear models subject to linear parametric constraints. *J. Statist. Comput. Simul.* *3*, 283-296.

Goldberger, A. S. (1962). Best linear unbiased prediction in the generalized linear regression model. *J. Amer. Statist. Assoc.* *57*, 369-375.

Golub, G., Heath, M. and Wahba, G. (1979). Generalized cross-validation as a method for choosing a good ridge parameter. *Technometrics* *21*, 215-223.

Golub, G. and Pereyra, V. (1973). The differentiation of pseudo-inverses and nonlinear least squares problems whose variables separate. *SIAM J. Numer. Anal.* *10*, 413-432.

Good, I. J. and Gaskins, R. A. (1971). Nonparametric roughness penalties for probability densities. *Biometrika* *58*, 255-277.

Graybill, F. A. (1976). *Theory and Application of the Linear Model.* North Scituate, Mass: Duxbury Press.

Greblicki, W., Krzyżak, A. and Pawlak, M. (1984). Distribution-free pointwise consistency of kernel regression estimate. *Ann. Statist. 12*, 1570-1575.

Green, P. (1985). Linear models for field trials, smoothing and cross-validation. *Biometrika 72*, 527-537.

Green, P., Jennison, C. and Seheult, A. (1985). Analysis of field experiments by least squares smoothing. *J. Roy. Statist. Soc. B 47*, 299-315.

Greville, T.N.E. (1969). Introduction to spline functions. In *Theory and Application of Spline Function* (T.N.E. Greville, ed.), 1-35. New York: Academic Press.

Gunst, R. F. and Mason, R. L. (1980). *Regression Analysis and Its Application: A Data-Oriented Approach.* New York: Marcel-Dekker.

Guttman, I., Pereyra, V.and Scolnik, H. D. (1973). Least squares estimation for a class of non-linear models. *Technometrics 15*, 209-218.

Hall, P. (1984). Integrated square error properties of kernel estimators of regression functions. *Ann. Statist. 12*, 241-260.

Hall, P. and Titterington, D. M. (1985). On the structure of smoothing techniques in regression problems. Manuscript.

Halperin, M. (1963). Confidence interval estimation in nonlinear regression. *J. R. Statist. Soc. B 25*, 330-333.

Hannan, E. J. and Quinn, B. G. (1979). The determination of the order of an autoregression. *J. Roy. Statist. Soc. B 41*, 190-195.

Härdle, W. (1984a). A law of the iterated logarithm for non-parametric regression function estimators. *Ann. Statist. 12*, 624-635.

_____ (1984b). Robust regression function estimation. *J. Multivariate Anal. 14*, 169-180.

Härdle, W., Hall, P. and Marron, S. (1985). How far are automatically chosen regression smoothing parameters from their optimum. Mimeo Series #1589, Department of Statistics, North Carolina State Univ.-Chapel Hill.

Härdle, W. and Gasser, Th. (1984). Robust non-parametric function fitting. *J. Roy. Statist. Soc. B 46*, 42-51.

_____ (1985). On robust kernel estimation of derivatives of regression functions. *Scand. J. Statist. 12*, 233-240.

Härdle, W. and Luckhaus, S. (1984). Uniform consistency of a class of regression function estimators. *J. Multivariate Anal. 12*, 612-623.

Härdle, W. and Marron, J. S. (1985a). Asymptotic nonequivalence of some bandwidth selectors in nonparametric regression. *Biometrika 72*, 481-484.

_____ (1985b). Optimal bandwidth selection in nonparametric regression function estimation. *Ann. Statist. 13*, 1465-1481.

Hart, J. (1985). On the choice of a truncation point in Fourier series density estimation. *J. Statist. Comput. Simul. 21*, 95-116.

Hart, J. and Wehrly, T. (1986). Kernel regression estimation using repeated measurements data. *J. Amer. Statist. Assoc. 81*, 1080-1088.

Hartley, H. O. (1961). The modified Gauss-Newton method for the fitting of non-linear regression functions by least squares. *Technometrics 3*, 269-280.

_____ (1964). Exact confidence regions for the parameters in non-linear regression laws. *Biometrika 51*, 347-353.

Hartley, H. O. and Booker, A. (1965). Non-linear least squares estimation. *Ann. Math. Statist. 36*, 638-650.

Hastie, T. and Tibshirani, R. (1986). Generalized additive models. *Statistical Science 1*, 297-318.

Heckman, N. E. (1986). Spline smoothing in a partly linear model. *J. R. Statist. Soc. B, 48*, 244-248.

_____ (1986). Minimax estimates in a semiparametric model. Tech. Rep. No. 41, Dept. of Statist., Univ. of British Columbia.

Hocking, R. R. (1976). The analysis and selection of variables in linear regression. *Biometrics 32*, 1-49.

Hoerl, A. E. and Kennard, R. W. (1970a). Ridge regression: biased estimation for nonorthogonal problems. *Technometrics 12*, 55-67.

_____ (1970b). Ridge regression: applications to nonorthogonal problems. *Technometrics 12*, 69-82.

Huber, P. J. (1964). Robust estimation of a location parameter. *Ann. Math. Statist. 35*, 73-101.

_____ (1979). Robust smoothing. In *Robustness in Statistics* (Launer and Wilkinson, eds.), 33-47. New York: Academic Press.

_____ (1981). *Robust Statistics.* New York: John Wiley.

_____ (1985). Projection pursuit (with discussion). *Ann. Statist. 13*, 435-525.

Hutchinson, M. F. and Bischof, R. J. (1983). A new method for estimating the spatial distribution of mean seasonal and annual rainfall applied to the Hunter Valley, New South Wales. *Aust. Met. Mag. 31*, 179-184.

Hutchinson, M. F. and deHoog, F. R. (1985). Smoothing noisy data with spline functions. *Numer. Math. 47*, 99-106.

_____ (1987). An efficient method for calculating smoothing splines using orthogonal transformations. *Numer. Math. 50*, 311-319.

Ibragimov, I. A. and Hasminski, R. Z. (1981). *Statistical Estimation: Asymptotic Theory.* New York: Springer Verlag.

Irvine, L. D., Marin, S. P. and Smith, P. W. (1985). Constrained interpolation and smoothing. Res. Pub. GMR-5102, General Motors Research Lab., Warren, Michigan.

Johnson, R. A. and Wichern, D. W. (1982). *Applied Multivariate Statistical Analysis.* Englewood Cliffs, N.J.: Prentice Hall.

Johnston, G. J. (1982). Probabilities of maximal deviations for nonparametric regression function estimates. *J. Multivariate Anal. 12*, 402-414.

Jupp, D.L.B. (1972). Curve fitting by splines as an application of unconstrained optimization. In *Optimization* (R. S. Anderssen, L. S. Jennings and D. M. Ryan, eds.), 49-59. St. Lucia, Queensland: University of Queensland Press.

_____ (1975). The "lethargy" theorem - a property of approximation by $\gamma$-polynomials. *J. Approx. Theor. 14*, 204-217.

_____ (1978). Approximation to data by splines with free knots. *SIAM J. Numer. Anal. 15*, 328-343.

Kennedy, W. and Gentle, J. (1980). *Statistical Computing*. New York: Marcel-Dekker.

Kimeldorf, G. S. and Wahba, G. (1970a). A correspondence between Bayesian estimation on stochastic processes and smoothing by splines. *Ann. Math. Statist. 2*, 495-502.

_____ (1970b). Spline functions and stochastic processes. *Sankhyā A 32*, 173-180.

_____ (1971). Some results on Tchebycheffian spline functions. *J. Math. Anal. Appl. 33*, 82-95.

Knafl, G., Sacks, J. and Ylvisaker, D. (1982). Model robust confidence intervals. *J. Statist. Plann. Infer. 6*, 319-334.

_____ (1985). Confidence bands for regression functions. *J. Amer. Statist. Assoc. 80*, 683-691.

Konakov, V. D. (1977). On a global measure of deviation for an estimate of the regression line. *Theor. Probab. Appl. 22*, 858-868.

Kronmal, R. A. and Tarter, M. E. (1968). The estimation of probability densities and cumulatives by Fourier series methods. *J. Amer. Statist. Assoc. 63*, 925-952.

Kshirsagar, A. (1983). *A Course in Linear Models*. New York: Marcel-Dekker.

Kuks, J. A. and Olman, V. (1972). A minimax linear estimator of regression coefficients (in Russian). *Izv. Akad. Nauk. Eston. SSR 21*, 66-72.

Largo, R. H., Gasser, Th., Prader, A., Stuetzle, W. and Huber, P. J. (1978). Analysis of the adolescent growth spurt using smoothing spline functions. *Ann. Human Biology 5*, 421-434.

Laurent, P. J. (1969). Construction of spline functions on a convex set. In *Approximation with Special Emphasis on Spline Functions* (I. J. Schoenberg, ed.), 415-446. New York: Academic Press.

Lenth, R. V. (1977). Robust splines. *Commun. Statist. A 6*, 847-854.

Li, K. C. (1982). Minimaxity of the method of regularization on stochastic processes. *Ann. Statist. 10*, 937-942.

_____ (1983). From Stein's unbiased risk estimates to the method of generalized cross-validation. Tech. Rept. No. 83-34, Dept. of Statist., Purdue University.

_____ (1984a). Regression models with infinitely many parameters: consistency of bounded linear functions. *Ann. Statist. 12*, 601-611.

_____ (1984b). Consistency for cross-validated nearest neighbor estimates in nonparametric regression. *Ann. Statist. 12*, 230-240.

_____ (1985). From Stein's unbiased risk estimates to the method of generalized cross validation. *Ann. Statist. 13*, 1352-1377.

_____ (1986). Asymptotic optimality of $C_L$ and generalized cross-validation in ridge regression with application to spline smoothing. *Ann. Statist. 14*, 1101-1112.

_____ (1987). Asymptotic optimality for $C_p$, $C_L$, cross-validation and generalized cross-validation: discrete index set. *Ann. Statist.*, to appear.

Li, K. C. and Hwang, J. (1984). The data-smoothing aspect of Stein estimates. *Ann. Statist. 12*, 887-897.

Luenberger, D. G. (1969). *Optimization by Vector Space Methods.* John Wiley: New York.

Lukas, M. A. (1980). Regularization. In *Numerical Solution of Integral Equations* (R. Anderssen, ed.), 152-182. Stockholm: Sijthoff and Noordoff.

Lyche, T. and Schumaker, L. L. (1973). Computation of smoothing and interpolating natural splines via local bases. *SIAM J. Numer. Anal. 10*, 1027-1038.

Mack, Y. and Silverman, B. (1982). Weak and strong uniform consistency of kernel regression estimates. *Z. Wahrsch. Verw. Gebiete 60*, 405-415.

Mallows, C. L. (1973). Some comments on $C_p$. *Technometrics 15*, 661 -675.

Marquardt, D. W. (1963). An algorithm for least-squares estimation of nonlinear parameters. *J. Soc. Indust. Appl. Math. 11*, 431-441.

Marron, S. (1986). Personal communication.

McCullagh, P. and Nelder, J. A. (1983). *Generalized Linear Models.* London: Chapman and Hall.

Meinguet, J. (1979). Multivariate interpolation at arbitrary points made simple. *J. Appl. Math. and Physics (ZAMP) 30*, 292-304.

Merz, P. H. (1980). Determination of absorption energy distribution by regularization and a characterization of certain absorption isotherms. *J. Comput. Physics 38*, 64-85.

Messer, K. (1986). A comparison of a spline estimate to its "equivalent" kernel estimate. Manuscript.

Montgomery, D. C. and Peck, E. A. (1982). *Introduction to Linear Regression Analysis.* New York: John Wiley.

Müller, H. G. (1984a). Smooth optimum kernel estimators of densities, regression curves and modes. *Ann. Statist. 12*, 766-774.

_____ (1984b). Optimal designs for nonparametric kernel regression. *Statist. and Prob. Letters 2*, 285-290.

_____ (1987). Weighted local regression and kernel methods for nonparametric curve fitting. *J. Amer. Statist. Assoc. 82*, 231-238.

Nadaraya, E. A. (1964). On estimating regression. *Theor. Probab. Appl. 9*, 141-142.

_____ (1965). On nonparametric estimation of density functions and regression curves. *Theor. Probab. Appl. 10*, 186-190.

_____ (1970). Remarks on non-parametric estimates for density functions and regression curves. *Theor. Probab. Appl. 15*, 134-137.

Naimark, M. A. (1967). *Linear Differential Operators. Vol. I.* New York: Ungar.

Narula, S. C. (1979). Orthogonal polynomial regression. *Internat. Statist. Rev. 47*, 31-36.

Nelder, J. A. and Wedderburn, R.W.M. (1972). Generalized linear interactive models. *J. R. Statist. Soc. A 135*, 370-384.

Noda, K. (1976). Estimation of a regression function by the Parzen kernel-type density estimators. *Ann. Inst. Statist. Math. 28*, 221-234.

Nussbaum, M. (1985). Spline smoothing in regression models and asymptotic efficiency in $L_2$. *Ann. Statist. 13*, 984-997.

Nychka, D. (1986). The mean posterior variance of a smoothing spline and a consistent estimate of the mean squared error. Manuscript.

Nychka, D. and Cox, D. D. (1984). Convergence rates for regularized solutions of integral equations from discrete noisy data. Tech. Rep. No. 752, Dept. of Statist., Univ. of Wisconsin, Madison.

Nychka, D., Wahba, G., Goldfarb, S. and Pugh, T. (1983). Cross validated spline methods for the estimation of three dimensional tumor size distributions from observations on two dimensional cross sections. Tech. Rep. No. 711, Dept of Statist., University of Wisconsin-Madison.

_____ (1984). Cross-validated spline methods for the estimation of three-dimensional tumor size distributions from observations on two-dimensional cross sections. *J. Amer. Statist. Assoc. 79*, 832-846.

O'Hagan, A. (1978). Curve fitting and optimal design for prediction (with discussion). *J. R. Statist. Soc. B 40*, 1-42.

O'Sullivan, F. (1983). The analysis of some penalized likelihood schemes. Tech. Rep. No. 726, Dept. of Statist., Univ. of Wisconsin-Madison.

_____ (1986a). A statistical perspective on ill-posed inverse problems. *Statist. Science 1*, 502-527.

_____ (1986b). Estimation of densities and hazards by the method of penalized likelihood. Tech. Rep. No. 58, Dept. of Statist., Univ. of California-Berkeley.

O'Sullivan, F. and Wahba, G. (1985). A cross validated Bayesian retrieval algorithm for nonlinear remote sensing experiments. *J. Comput. Physics 59*, 441-455.

O'Sullivan, F., Yandell, B. S. and Raynor, W. J. (1986). Automatic smoothing of regression functions in generalized linear models. *J. Amer. Statist. Assoc. 81*, 96-103.

Parzen, E. (1962a). *Stochastic Processes.* San Francisco: Holden-Day.

_____ (1962b). On estimation of a probability density function and mode. *Ann. Math. Statist. 33*, 1065-1076.

_____ (1974). Some recent advances in time series modeling. *IEEE Trans. Auto. Control. 19*, 723-730.

_____ (1977). Multiple time series: Determining the order of approximating autoregressive schemes. In *Multivariate Analysis IV* (P. Krishnaiah, ed.), 283-296. Amsterdam: North Holland.

Penrose, L. S. (1945). Discrimination between normal and psychotic subjects by revised examination. *M. Bull. Canad. Psychol. Assoc. 5*, 37-40.

Poirier, D. J. (1973). Piecewise regression using cubic splines. *J. Amer. Statist. Assoc. 68*, 515-524.

Prakasa Rao, B. L. S. (1983). *Nonparametric Functional Estimation*. Orlando: Academic Press.

Preece, M. A. and Baines, M. J. (1978). A new family of mathematical models describing the human growth curve. *Ann. Hum. Biol. 5*, 1-24.

Priestley, M. B. and Chao, M. T. (1972). Non-parametric function fitting. *J. Roy. Statist. Soc. B 34*, 385-392.

Rafajlowicz, E. (1987). Nonparametric orthogonal series estimators of regression: a class attaining the optimal convergence rate in $L_2$. *Statist. and Prob. Letters 5*, 219-224.

Ragozin, D. L. (1983). Error bounds for derivative estimates based on spline smoothing of exact or noisy data. *J. Approx. Theor. 37*, 335-355.

Randles, R. H. and Wolfe, D. A. (1979). *Introduction to the Theory of Nonparametric Statistics*. New York: John Wiley.

Raynor, W. J. and Bates, D. M. (1983). Smoothing generalized linear models with applications to logistic regression. Tech. Rep. No. 724, Dept. of Statist., Univ. of Wisconsin-Madison.

Reinsch, C. (1967). Smoothing by spline functions. *Numer. Math. 10*, 177-183.

_____ (1971). Smoothing by spline functions. II. *Numer. Math. 16*, 451-454.

Rice, J. (1982). An approach to peak area estimation. *J. Res. Nat. Bur. Standards 87*, 53-65.

_____ (1984a). Boundary modification for kernel regression. *Commun. Statist. A 13*, 893-900.

_____ (1984b). Bandwidth choice for nonparametric regresssion. *Ann. Statist. 12*, 1215-1230.

_____ (1986a). Convergence rates for partially splined models. *Statist. and Prob. Letters 4*, 203-208.

_____ (1986b). Bandwidth choice for differentiation. *J. Multivariate Anal. 19*, 251-264.

_____ (1986c). Choice of smoothing parameter in deconvolution problems. *Contemp. Math. 59*, 137-151.

Rice, J. and Rosenblatt, M. (1981). Integrated mean squared error of a smoothing spline. *J. Approx. Theor. 33*, 353-369.

_____ (1983). Smoothing splines: regression, derivatives and deconvolution. *Ann. Statist. 11*, 141-156.

Rosenblatt, M. (1956). Remarks on some nonparametric estimates of a density function. *Ann. Math. Statist. 27*, 832-835.

_____ (1969). Conditional probability density and regression estimators. In *Multivariate Analysis II* (P. Krishnaiah, ed.), 25-31. New York: Academic Press.

_____ (1970). Density estimates and Markov sequences. In *Nonparametric Techniques in Statistical Inference* (M. L. Puri, ed.), 199-210. Oxford: Cambridge University Press.

_____ (1971). Curve estimates. *Ann. Math. Statist. 42*, 1815-1842.

Roussas, G. (1969). Nonparametric estimation of the transition distribution function of a Markov process. *Ann. Math. Statist. 40*, 1386-1400.

Royall, R. M. (1966). *A Class of Nonparametric Estimators of a Smooth Regression Function*. Ph.D. Dissertation, Dept. of Statist., Stanford Univ.

Rutkowski, L. (1982). On system identification by nonparametric function fitting. *IEEE Trans. Automat. Control 27*, 225-227.

Sacks, J. and Ylvisaker, D. (1968). Designs for regression problems with correlated errors: many parameters. *Ann. Math. Statist. 39*, 40-69.

_____ (1978). Linear estimation for approximately linear models. *Ann. Statist. 6*, 1122-1137.

Schoenberg, I. J. (1964). Spline functions and the problem of graduation. *Proc. Nat. Acad. Sci. USA 52*, 947-950.

Schucany, W. R. and Sommers, J. P. (1977). Improvement of kernel type density estimators. *J. Amer. Statist. Assoc. 72*, 420-423.

Schumaker, L. L. (1980). *Spline Functions: Basic Theory*. New York: John Wiley.

Schuster, E. F. (1972). Joint asymptotic distribution of the estimated regression function at a finite number of distinct points. *Ann. Math. Statist. 43*, 84-88.

Schuster, E. and Yakowitz, S. (1979). Contributions to the theory of nonparametric regression, with application to system identification. *Ann. Statist. 7*, 139-149.

Schwarz, G. (1978). Estimating the dimension of a model. *Ann. Statist. 6*, 461-464.

Searle, S. R. (1971). *Linear Models*. New York: John Wiley.

Serfling, R. J. (1980). *Approximation Theorems of Mathematical Statistics*. New York: John Wiley.

Shepp, L. A. (1966). Radon-Nickodym derivatives of Gaussian measures. *Ann. Math. Statist. 37*, 321-354.

Shiau, J.-J. (1985). Smoothing spline estimation of functions with discontinuities. Tech. Rep. No. 768, Dept. of Statist., Univ. of Wisconsin-Madison.

Shiau, J.-J., Wahba, G. and Johnson, D. R. (1985). Partial spline models for the inclusion of tropopause and frontal boundary information in otherwise smooth two and three dimensional objective analysis. Tech. Rep. No. 777, Dept. of Statist., Univ. of Wisconsin-Madison.

Shibata, R. (1981). An optimal selection of regression variables. *Biometrika 68*, 45-54.

Shiller, R. J. (1984). Smoothness priors and nonlinear regression. *J. Amer. Statist. Assoc. 79*, 609-615.

Silverman, B. W. (1982). On the estimation of a probability density function by the maximum penalized likelihood method. *Ann. Statist. 10*, 795-810.

_____ (1984a). A fast and efficient cross-validation method for smoothing parameter choice in spline regression. *J. Amer. Statist. Assoc. 79*, 584-589.

_____ (1984b). Spline smoothing: the equivalent variable kernel method. *Ann. Statist. 12*, 898-916.

_____ (1985). Some aspects of the spline smoothing approach to non-parametric regression curve fittting (with discussion). *J. R. Statist. Soc. B 47*, 1-52.

_____ (1986). *Density Estimation for Statistics and Data Analysis.* New York: Chapman and Hall.

Smith, C. A. B. (1947). Some examples of discrimination. *Ann. of Eugenics 13*, 272-282.

Smith, P. L. (1979). Splines as a useful and convenient statistical tool. *Amer. Statist. 33*, 57-62.

_____ (1983). Curve fitting and modeling with splines using statistical variable selection techniques. Manuscript.

Speckman, P. (1979). Minimax estimates of linear functionals in a Hilbert space. Manuscript.

_____ (1981). The asymptotic integrated mean square error for smoothing noisy data by splines. *Numer. Math.*, to appear.

_____ (1982). Efficient nonparametric regression with cross-validated smoothing splines. Manuscript.

_____ (1985). Spline smoothing and optimal rates of convergence in nonparametric regression models. *Ann. Statist. 13*, 970-983.

_____ (1986). Regression analysis for partially linear models. Manuscript.

Staniswalis, J. G. (1986). Local bandwidth selection for kernel estimates. Manuscript.

Stein, C. (1981). Estimation of the mean of a multivariate normal distribution. *Ann. Statist. 9*, 1135-1155.

Steinberg, D. (1983). Bayesian models for response surfaces of uncertain functional form. MRC Tech. Rep. No. 2474, Math. Res. Center, Univ. of Wisconsin-Madison.

_____ (1985). Model robust response surface designs: scaling two-level factorials. *Biometrika 72*, 513-526.

Stone, C. J. (1975). Nearest neighbor estimators of a nonlinear regression function. In *Proc. Computer Sci. Statist. 8th Ann. Symp. Interface*, 413-418.

_____ (1977). Consistent nonparametric regression (with discussion). *Ann. Statist. 5*, 549-645.

_____ (1980). Optimal rates of convergence for nonparametric estimators. *Ann. Statist. 8*, 1348-1360.

_____ (1982). Optimal global rates of convergence for non-parametric regression. *Ann. Statist. 10*, 1040-1053.

_____ (1985). Additive regression and other nonparametric models. *Ann. Statist. 13*, 689-705.

Stone, M. (1974). Cross-validatory choice and assessment of statistical predictions (with discussion). *J. Roy. Statist. Soc. B 36*, 111-147.

_____ (1977). An asymptotic equivalence of choice of model by cross-validation and Akaike's criterion. *J. Roy. Statist. Soc. B, 39*, 44-47.

_____ (1979). Comments on model selection criteria of Akaike and Schwarz. *J. Roy. Statist. Soc. B 41*, 276-278.

Stroud, A. H. and Secrest, D. (1966). *Gaussian Quadrature Formulas*. New Jersey: Prentice-Hall.

Stuetzle, W. and Mittal, Y. (1979). Some comments on the asymptotic behavior of robust smoothers. In *Smoothing Techniques for Curve Estimation* (Th. Gasser and M. Rosenblatt, eds.), 191-195. Lecture Notes in Mathematics, No. 757. Heidelberg: Springer-Verlag.

Suess, H. E. (1970). Bristlecone pine calibration of the radiocarbon time scale from 5400 BC to the present. In *Radiocarbon Variations and Absolute Chronology* (I.U. Olsson, ed.), 303-312. New York: John Wiley.

Sugiura, N. (1978). Further analysis of the data by Akaike's information criterion and the finite corrections. *Commun. Statist. A 7*, 13-26.

Szegö, G. (1959). *Orthogonal Polynomials*. American Mathematical Society Colloquium Publications, Vol. XXIII. New York: American Mathematical Society.

Tapia, R. A. and Thompson, J. R. (1978). *Nonparametric Probability Density Estimation*. Baltimore: Johns Hopkins Press.

Tarter, M. E. and Kronmal, R. A. (1976). An introduction to the implementation and theory of nonparametric density estimation. *Amer. Statist. 30*, 105-112.

Taylor, A. E. and Mann, W. R. (1972). *Advanced Calculus*. Lexington Mass: Xerox College Publishing.

Thompson, M. L. (1978). Selection of variables in multiple regression. *Internat. Statist. Rev. 46*, 1-49.

Tikhonov, A. N. (1963a). Solution of incorrectly formulated problems and the regularization method. *Soviet Math. Dokl. 4*, 1035-1038.

_____ (1963b). Regularization of incorrectly posed problems. *Soviet Math. Dokl. 4*, 1624-1627.

Tikhonov, A. N. and Arsenin, V. Y. (1977). *Solutions of Ill-Posed Problems*. New York: John Wiley.

Titterington, D. M. (1985). Common structure of smoothing techniques in statistics. *Internat. Statist. Rev. 53*, 141-170.

Utreras, F. (1979). Cross-validation techniques for smoothing spline functions in one or two dimensions. In *Smoothing Techniques for Curve Estimation* (T. Gasser and M. Rosenblatt, eds.), 196-232. Lecture Notes in Math. No. 757. New York: Springer-Verlag.

_____ (1981a). On computing robust splines and applications. *SIAM J. Sci. Statist. Comput. 2*, 153-163.

_____ (1981b). Optimal smoothing of noisy data using spline functions. *SIAM J. Sci. Statist. Comput. 2*, 349-362.

_____ (1983). Natural spline functions, their associated eigenvalue problem. *Numer. Math. 42*, 107-117.

_____ (1984). Positive thin plate splines. CAT Rep. No. 68, Dept. of Math., Texas A & M Univ.

_____ (1985a). Smoothing noisy data under monotonicity constraints: existence, characterization and convergence rates. *Numer. Math. 47*, 611-625.

_____ (1985b). On generalized cross-validation for multivariate smoothing spline functions. CAT Rep. No. 81, Dept. of Math., Texas A & M Univ.

_____ (1986a). Boundary effects on convergence rates for Tikhonov regularization. *J. Approx. Theor.*, to appear.

_____ (1986b). Convergence rates for multivariate smoothing spline functions. *J. Approx. Theor.*, to appear.

Velleman, P. F. and Welsch, R. E. (1981). Efficient computing of regression diagnostics. *Amer. Statist. 35*, 234-242.

Villalobos, M. (1983). Estimation of posterior probabilities using multivariate smoothing splines and generalized cross-validation. Tech. Rep. No. 725, Dept. of Statist., Univ. of Wisconsin-Madison.

Villalobos, M. and Wahba, G. (1983). Multivariate thin plate spline estimates for the posterior probabilities in the classification problem. *Commun. Statist. A 12*, 1449-1479.

_____ (1985). Inequality constrained multivariate smoothing splines with application to the estimation of posterior probabilities. Tech. Rep. No. 756, Dept. of Statist., Univ. of Wisconsin-Madison.

_____ (1987). Inequality-constrained multivariate smoothing splines with application to the estimation of posterior probabilities. *J. Amer. Statist. Assoc. 82*, 239-248.

Wahba, G. (1973). On the minimization of a quadratic functional subject to a continuous family of linear inequality constraints. *SIAM J. Control 11*, 64-79.

_____ (1975). Smoothing noisy data with spline functions. *Numer. Math. 24*, 383-393.

_____ (1977a). Practical approximate solutions to linear operator equations when the data are noisy. *SIAM J. Numer. Anal. 14*, 651-667.

_____ (1977b). Optimal smoothing of density estimates. In *Classification and Clustering* (J. Van Ryzin, ed.), 423-458. New York: Academic Press.

_____ (1977c). A survey of some smoothing problems and the method of generalized cross-validation for solving them. In *Applications of Statistics* (P. R. Krishnaiah, ed.), 507-523. Amsterdam: North Holland.

_____ (1978). Improper priors, spline smoothing and the problem of guarding against model errors in regression. *J. R. Statist. Soc. B 40*, 364-372.

_____ (1979). Convergence rates of "thin plate" smoothing splines when the data are noisy. In *Smoothing Techniques for Curve Estimation* (T. Gasser and M. Rosenblatt, eds.), 233-245. Lecture Notes in Math. No. 757. New York: Springer-Verlag.

_____ (1980a). Ill posed problems: numerical and statistical methods for mildly, moderately and severely ill posed problems with noisy data. Tech. Rep. No. 595, Dept. of Statist., Univ. of Wisconsin-Madison. To appear in the Proceedings of the *International Conference on Ill Posed Problems*, M. Z. Nashed, ed.

_____ (1980b). Automatic smoothing of the log periodogram. *J. Amer. Statist. Assoc. 75*, 122-132.

_____ (1981a). Numerical experiments with the thin plate histospline. *Commun. Statist. A 10*, 2475-2514.

_____ (1981b). Spline interpolation and smoothing on the sphere. *SIAM J. Sci. Statist. Comput. 2*, 5-16. Erratum 3, 385-386.

_____ (1981c). Data-based optimal smoothing of orthogonal series density estimates. *Ann. Statist. 9*, 146-156.

_____ (1982). Constrained regularization for ill-posed linear operator equations, with applications in meteorology and medicine. In *Statistical Decision Theory and Related Topics III* (J. O. Berger and S. S. Gupta, eds.), 383-418. New York: Academic Press.

_____ (1983). Bayesian "confidence intervals" for the cross-validated smoothing spline. *J. R. Statist. Soc. B 45*, 133-150.

_____ (1984a). Cross-validated spline methods for the estimation of multivariate functions from data on functionals. In *Statistics: An Appraisal, Proceeding 50th Anniversary Conference Iowa State Statistical Laboratory* (H. A. David and H. T. David, eds.), 205-235. Ames: Iowa State Univ. Press.

_____ (1984b). Partial spline models for the semi-parametric estimation of functions of several variables. In *Statistical Analysis of Time Series*, 319-329. Tokyo: Institute of Statistical Mathematics.

_____ (1985). A comparison of GCV and GML for choosing the smoothing parameter in the generalized spline smoothing problem. *Ann. Statist. 13*, 1378-1402.

_____ (1986). Partial and interaction splines for semiparametric estimation of functions of several variables. Tech. Rep. No. 784, Dept. of Statist., Univ. of Wisconsin-Madison.

Wahba, G. and Wendelberger, J. G. (1980). Some new mathematical methods for variational objective analysis using splines and cross validation. *Monthly Weather Review 108*, 1122-1143.

Wahba, G. and Wold, S. (1975a). A completely automatic French curve: fitting spline functions by cross validation. *Commun. Statist. 4*, 1-17.

_____ (1975b). Periodic splines for spectral density estimation: the use of cross validation for determining the degree of smoothing. *Commun. Statist. 4*, 125-141.

Watson, G. S. (1964). Smooth regression analysis. *Sankhyā A 26*, 359-372.

Wecker, W. E. and Ansley, C. F. (1983). The signal extraction approach to nonlinear regression and spline smoothing. *J. Amer. Statist. Assoc. 78*, 81-89.

Wegman, E. J. (1972). Nonparametric probability density estimation: I. A summary of available methods. *Technometrics 14*, 513-546.

_____ (1982). Optimal nonparametric function estimation. *Proc. Conf. on Nonparametrics* (Z. Govindarajulu, ed.).

Wegman, E. J. and Wright, I. W. (1983). Splines in statistics. *J. Amer. Statist. Assoc. 78*, 351-365.

Weinberger, H. F. (1974). *Variational Methods for Eigenvalue Approximation*. CBMS-NSF Monograph No. 15. Philadelphia: SIAM.

Weinert, H. L. (1978). Statistical methods in optimal curve fitting. *Commun. Statist. B 7*, 417-435.

Weinert, H. L., Byrd, R. H. and Sidhu, G. S. (1980). A stochastic framework for recursive computation of spline functions: part II, smoothing splines. *J. Optim. Theory Appl. 30*, 255-268.

Weinert, H. L., Desai, U. B. and Sidhu, G. S. (1979). ARMA splines, system inverses, and least-squares estimates. *SIAM J. Control Optim. 17*, 525-536.

Weinert, H. L. and Kailath, T. (1974). Stochastic interpretations and recursive algorithms for spline functions. *Ann. Statist. 2*, 787-794.

Weinert, H. L. and Sidhu, G. S. (1978). A stochastic framework for recursive computation of spline functions: part I, interpolating splines. *IEEE Trans. Inform. Theory IT 24*, 45-50.

Wendelberger, J. G. (1981). The computation of Laplacian smoothing splines with examples. Tech. Rep. No. 648, Dept. of Statist., Univ. of Wisconsin-Madison.

_____ (1982). Estimation of divergence and vorticity using multidimensional smoothing splines. In *Proc. Workshop on Density Estimation and Smoothing* (E. Parzen, ed.).

Wertz, W. and Schneider, B. (1979). Statistical density estimation: a bibliography. *Internat. Statist. Rev. 47*, 155-175.

Whitney, P. (1986). Consistent least squares nonparametric regression. *J. Statist. Plan. Infer.*, to appear.

Whittaker, E. T. (1923). On a new method of graduation. *Proc. Edinburgh Math. Soc. 41*, 63-75.

Wold, S. (1974). Spline functions in data analysis. *Technometrics 16*, 1-11.

Wong, W. H. (1983). On the consistency of cross-validation in kernel nonparametric regression. *Ann. Statist. 11*, 1136-1141.

_____ (1984). On constrained multivariate splines and their approximations. *Numer. Math. 43*, 141-152.

Wright, I. W. and Wegman, E. J. (1980). Isotonic, convex and related splines. *Ann. Statist. 8*, 1023-1035.

Yakowitz, S. J. (1985). Nonparametric density estimation, prediction, and regression for markov sequences. *J. Amer. Statist. Assoc. 80*, 215-221.

Yakowitz, S. J. and Szidarovszky, F. (1985). A comparison of kriging with nonparametric regression methods. *J. Multivariate Anal. 16*, 21-53.

Young, A. S. (1977). A Bayesian approach to prediction using polynomials. *Biometrika 64*, 309-317.

______ (1982). The bivar criterion for selecting regressors. *Technometrics 24*, 181-189.

# Index